Springer Complexity

Springer Complexity is an interdisciplinary program publishing the best research and academic-level teaching on both fundamental and applied aspects of complex systems – cutting across all traditional disciplines of the natural and life sciences, engineering, economics, medicine, neuroscience, social and computer science.

Complex Systems are systems that comprise many interacting parts with the ability to generate a new quality of macroscopic collective behavior the manifestations of which are the spontaneous formation of distinctive temporal, spatial or functional structures. Models of such systems can be successfully mapped onto quite diverse "real-life" situations like the climate, the coherent emission of light from lasers, chemical reaction-diffusion systems, biological cellular networks, the dynamics of stock markets and of the internet, earthquake statistics and prediction, freeway traffic, the human brain, or the formation of opinions in social systems, to name just some of the popular applications.

Although their scope and methodologies overlap somewhat, one can distinguish the following main concepts and tools: self-organization, nonlinear dynamics, synergetics, turbulence, dynamical systems, catastrophes, instabilities, stochastic processes, chaos, graphs and networks, cellular automata, adaptive systems, genetic algorithms and computational intelligence.

The two major book publication platforms of the Springer Complexity program are the monograph series "Understanding Complex Systems" focusing on the various applications of complexity, and the "Springer Series in Synergetics", which is devoted to the quantitative theoretical and methodological foundations. In addition to the books in these two core series, the program also incorporates individual titles ranging from textbooks to major reference works.

Understanding Complex Systems

Founding Editor: J.A. Scott Kelso

Future scientific and technological developments in many fields will necessarily depend upon coming to grips with complex systems. Such systems are complex in both their composition – typically many different kinds of components interacting simultaneously and nonlinearly with each other and their environments on multiple levels – and in the rich diversity of behavior of which they are capable.

The Springer Series in Understanding Complex Systems series (UCS) promotes new strategies and paradigms for understanding and realizing applications of complex systems research in a wide variety of fields and endeavors. UCS is explicitly transdisciplinary. It has three main goals: First, to elaborate the concepts, methods and tools of complex systems at all levels of description and in all scientific fields, especially newly emerging areas within the life, social, behavioral, economic, neuro- and cognitive sciences (and derivatives thereof); second, to encourage novel applications of these ideas in various fields of engineering and computation such as robotics, nano-technology and informatics; third, to provide a single forum within which commonalities and differences in the workings of complex systems may be discerned, hence leading to deeper insight and understanding.

UCS will publish monographs, lecture notes and selected edited contributions aimed at communicating new findings to a large multidisciplinary audience.

Santo Banerjee • Mala Mitra • Lamberto Rondoni
Editors

Applications of Chaos and Nonlinear Dynamics in Engineering – Vol. 1

Springer

Editors
Santo Banerjee
Department of Mathematics
Politechnico di Torino
Corso Duca degli Abruzzi 24
10129 Torino
Italy
santoban@gmail.com
santo.banerjee@polito.it

Mala Mitra
Department of Physics
Camellia School of Engineering
and Technology
Nadibag PO Kajipara Barasat
700124 Kolkata
India
malamitra0@gamil.com

Lamberto Rondoni
Dipartimento di Matematica
Politecnico di Torino
Corso Duca degli Abruzzi 24
10129 Torino
Italy
lamberto.rondoni@polito.it

INFN, Sezione di Torino
Via P. Giuria 1
10125 Torino
Italy

ISSN 1860-0832 e-ISSN 1860-0840
ISBN 978-3-642-21921-4 e-ISBN 978-3-642-21922-1
DOI 10.1007/978-3-642-21922-1
Springer Heidelberg Dordrecht London New York

Library of Congress Control Number: 2011936888

Printed on acid-free paper

Springer is part of Springer Science+Business Media (www.springer.com)

Preface

In the past 60 years, the terms nonlinear dynamics and chaos have become familiar in the technical vocabulary of most sciences and technology. Indeed, the mathematical formulation of the vast majority of phenomena evolving in time, which has been given so far, consists of nonlinear ordinary or partial differential equations, or of nonlinear space and time discrete iterative processes. The typical situation, then, presents a kind of propagation of uncertainties which is exponential in time, is known as sensitive dependence on initial conditions, and is concisely and suggestively called chaos. Nonlinearities appear in feedback phenomena and generically in the evolution equations of most systems consisting of interacting parts or interacting with an external environment, which is itself affected by the interaction. Because of the mutual interactions, a given perturbation or action performed on the system of interest is shared by its elementary constituents in different ways, and the resulting effect is hardly predictable, especially in the long term, when the concatenation of causes and effects amounts to a long list. This concatenation may have constructive or destructive outcomes, with respect to the initial perturbations, and, typically, the consequent response of the object under investigation to the external actions will not be merely proportional to the size of the imposed perturbations. This is what nonlinearities mean, in general terms, and it is evident that observable natural phenomena and human artifacts commonly behave in nonlinear fashions.

Long-term predictions of the response of nonlinear dynamics to perturbations are usually problematic, but unpredictability is not an exclusive feature of nonlinear dynamics; even the simplest idealization of motion in space, the uniform motion of a single body subjected to no forces, enhances in time the initial uncertainty on its initial condition. Therefore, it is impossible to even predict whether an ideal arrow, which moves strictly along a straight line, will hit its target, if the direction of the motion is affected by some uncertainty and the target is sufficiently far. In this case, uncertainties grow linearly in time: to double the accuracy of predictions, it suffices to halve the uncertainty on the initial conditions. Beyond a certain limit, however, that may be practically impossible to achieve. Then, given the fact that any measurement one may perform, like any estimate of the initial state of any material object, is bound to be affected by uncertainties, one concludes that some degree of unpredictability is intrinsic, in practice as well as in principle, to our descriptions of all time dependent phenomena.

Nonlinearities commonly result in more serious difficulties as far as predictions, and hence control of the phenomena of interest is concerned. The limiting situation, known as chaos, is obtained when uncertainties are enhanced at an exponential rate. This situation is qualitatively, not simply quantitatively, different from those that enjoy linear or polynomial growths of perturbations. The striking fact is that even simple devices, such as the double pendulum, and not only very complex phenomena, like the climate, follow this kind of dynamics, in which there are no chances of reliable predictions beyond quite short times. This has then pushed researchers to develop the statistical approach to nonlinear dynamics, an approach that performs surprisingly well, in many circumstances, particularly when the degree of chaos, measured, e.g., by Lyapunov exponents, is quite high, or when the number of interacting elementary constituents of the system of interest is very large.

In the past decades, the growing understanding of these concepts has turned in practical applications, of particular interest in the development of present day technology. Time is therefore ripe for a review of the applications of chaos and nonlinear dynamics in engineering. As a matter of fact, numerous books are devoted to this purpose, but they are typically quite theoretical in nature. Therefore, the present collection of articles takes a more practical stand by addressing the various issues in the form of self-contained tutorials, which will guide step by step even the inexperienced reader to an informed use of the existing mathematical tools and softwares.

The first contribution, by S. Lynch, introduces the very popular and powerful Matlab software, by considering examples drawn from mechanical and electrical engineering applications, like Leon Chua's circuit, relevant for the problem of synchronization. M.F. Alves and Z.M. Assis Peixoto then tackle the important phenomenon of voltage flicker in electrical networks, which may be produced by an arc furnace operation. A case study concerning a 30 MVA arc furnace plant is considered in detail. Chapter 3, by S. Lynch and A. Steel, concerns the nonlinear dynamics and the stability properties of optical resonators, which are, for instance, one fundamental component of lasers. Chapter 4 is devoted to the problem of turbulence, and to the control of some of its aspects which are of primary technological importance, like turbulent mixing, which concerns the emissions of carbon dioxide and the spreading of pollutants, among its very many manifestations. This chapter is coauthored by B.S.V. Patnaik and S. Muddada. Vibration-based damage detection methods are widely used to identify hidden damages in beam and structural components. This application of nonlinear dynamics is considered by C.D. Dubey and V. Kapila in Chap. 5. The authors of Chap. 6, B. Kaygisiz, M. Karahan, A.M. Erkmen, and I. Erkmen, address another aspect of modern technology, that of the design of robots. This technology faces strong challenges, including those posed by vibrations, noisy sensing, and robot/irregular-environment interactions. The last four chapters are devoted to the evergrowing field of telecommunications and communications protocols, a number of which are based on chaos techniques. Chapter 7, by Jose M.V. Grzybowski, M. Eisencraft, and E.E.N. Macau, addresses, among other issues, the possibility of chaos-based ultra-fast communication. Chapter 8, by R. Martínez-Guerra, J.L. Mata-Machuca, A. Rodríguez-Bollain, and R. Aguilar-López, concerns secure transmission of information. In Chap. 9, Yu. Andreyev, A. Dmitriev, A.N. Miliou, and A.N. Anagnostopoulos describe an efficient method for storing, retrieving and processing information. Chapter 10, by S. Banerjee and S. Mukhopadhyay, concludes this book with another presentation of chaos-based secure communication methods.

This volume is the first of two, devoted to applications of chaos and nonlinear dynamics in engineering. The ten chapters present are organized in five parts, each concerning one of the most active fields of present-day engineering:

I. Nonlinearity and Computer simulations
II. Chaos and Nonlinear Dynamics in Electrical Engineering
III. Chaos and Nonlinear Dynamics in Building Mechanism and Fluid Dynamics
IV. Chaos in Robotics
V. Chaos and Nonlinear Dynamics in Communication.

We wish that this collection of essays, with their tutorial character, provide valuable help to those who intend to familiarize with both the theoretical and mathematical aspects of engineering applications of the fascinating modern theory of dynamical systems. As proved by the applications discussed in the following ten chapters, and those of the second volume which will follow, many applications of this theory, which are of high technological interest, have already been realized. But given the pace at which new ideas and new techniques are being developed, we anticipate that many more applications will be developed in the coming years, which makes of tantamount importance for scientists as well as engineers to familiarize with issues such as those addresses in this book.

Torino

Santo Banerjee
Mala Mitra
Lamberto Rondoni

Contents

Part I
Nonlinearity and Computer Simulations

Chapter 1
MATLAB Programming for Engineers

S. Lynch

1.1 Introduction

One of the many ventures in the UK National Higher Education Science, Technology, Engineering and Mathematics (HE STEM) programme is the Supporting MATLAB Automated assessment to Reinforce Teaching (SMART) project. Some staff involved include the author (Manchester Metropolitan University and the Open University), Professor Alan Irving (Liverpool University) and Dr Adam Crawford (Loughborough University). Some of the aims include developing students programming skills, engaging students in contextualised real world problems and developing a broader STEM community where academics can share expertise. Some of the objectives include developing an automated assessment system to give individual feedback to large cohorts of students in a short space of time, to disseminate findings through the Institute for Maths and its Applications (IMA) and Royal Academy of Engineers as HE STEM partners and to encourage input and feedback from international colleagues by means of an open educational resource channel. It is hoped that these resources will be freely available within the next few years.

It is assumed that the reader is familiar with either the *Windows* or *UNIX* platform. This chapter was prepared using MATLAB (R2011a) but most programs should work under earlier and later versions of the package.

A tutorial introduction to MATLAB can be downloaded from the MathWorks Central File Exchange. The commands and programs listed have been chosen to allow the reader to become familiar with MATLAB and the Symbolic Math Toolbox within a few hours. They provide a concise summary of the type of commands

S. Lynch (✉)
School of Computing, Mathematics and Digital Technology, Manchester Metropolitan University, Manchester M1 5GD, UK
e-mail: s.lynch@mmu.ac.uk

S. Banerjee et al. (eds.), *Applications of Chaos and Nonlinear Dynamics in Engineering – Vol. 1*, Understanding Complex Systems, DOI 10.1007/978-3-642-21922-1_1,

that will be used throughout this chapter. New users should be able to start on their own problems after completing this tutorial, and experienced users should find this chapter an excellent source of reference. Of course, there are many MATLAB textbooks on the market for those who require further applications or more detail. Note that Version 5 of the Symbolic Math Toolbox is powered by the MuPAD symbolic engine and nearly all Symbolic Math Toolbox functions work the same way as in previous versions. The Symbolic Math Toolbox provides tools for solving and manipulating symbolic math expressions and performing variable-precision arithmetic. The toolbox contains hundreds of MATLAB symbolic functions that leverage the MuPAD engine for tasks such as differentiation, integration, simplification, transforms, and equation solving. See the online MATLAB help pages if you require further information.

If you experience any problems there are several options for you to take. There is an excellent index within MATLAB, there are thousands of textbooks, see references [1, 4, 5, 9, 12, 13, 15, 17], for example, and MATLAB program files (or M-files) can be downloaded from the MATLAB Central File Exchange at the link: http://www.mathworks.com/matlabcentral/fileexchange/.

Download the zipped M-files and **Extract** the relevant M-files from the archive onto your computer. Since 2002, the authors files associated with his MATLAB book [12] have been downloaded over 20 000 times by users from all over the world.

To start MATLAB, simply double-click on the MATLAB icon. In the Unix environment, one types matlab as a shell command. The author has used the Windows platform in the preparation of this material. When MATLAB starts up, by default four windows are displayed entitled Current Folder, Command Window, Workspace and Command History. Figure 1.1 shows a typical screen when running MATLAB.

Each of the windows appearing in Fig. 1.1 will now be described:

Current Folder. This window shows the current folder where files are listed. Note that when typing in the Command Window, the command pwd also displays the MATLAB current folder.

Command Window. This is where the user types in MATLAB commands followed by the RETURN or ENTER key on the keyboard. If you do not want to see the result (for example, you would not want to see a list of numbers from 1 to 1000) then use the semi-colon delimeter.

Workspace. The workspace consists of the set of variables built up during a session of using the MATLAB software and stored in memory.

Command History. The Command History window displays a log of the statements most recently run in the Command Window.

The Command Window is where the MATLAB commands are typed and executed and this is where the user can use MATLAB just like a graphing calculator. When in the Command Window, command lines may be returned to and edited using the up and down arrow keys. If you do make a typing error, MATLAB will

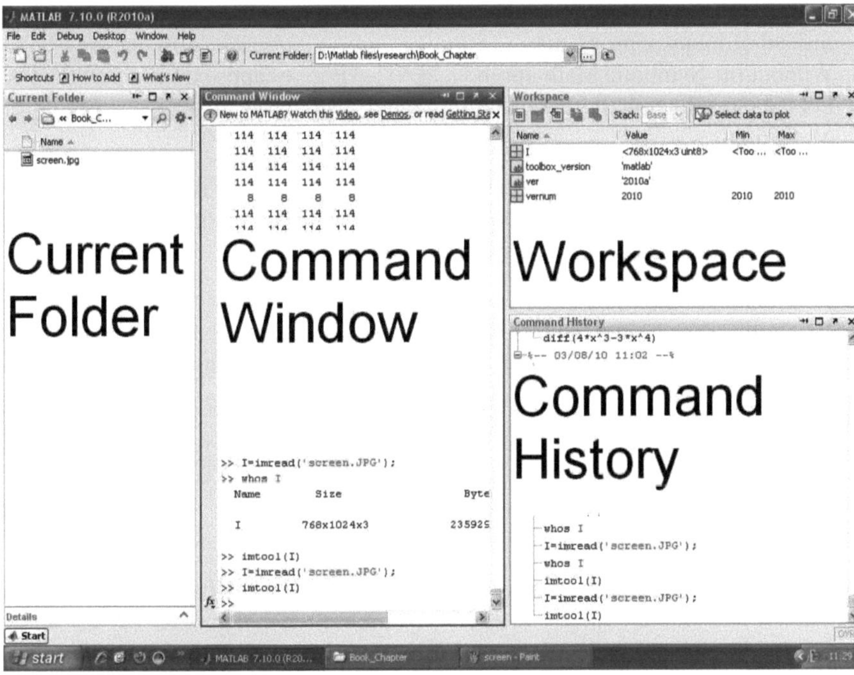

Fig. 1.1 The MATLAB user interface

give an **Error: message** and even point to the mistake in the code. Do not re-type the line, simply use the up arrow key and edit your mistake. In this chapter, each MATLAB program is displayed between horizontal lines and kept short to aid in understanding; the output is also included. Take care, M-files will only run if they are in the current working directory. To edit a file in MATLAB simply type edit filename.m in the Command Window or select **File**, **New** or **Open** in the MATLAB desktop or any desktop tool. Ensure that the files you are working with are in your Current Folder.

Some Well-Known Pitfalls of MATLAB

1. Array indices start with 1 and not zero. In some programs a translation may be required to start indices with zero.
2. MATLAB allows use of variable names of pre-defined functions. For example, you can set rand = 1 which means that you can not subsequently use the rand command. Readers should use the which command to check variable names.
3. Take care with element-wise and matrix matrix multiplication.
4. Some functions such as max, min, sort, sum, mean etc behave differently for complex and real data.

5. In MATLAB it is better (though not easier) to use matrix or vector operations instead of loops in programs.
6. Without the Symbolic Math Toolbox, MATLAB gives approximate answers.
7. For advanced symbolic computation readers will need to purchase Maple or Mathematica.

The reader should also be aware that even when the programs run successfully the output may not be correct. The following phenomena can greatly affect the results obtained for nonlinear dynamical systems:

1. Chaos and sensitivity to initial conditions.
2. Using the correct numerical solver for your particular problem.
3. Feedback, especially in the case of the second iterative method which is defined later in this chapter.

Any results obtained with MATLAB should be verified by experiment where possible.

1.2 Chemical Kinetics and Autocatalytic Reactions

Even the simplest chemical reactions can be highly complex and difficult to model. Physical parameters such as temperature, texture, pressure, and mixing, for example, are ignored in this chapter, and differential equations are constructed that are dependent only on the concentrations of the chemicals involved in the reaction. This is potentially a very difficult subject and some assumptions have to be made to make progress.

The Chemical Law of Mass Action. The rates at which the concentrations of the various chemical species change with time are proportional to their concentrations.

Consider the simple chemical reaction

$$A + \beta B \overset{k_1}{\rightleftharpoons} \gamma C,$$

where β and γ are the stoichiometric coefficients, A and B are the reactants, C is the product, and k_1 is the rate constant of the equation. The rate of reaction, say, r, is given by

$$r = \frac{\text{change in concentration}}{\text{change in time}}.$$

For this simple example,

$$r = k_1[A][B] = -\frac{d[A]}{dt} = -\frac{1}{\beta}\frac{d[B]}{dt} = \frac{1}{\gamma}\frac{d[C]}{dt},$$

Table 1.1 One of the possible reaction rate equations for each chemical reaction

Chemical reaction	The reaction rate equation for one species may be expressed as follows:
$2A \underset{k_r}{\overset{k_f}{\rightleftharpoons}} B$	$\dot{b} = k_f(a_0 - 2b)^2 - k_r b$
$A \underset{k_r}{\overset{k_f}{\rightleftharpoons}} 2B$	$\dot{b} = k_f\left(a_0 - \frac{b}{2}\right) - k_r b^2$
$A \underset{k_r}{\overset{k_f}{\rightleftharpoons}} B{+}C$	$\dot{c} = k_f(a_0 - c) - k_r(b_0 + c)(c_0 + c)$
$A{+}B \underset{k_r}{\overset{k_f}{\rightleftharpoons}} C$	$\dot{c} = k_f(a_0 - c)(b_0 - c) - k_r c$
$A{+}B \underset{k_r}{\overset{k_f}{\rightleftharpoons}} C{+}D$	$\dot{c} = k_f(a_0 - c)(b_0 - c) - k_r(c_0 + c)(d_0 + c)$

where $[A], [B]$, and $[C]$ represents the concentrations of A, B, and C, respectively. By adding a second chemical equation, a slightly more complex system is produced,

$$\alpha A \overset{k_2}{\rightleftharpoons} \delta D,$$

where k_2 is the rate constant of the second equation and α and δ are the stoichiometric coefficients. Two of the possible reaction rate equations for this system now become

$$\frac{d[A]}{dt} = -k_1\beta[A][B]^\beta - k_2\alpha[A]^\alpha, \quad \frac{d[D]}{dt} = k_2\delta[A]^\alpha.$$

Suppose that species A, B, C, and D have concentrations $a(t), b(t), c(t)$, and $d(t)$ at time t and initial concentrations a_0, b_0, c_0, and d_0, respectively. Table 1.1 lists some reversible chemical reactions and one of the corresponding reaction rate equations, where k_f and k_r are the forward and reverse rate constants, respectively.

Example 1.1. A reaction equation for sulphate and hydrogen ions to form bisulphite ions is given by

$$SO_3^{2-} + H^+ \underset{k_r}{\overset{k_f}{\rightleftharpoons}} HSO_3^-,$$

where k_f and k_r are the forward and reverse rate constants, respectively. Denote the concentrations by $a = [SO_3^{2-}]$, $b = [H^+]$, and $c = [HSO_3^-]$, and let the initial concentrations be a_0, b_0, and c_0. Assume that there is much more of species H^+ than the other two species, so that its concentration b can be regarded as constant. The reaction rate equation for $c(t)$ is given by

$$\frac{dc}{dt} = \dot{c} = k_f(a_0 - c)b - k_r(c_0 + c).$$

Find a general solution for $c(t)$.

Solution. The differential equation is separable and

$$\int \frac{dc}{k_f(a_0-c)b-k_r(c_0+c)} = \int dt.$$

Integration yields

$$c(t) = \frac{k_f a_0 b - k_r c_0}{k_f b + k_r} - \frac{k_r c_0}{k_f b + k_r} + Ae^{(-k_f a_0 - k_r)t},$$

where A is a constant. Using the MATLAB Command Window:

MATLAB Commands **Comments**

```
>> dsolve('Dc=kf*(a0-c)*b-kr*(c0+c)')    % Solve the ODE.
```

Example 1.2. The chemical equation for the reaction between nitrous oxide and oxygen to form nitrogen dioxide at 25°C,

$$2NO + O_2 \rightarrow 2NO_2$$

obeys the law of mass action. The rate equation is given by

$$\frac{dc}{dt} = k(a_0-c)^2\left(b_0 - \frac{c}{2}\right),$$

where $c = [NO_2]$ is the concentration of nitrogen dioxide, k is the rate constant, a_0 is the initial concentration of NO, and b_0 is the initial concentration of O_2. Find the concentration of nitrogen dioxide after time t given that $k = 0.00713 l^2 M^{-2} s^{-1}$, $a_0 = 4Ml^{-1}$, $b_0 = 1Ml^{-1}$, and $c(0) = 0Ml^{-1}$.

Solution. The differential equation is separable and

$$\int \frac{dc}{(4-c)^2(1-c/2)} = \int k\,dt.$$

Integrating using partial fractions gives

$$kt = \frac{1}{c-4} + \frac{1}{2}\ln|c-4| - \frac{1}{2}\ln|c-2| + \frac{1}{4} - \frac{1}{2}\ln 2.$$

It is not possible to obtain $c(t)$ explicitly, so numerical methods are employed using MATLAB. The concentration of nitrogen dioxide levels off at two moles per liter as time increases, as depicted in Fig. 1.2. The MATLAB program used to plot Fig. 1.2 is listed below.

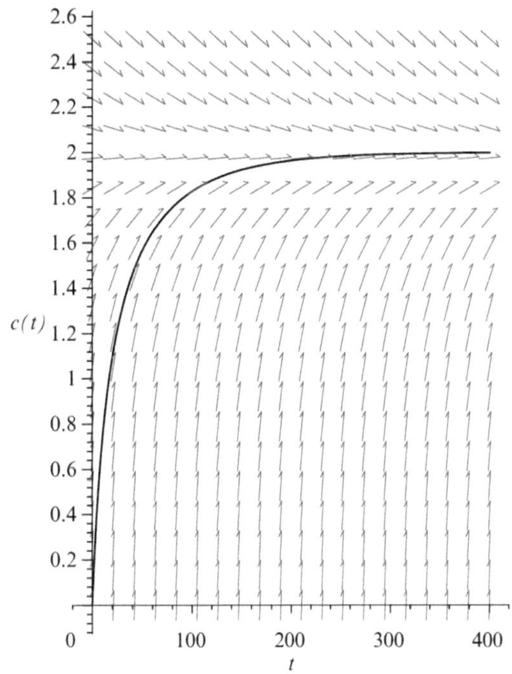

Fig. 1.2 The concentration of NO_2 in moles per liter against time in seconds

---

```
% Plotting the solution to an ODE.
deqn=inline('.00713*(4-c(1))^2*(1-c(1)/2)','t','c');
[t,ca]=ode45(deqn,[0 400],0);
plot(t,ca(:,1))
axis([0 400 0 3])
fsize=15;
set(gca,'xtick',[0:100:400],'FontSize',fsize)
set(gca,'ytick',[0:1:3],'FontSize',fsize)
xlabel('t','FontSize',fsize)
ylabel('c(t)','FontSize',fsize)
% End of Program.
```

There may be a wide range of time scales involved in chemical reactions and this can lead to stiff systems. Loosely speaking, a stiff system of differential equations is one in which the velocity or magnitude of the vector field changes rapidly in phase space. The final example illustrates how certain stiff systems can be solved using special solvers in MATLAB.

Example 1.3. The chemical rate equations for the Chapman cycle modeling the production of ozone are

$$O_2 + hv \rightarrow O + O, \qquad \text{Rate} = k_1,$$

$$O_2 + O + M \rightarrow O_3 + M, \quad \text{Rate} = k_2,$$

$$O_3 + \text{hv} \rightarrow O_2 + O, \quad \text{Rate} = k_3,$$

$$O + O_3 \rightarrow O_2 + O_2, \quad \text{Rate} = k_4,$$

where O is a singlet, O_2 is oxygen, and O_3 is ozone. The reaction rate equations for species $x = [\text{O}]$, $y = [\text{O}_2]$, and $z = [\text{O}_3]$ are

$$\dot{x} = 2k_1 y + k_3 z - k_2 xy[\text{M}] - k_4 xz,$$

$$\dot{y} = k_3 z + 2k_4 xz - k_1 y - k_2 xy[\text{M}],$$

$$\dot{z} = k_2 xy[\text{M}] - k_3 z - k_4 xz.$$

This is a stiff system of differential equations. Given that $[\text{M}] = 9e17$, $k_1 = 3e{-}12$, $k_2 = 1.22e{-}33$, $k_3 = 5.5e{-}4$, $k_4 = 6.86e{-}16$, $x(0) = 4e16$, $y(0) = 2e16$, and $z(0) = 2e16$, show that the steady-state reached is $[\text{O}] = 4.6806e7$, $[\text{O}_2] = 6.999e16$, and $[\text{O}_3] = 6.5396e12$.

Solution. The MATLAB commands used for solving this system are listed below. The reader should look at the Help pages in MATLAB for further information on stiff solvers.

```
% Solving a stiff system of ODEs. See Example 3.
% x=[O], y=[O_2], z=[O_3].
% Simple model of ozone production.
clear
deq=inline('[2*3e-12*x(2)+5.5e-4*x(3)-1.22e-33*x(1)*x(2)*9e17
   -6.86e-16*x(1)*x(3);
5.5e-4*x(3)+2*6.86e-16*x(1)*x(3)-3e-12*x(2)-1.22e-33*x(1)*x(2)
   *9e17;
1.22e-33*9e17*x(1)*x(2)-5.5e-4*x(3)-6.86e-16*x(1)*x(3)]','t','x');
[t,xa]=ode23s(deq,[0 1e8],[4e16,2e16,2e16]);
last=size(xa,1)
O=xa(last,1)
O2=xa(last,2)
O3=xa(last,3)
% End of Program.
```

Autocatalytic Chemical Reactions. Autocatalytic chemical reactions are those in which at least one of the reactants is also a product. One of the simplest autocatalytic reactions can be written

$$A + 2B \underset{}{\overset{k_1}{\rightleftharpoons}} 3B,$$

with rate equations

$$\frac{d[A]}{dt} = -2k_1[A][B]^2, \quad \frac{d[B]}{dt} = 3k_1[A][B]^2,$$

where k_1 is the rate constant.

These rate equations are nonlinear and a richer variety of behaviour is possible including emergent temporal order and oscillatory states, multiple steady states and hysteresis, emergent spatial order and self-organization, and chaos. The following references contain a rich variety of examples and applications in the real world [6, 8, 16].

As a particular example of emergent temporal order, consider the Brusselator model characterized by the chemical equations

$$A \to X,$$
$$2X + Y \to 3X,$$
$$B + X \to Y + C,$$
$$X \to D,$$

with rate equations given by

$$\frac{d[X]}{dt} = [A] + [X]^2[Y] - [B][X] - [X],$$
$$\frac{d[Y]}{dt} = [B][X] - [X]^2[Y],$$

where the rate constants have all been scaled to one, and the amount of species A and B is assumed much higher than species X and Y, so their concentrations can be regarded as constant. For mathematical convenience, suppose that $[X] = x$, $[Y] = y$, $[A] = a$, and $[B] = b$. Then the rate equations become

$$\dot{x} = a + x^2y - bx - x, \quad \dot{y} = bx - x^2y,$$

where a and b are constant. This is a system of ordinary differential equations and it is possible to carry out a phase plane analysis.

The critical points of the system are found by solving the equations, $\dot{x} = \dot{y} = 0$. These equations are easily solved and there is a unique critical point at $\left(a, \frac{b}{a}\right)$. To determine the stability of the critical point the Jacobian matrix is used to compute the eigenvalues.

MATLAB Commands **Comments**

```
>> syms a b                                  % The parameters.
>> P=a+x^2*y-b*x-x;Q=b*x-x^2*y;              % Define x dot and y dot.
>> J=[diff(P,x) diff(P,y);diff(Q,x)
   diff(Q,y)]                                % The Jacobian matrix.
>> x=a;y=b/a;                                % The critical point.
>> eig(J)                                    % The eigenvalues.
```

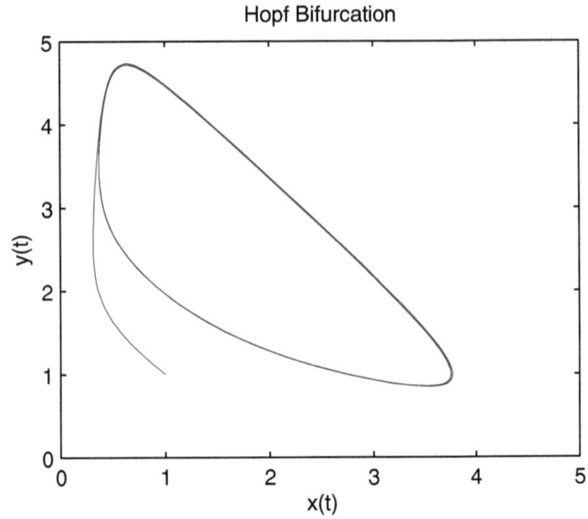

Fig. 1.3 Limit cycle of the Brusselator model when $a = 1$ and $b = 3$. Note that the MATLAB program produces an animation of a Hopf bifurcation which takes place when $b = 1 + a^2$

The Jacobian matrix at the critical point $(a, \frac{b}{a})$ is given by

$$A = \begin{pmatrix} b-1 & a^2 \\ -b & -a^2 \end{pmatrix}.$$

The determinant of J is $det(J) = a^2$ and the trace of J is $trace(J) = b - a^2 - 1$. Thus the critical point is stable when $b < 1 + a^2$ and unstable when $b > 1 + a^2$. The critical point goes through a bifurcation at $b = 1 + a^2$, and at this point the critical point has pure imaginary eigenvalues. A *Hopf bifurcation* or more accurately a *Poincaré-Andronov-Hopf bifurcation* arises when these two eigenvalues cross the imaginary axis because of a variation of the system parameters. For a mathematical introduction to bifurcation theory the reader is directed to [11, 12]. The MATLAB program listed below can be used to produce an animation of a Hopf bifurcation for the Brusselator model, and Fig. 1.3 shows the limit cycle when $a = 1$ and $b = 2.5$. For an introduction to limit cycles see reference [12]. When watching the animation, imagine the critical point blowing a smoke ring.

```
% Save the file as Hopf_System.m.
function sys=Hopf_System(t,x);
global b
X=x(1,:);
Y=x(2,:);
% Define the system.
P=1+Y.*X.^2-b*X-X;
Q=b*X-X.^2.*Y;
sys=[P;Q];
```

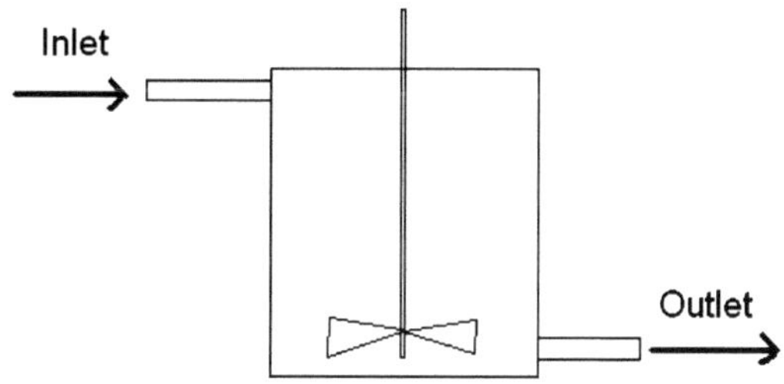

Fig. 1.4 A perfectly mixed continuously stirred tank reactor

```
% Animation of Hopf bifurcation of a limit cycle from a critical
    point.
% Save the file as Hopf_bif.m.
% NOTE: Hopf_System must be in the same directory as Hopf_bif.m.
clear
Max=120;global b;
for j=1:Max
    F(j)=getframe;
    b=j/40;    % b goes from 0 to 3.
    options=odeset('RelTol',1e-4,'AbsTol',1e-4);
    x0=1;y0=1;
    [t,x]=ode45(@Hopf_System,[0 100],[x0 y0],options);
    plot(x(:,1),x(:,2),'b');
    axis([0 5 0 5])
    fsize=15;
    set(gca,'xtick',[0:1:5],'FontSize',fsize)
    set(gca,'ytick',[0:1:5],'FontSize',fsize)
    xlabel('x(t)','FontSize',fsize)
    ylabel('y(t)','FontSize',fsize)
    title('Hopf Bifurcation','FontSize',15);
    F(j)=getframe;
end
movie(F,5)
% End of Program.
```

As a particular example of multiple steady states and hysteresis, consider a simple autocatalytic reaction in an isothermal continuously stirred tank reactor (CSTR). Select a volumetric control $V(t)$ as indicated in Fig. 1.4. The autocatalytic reaction is modeled by the following chemical equation:

$$A + 2B \underset{}{\overset{k_1}{\rightleftharpoons}} 3B,$$

where the reaction is taken to be elementary and obeys the chemical law of mass action. Applying rate equations and macroscopic balance equations, the differential equations modeling this system are written as

$$V\frac{da}{dt} = Q(a_0 - a) - Vk_1ab^2,$$

$$V\frac{db}{dt} = Q(b_0 - b) + Vk_1ab^2, \qquad (1.1)$$

where a and b are the concentrations of A and B, respectively, a_0 and b_0 are initial concentrations, V is the control volume, k_1 is the rate constant, and $\tau = \frac{V}{Q} = \frac{1}{k_f}$, is the residence time for the reactor. The equations provide a simple model of an autocatalytic reaction in an isothermal CSTR with perfect mixing. Interested readers should see reference [18] for an example of non-ideal mixing in an isothermal CSTR with two unpremixed feeds. Adding the equations in (1.1) and noting that conservation of mass gives $a(t)+b(t)=a_0+b_0$, the term b can be eliminated to give

$$\frac{da}{dt} = k_f\left(a_0 - a\right) - k_1 a\left(a_0 + b_0 - a\right)^2. \tag{1.2}$$

The differential equation (1.2) can have one, two or three critical points. Critical points are found by solving the equation

$$k_f\left(a_0 - a\right) = k_1 a\left(a_0 + b_0 - a\right)^2.$$

Using a tangency condition, it is not difficult to show that this equation has more than one solution if $a_0 > 8b_0$. This is left as an exercise for the reader. Taking the transformations

$$\gamma = 1 + \frac{b_0}{a_0}, \quad \beta = \frac{k_1}{k_f}a_0^2, \quad x = \frac{a}{a_0},$$

equation (1.2) simplifies to

$$\frac{dx}{dt} = (1-x) - \beta x(\gamma - x)^2, \tag{1.3}$$

where β is related to volumetric rate and γ is the catalyst concentration. By changing the volumetric rate in the tank reactor it is possible to obtain a hysteresis curve. A bifurcation diagram can be plotted using MATLAB and a feedback mechanism has to be incorporated. There are two well-known ways of plotting bifurcation diagrams which the author has labeled the first and second iterative methods.

The First Iterative Method. A parameter is fixed and one or more initial points are iterated forward. Transients are ignored and a number of the final iterates are plotted. The parameter is then increased by a suitable step length and the process is repeated. There are many points plotted for each value of the parameter.

The Second Iterative Method. A parameter is varied and the solution to the previous iterate is used as the initial condition for the next iterate. In this way, a feedback mechanism is introduced. In this case, there is a history associated with the process and only one point is plotted for each value of the parameter.

The MATLAB program listed below uses the second iterative method to produce a bifurcation diagram for (1.3) as the volumetric rate β is increased and then decreased.

```
% The differential equation.
% Save file as Chcm.m.
function xdot=Chem(t,x);
global beta;
xdot(1)=(1-x(1))-beta*x(1)*(1.1-x(1))^2;
xdot=[xdot(1)];
```

```
% Bifurcation diagram using the second iterative method.
% Ensure that Chem.m is in the same directory.
clear
figure
global beta;
endt=1; % Time span between each iteration.
Max=300;step=0.01;interval=Max*step;a=0.99;
betaup=zeros(Max);betadown=zeros(Max);
start=1;final=4;
% Ramp the volumetric rate up.
for n=1:Max
    beta=start+step*n;
    [t,xa]=ode45('Chem',[0 endt],a);
    a=xa(size(xa,1));
    betaup(n)=xa(size(xa,1));
end
% Ramp the volumetric rate down.
for n=1:Max
    beta=final-step*n;
    [t,xa]=ode45('Chem',[0 endt],xa(size(xa,1)));
    a=xa(size(xa,1));
    betadown(n)=xa(size(xa,1));
end
hold on
rr=step+start:step:final;
plot(rr,betaup)
plot(final+start-rr,betadown)
hold off
fsize=15;
axis([start final 0 1])
xlabel('\beta','FontSize',fsize)
ylabel('x','FontSize',fsize)
% End of Program.
```

Figure 1.5 shows the bifurcation diagram for (1.3) when $\gamma = 1.1$ and the volumetric rate β is increased from $\beta = 1$ to $\beta = 4$ and then decreased again. A hysteresis cycle is clearly visible.

1.3 The Duffing Equation

The existence and uniqueness theorems hold for autonomous systems of differential equations. This means that trajectories cannot cross, and the Poincaré-Bendixson theorem further implies that there is no chaos in two dimensions. However, chaos

Fig. 1.5 Hysteresis in a perfectly mixed isothermal CSTR

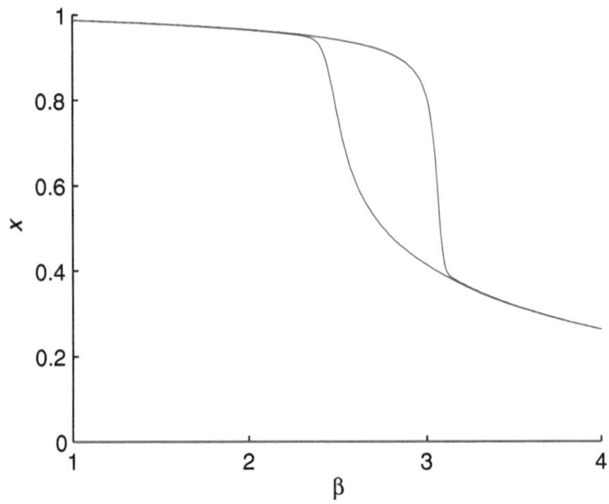

can be displayed in three-dimensional autonomous systems where various strange attractors can be plotted. This section is concerned with *nonautonomous* (or *forced*) systems of differential equations of the form

$$\ddot{x} = f(x, \dot{x}, t),$$

where the function f depends explicitly on t. There is no longer uniqueness of the solutions, and trajectories can cross in the phase plane. For certain parameter values, the phase portrait can become entangled with trajectories criss-crossing one another. By introducing a Poincaré map, it becomes possible to observe the underlying structure of the complicated flow.

As a particular example, consider the *Duffing equation* given by

$$\ddot{x} + k\dot{x} + (x^3 - x) = \Gamma\cos(\omega t),$$

where, in physical models, k is a damping coefficient, Γ represents a driving amplitude, and ω is the frequency of the driving force. Let $\dot{x} = y$; then the Duffing equation can be written as a system of the form

$$\dot{x} = y, \quad \dot{y} = x - ky - x^3 + \Gamma\cos(\omega t). \tag{1.4}$$

The equations may be used to model a periodically forced pendulum that has a cubic restoring force, where $x(t)$ represents displacement and $\dot{x}$ the speed of a simple mass, see Fig. 1.6. The equation can also be used to model periodically forced resistor-inductor-capacitor circuits with nonlinear circuit elements, where $x(t)$ would represent the charge oscillating in the circuit at time t.

Systems of the form (1.4) have been studied extensively in terms of, for example, stability, harmonic solutions, subharmonic solutions, transients, chaotic output, chaotic control, and Poincaré maps. The work here will be restricted to

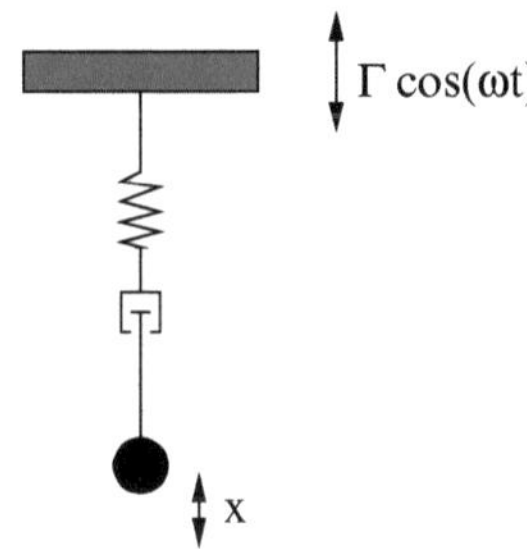

Fig. 1.6 A periodically driven pendulum

considering the Poincaré maps and bifurcation diagrams for system (1.4) as the driving amplitude Γ varies when $k = 0.3$ and $\omega = 1.25$ are fixed.

It is interesting to apply quasiperiodic forcing to nonlinear systems, as in [14], where nonchaotic attractors appear for a quasiperiodically forced van der Pol system.

Any periodically forced nonautonomous differential equation can be represented in terms of an autonomous flow in a torus. To achieve this transformation, simply introduce a third variable $\theta = \omega t$. System (1.4) then becomes a three-dimensional autonomous system given by

$$\dot{x} = y, \quad \dot{y} = x - ky - x^3 + \Gamma\cos(\theta), \quad \dot{\theta} = \omega. \tag{1.5}$$

A flow in this state space corresponds to a trajectory flowing around a torus with period $\frac{2\pi}{\omega}$. This naturally leads to a Poincaré mapping of a $\theta = \theta_0$ plane to itself.

When $\Gamma = 0$, system (1.4) becomes the autonomous system

$$\dot{x} = y, \quad \dot{y} = x - ky - x^3. \tag{1.6}$$

MATLAB can be used to compute the location and type of critical points for this system. Critical points are determined by solving the equations $\dot{x} = \dot{y} = 0$.

MATLAB Commands	**Comments**
`>> syms x y`	`% Symbolic objects.`
`>> [x,y]=solve('y','x-k*y-x^3')`	`% Solve the simultaneous equations.`

Thus there three critical points at $M = (-1,0)$, $N = (1,0)$, and $O = (0,0)$. The next step is to determine the stability of these critical points using the Jacobian matrix.

MATLAB Commands	**Comments**
`>> P=y;Q=x-k*y-x^3;`	`% Define x dot and y dot.`
`>> J=[diff(P,x) diff(P,y);diff(Q,x) diff(Q,y)]`	`% The Jacobian matrix.`
`>> eig([0 1;1 -k])`	`% The eigenvalues for (0,0).`
`>> eig([0 1;-2 -k])`	`% The eigenvalues for (-1,0) and (1,0).`

Thus the Jacobian matrix is

$$J = \begin{pmatrix} \frac{\partial P}{\partial x} & \frac{\partial P}{\partial y} \\ \frac{\partial Q}{\partial x} & \frac{\partial Q}{\partial y} \end{pmatrix} = \begin{pmatrix} 0 & 1 \\ 1-3x^2 & -k \end{pmatrix}.$$

The eigenvalues of O are $\lambda_{1,2} = \frac{-k \pm \sqrt{k^2+4}}{2}$ and the eigenvalues of M and N are $\lambda_{1,2} = \frac{-k \pm \sqrt{k^2-8}}{2}$. The point O is a saddle point and M and N are stable foci as long as $0 < k < 2\sqrt{2}$.

As Γ is increased from zero (now a nonautonomous system), stable periodic cycles appear from M and N and there are bifurcations of subharmonic oscillations. The system can also display chaotic behavior for certain values of Γ.

Only periodic cycles initially appearing from the critical point N will be considered here. A gallery of phase portraits along with their respective Poincaré return maps are presented in Figs. 1.7 and 1.8.

The MATLAB programs for plotting trajectories and Poincaré maps for the nonautonomous Duffing system will now be listed. In the programs, $k = 0.3$, $\omega = 1.25$ and $\Gamma = 0.5$.

```
% Phase portrait for a nonautonomous system of ODEs.
% The Duffing system.
deq=inline('[x(2);x(1)-0.3*x(2)-(x(1))^3+0.5*cos(1.25*t)]',
    't','x');
% Set tolerances.
options=odeset('RelTol',1e-4,'AbsTol',1e-4);
% Use ode45, 0<t<200, initial value (x,y)=(1,0).
[t,xx]=ode45(deq,[0 200],[1,0],options);
plot(xx(:,1),xx(:,2))
% Plot the trajectory.
fsize=15;
axis([-2 2 -2 2])
xlabel('x','FontSize',fsize)
ylabel('y','FontSize',fsize)
% End of Program.
```

```
% Poincare section for the Duffing system.
clear
deq=inline('[x(2);x(1)-0.3*x(2)-(x(1))^3+0.5*cos(1.25*t)]',
    't','x');
options=odeset('RelTol',1e-4,'AbsTol',1e-4);
% Solve the ODE at specific values of time.
% Multiples of 2*pi take you to the Poincare section.
[t,xx]=ode45(deq,[0:(2/1.25)*pi:(4000/1.25)*pi],[1,0]);
% Plot the chaotic attractor.
plot(xx(:,1),xx(:,2),'.','MarkerSize',1)
fsize=15;
axis([-2 2 -2 2])
xlabel('x','FontSize',fsize)
```

```
ylabel('y','FontSize',fsize)
title('Poincare Section of the Duffing System')
% End of Program.
```

When $\Gamma = 0.2$, there is a period-one harmonic solution of period $\frac{2\pi}{\omega}$, which is depicted as a closed curve in the phase plane and as a single point in the $\theta = 0$ plane (see Fig. 1.7a). When $\Gamma = 0.3$, a period-two cycle of period $\frac{4\pi}{\omega}$ appears; this is a subharmonic of order $\frac{1}{2}$. A period-two cycle is represented by two points in the Poincaré section (see Fig. 1.7b); note that the trajectory crosses itself in this case. A period-four cycle of period $\frac{8\pi}{\omega}$ is present when $\Gamma = 0.31$ (see Fig. 1.7c). When $\Gamma = 0.37$, there is a period-five cycle that is centered at O and also surrounds both M and N (see Fig. 1.8a). When $\Gamma = 0.5$, the system becomes chaotic. A single trajectory plotted in the phase plane intersects itself many times, and the portrait soon becomes very messy. However, if one plots the first returns on the Poincaré section, then a strange attractor is formed that demonstrates some underlying structure (see Fig. 1.8b). It must be noted that the chaotic attractor will have different forms on different Poincaré sections. This strange (or chaotic) attractor has fractal structure. At $\Gamma = 0.8$, there is once more a stable period-one solution. However, it is now centered at O (see Fig. 1.8c).

Figures 1.7 and 1.8 display some of the behavior possible for the Duffing equation for specific values of the parameter Γ. Of course, it would be far better to summarize all of the possible behaviors as the parameter Γ varies on one diagram. To achieve this goal, one must plot bifurcation diagrams. There are basically two ways in which bifurcation diagrams may be produced; one involves a feedback mechanism, the other does not. The first and second iterative methods are described in the second section of this chapter.

The MATLAB program to plot the bifurcation diagram in Fig. 1.10 will now be listed.

```
% Define a function. Save file as Duffing.m.
function xdot=Duffing(t,x)
% The Duffing System.
global Gamma;
xdot(1)=x(2);
xdot(2)=x(1)-0.1*x(2)-(x(1))^3+Gamma*cos(1.25*t);
xdot=[xdot(1);xdot(2)];
% End of Duffing.
```

```
% Bifurcation diagram for the Duffing equation.
% Make sure Duffing.m is in your directory.
clear
global Gamma;
Max=120;step=0.001;interval=Max*step;a=1;b=0;
% Ramp the amplitude up.
for n=1:Max
    Gamma=step*n;
    [t,x]=ode45('Duffing',[0:(2*pi/1.25):(4*pi/1.25)],[a,b]);
    a=x(2,1);
```

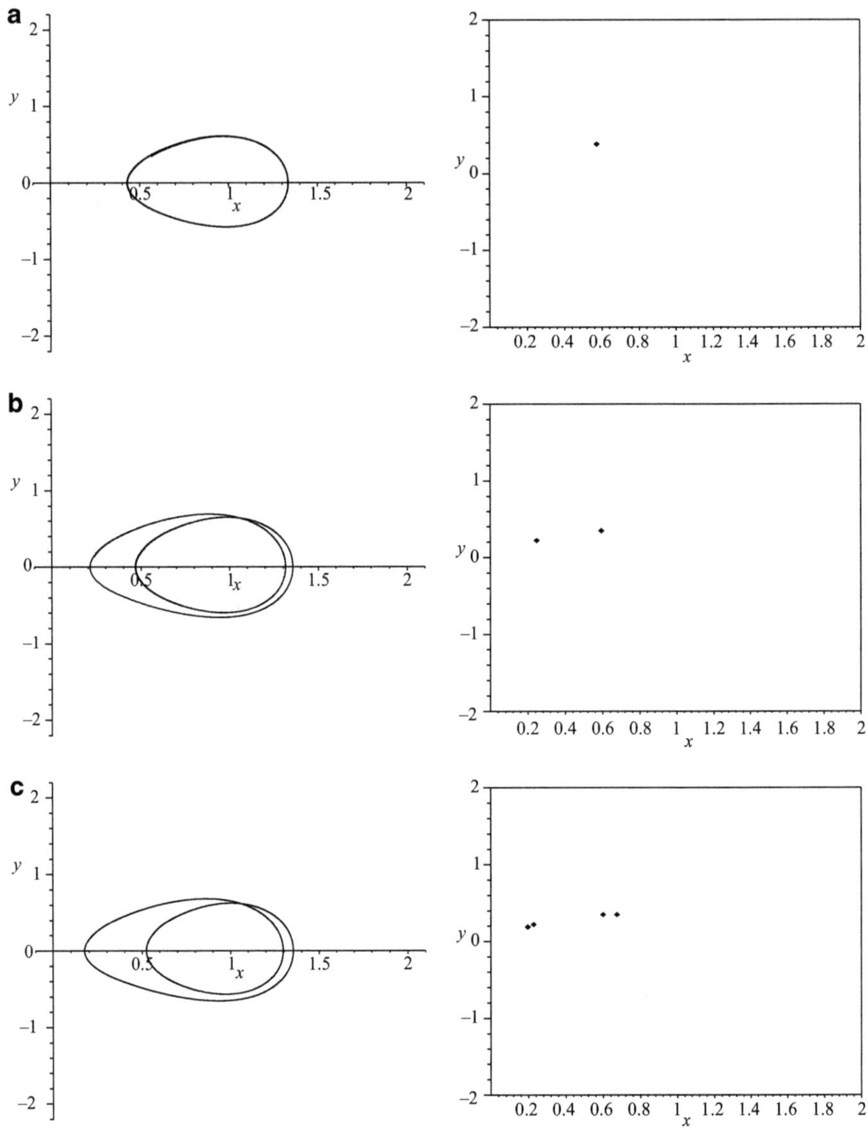

Fig. 1.7 A gallery of phase portraits and Poincaré maps for system (1.4) when $k = 0.3$ and $\omega = 1.25$: (**a**) $\Gamma = 0.2$ (forced period one), (**b**) $\Gamma = 0.3$ (a period-two subharmonic), (**c**) $\Gamma = 0.31$ (a period-four subharmonic)

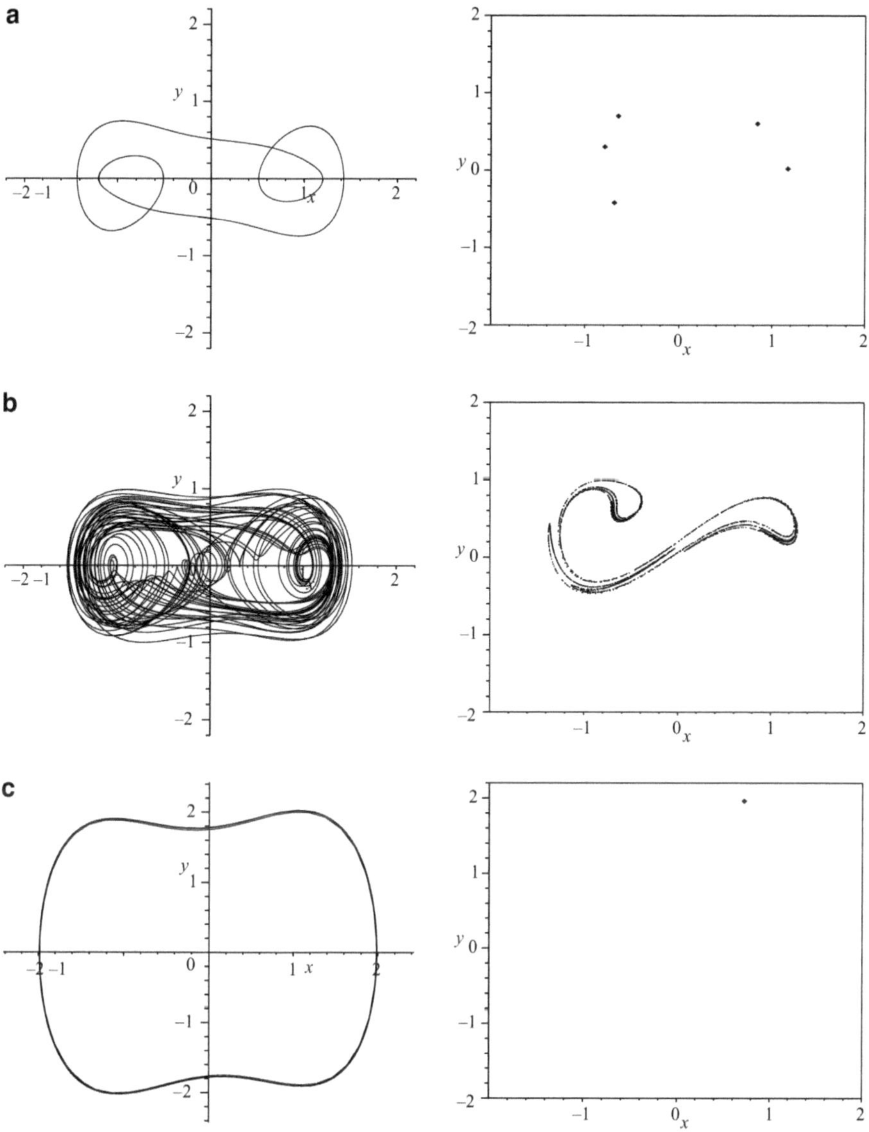

Fig. 1.8 A gallery of phase portraits and Poincaré maps for system (1.4) when $k = 0.3$ and $\omega = 1.25$: (**a**) $\Gamma = 0.37$ (a period-five subharmonic); (**b**) $\Gamma = 0.5$ (chaos), 4,000 points are plotted; (**c**) $\Gamma = 0.8$ (forced period one)

```
    b=x(2,2);
    rup(n)=sqrt((x(2,1))^2+(x(2,2))^2);
end
% Ramp the amplitude down.
for n=1:Max
    Gamma=interval-step*n;
    [t,x]=ode45('Duffing',[0:(2*pi/1.25):(4*pi/1.25)],[a,b]);
    a=x(2,1);
    b=x(2,2);
    rdown(n)=sqrt((x(2,1))^2+(x(2,2))^2);
end
% Plot the bifurcation diagram.
hold on
rr=step:step:interval;
plot(rr,rup)
plot(interval-rr,rdown)
hold off
fsize=15;
axis([0 .12 0 2])
xlabel('\Gamma','FontSize',fsize)
ylabel('r','FontSize',fsize)
title('Bifurcation Diagram of the Duffing System')
% End of Program.
```

Figure 1.9 shows a possible bifurcation diagram for system (1.4) for forcing amplitudes in the range $0 < \Gamma < 0.4$ near the critical point at N. The vertical axis labeled r represents the distance of the point in the Poincaré map from the origin ($r = \sqrt{x^2+y^2}$). The first iterative method was employed in this case. For each value of Γ, the last 10 of 50 iterates were plotted, and the step length used in this case was 0.01. The initial values were chosen close to one of the existing periodic solutions. The diagram shows period-one behavior for $0 < \Gamma < 0.28$, approximately. For values of $\Gamma > 0.28$, there is period-two behavior, and then the results become a little obscure.

Figure 1.10 shows a possible bifurcation diagram produced using the second iterative method. The parameter Γ is increased from zero to 0.4 and then decreased from $\Gamma = 0.4$ back to zero. There were 2000 iterates used as Γ was increased and then decreased. The solid curve lying approximately between $0 \leq \Gamma < 0.32$ represents steady-state behavior. As Γ increases beyond 0.32, the system goes through a chaotic regime and returns to periodic behavior before $\Gamma = 0.4$. As the parameter Γ is decreased, the system returns through the periodic paths, enters a chaotic region, and period undoubles back to the steady-state solution at $\Gamma \approx 0.28$. Note that on the ramp-up part of the iterative scheme, the steady state overshoots into the region where the system is of period two, roughly where $0.28 < \Gamma < 0.32$.

Figure 1.11 shows a bifurcation diagram produced as Γ is increased from zero to 0.45 and then decreased back to zero. Once more as Γ is increased, there is

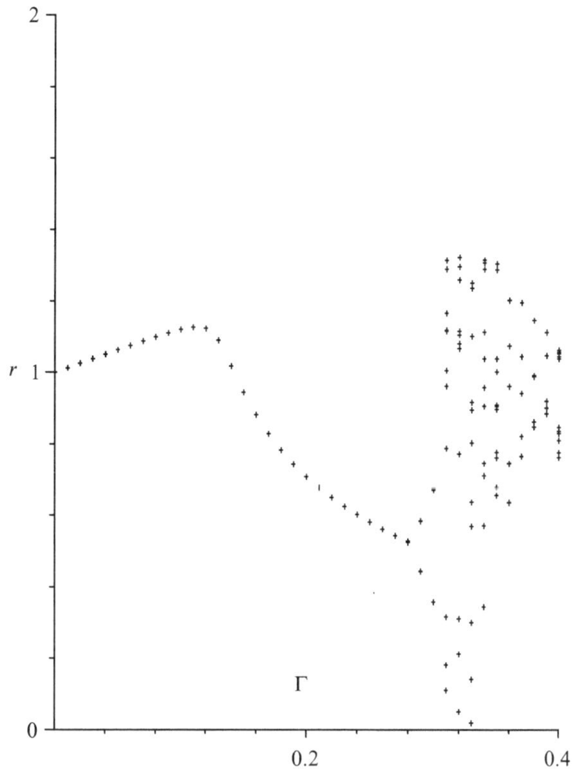

Fig. 1.9 A bifurcation diagram for system (1.4) produced using the first iterative method

steady-state behavior for Γ lying between zero and approximately 0.32. However, as the parameter is decreased a different steady state is produced and a large bistable region is present.

Note that there will also be steady-state behavior and bifurcations associated with the critical point at M.

1.4 Chua's Electronic Circuit

In the mid-1980's Chua modeled a circuit which was a simple oscillator exhibiting a variety of bifurcation and chaotic phenomena. Chua's electronic circuit is easy to construct in a physics laboratory which has made it a ubiquitous real-world example of a chaotic system. The circuit diagram is given in Fig. 1.12, and using Kirchhoff's laws, the circuit equations are given by

$$\frac{dv_1}{dt} = \frac{(G(v_2 - v_1) - f(v_1))}{C_1}, \frac{dv_2}{dt} = \frac{1(G(v_1 - v_2) + i)}{C_2}, \frac{di}{dt} = -\frac{v_2}{L}, \quad (1.7)$$

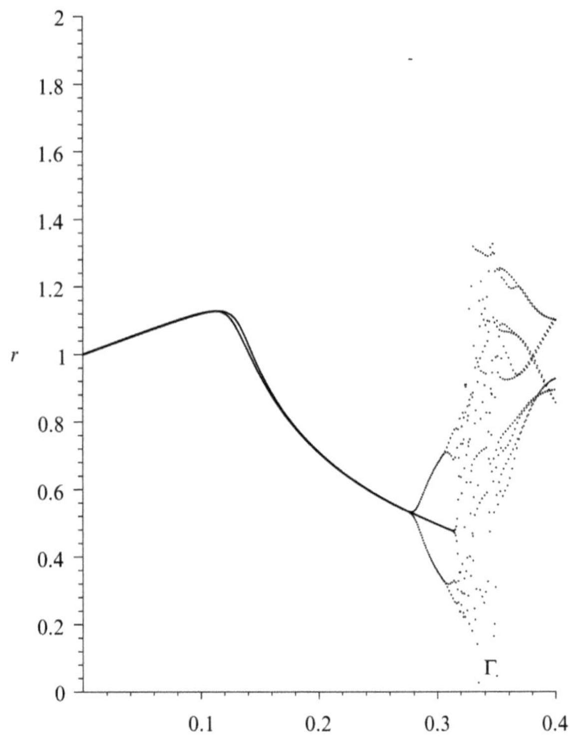

Fig. 1.10 A bifurcation diagram for system (1.4) produced using the second iterative method

where v_1, v_2, and i are the voltages across C_1, C_2 and the current through L, respectively. The characteristic of the nonlinear resistor N_R is given by

$$f(v_1) = G_b v_1 + 0.5(G_a - G_b)\left(|v_1 + B_p| - |v_1 - B_p|\right),$$

where $G = 1/R$. Typical parameters used are $C_1 = 10.1nF$, $C_2 = 101nF$, $L = 20.8mH$, $R = 1420\Omega, r = 63.8\Omega, G_a = -0.865mS, G_b = -0.519mS$, and $B_p = 1.85V$. In the simple case, Chua's equations can be written in the following dimensionless form:

$$\dot{x} = a(y - x - g(x)), \quad \dot{y} = x - y + z, \quad \dot{z} = -by, \tag{1.8}$$

where a and b are dimensionless parameters. The function $g(x)$ has the form

$$g(x) = cx + \frac{1}{2}(d - c)\left(|x + 1| - |x - 1|\right),$$

where c and d are constants.

To demonstrate that the output is chaotic one can use MATLAB to plot a time series plot for one of the variables against time.

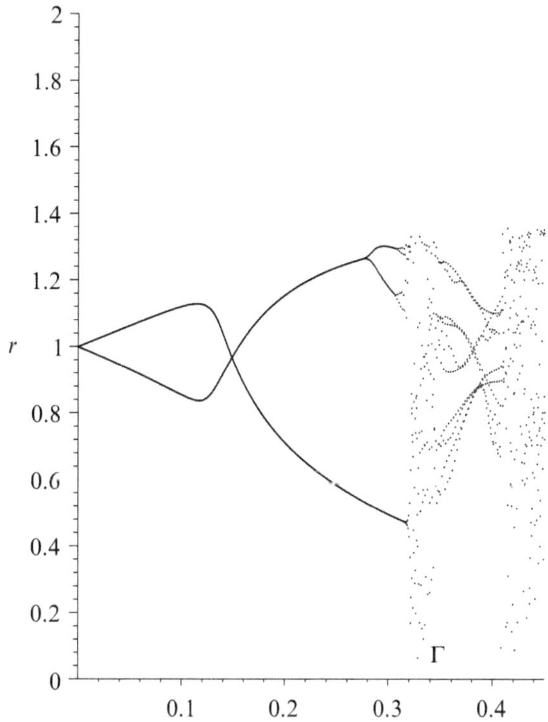

Fig. 1.11 A bifurcation diagram for system (1.4) produced using the second iterative method. There is a large bistable region

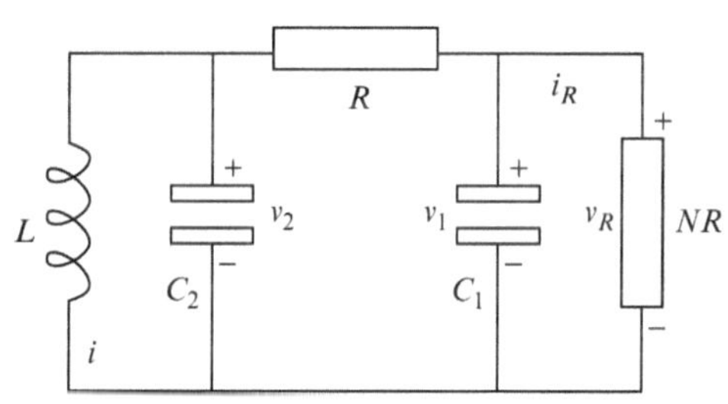

Fig. 1.12 Chua's electronic circuit

```
% Chua's circuit.
Chua=inline('[15*(x(2)-x(1)-(-(5/7)*x(1)+(1/2)*(-(8/7)-(-5/7))
      *(abs(x(1)+1)-abs(x(1)-1))));
x(1)-x(2)+x(3);-25.58*x(2)]','t','x');
options = odeset('RelTol',1e-4,'AbsTol',1e-4);
% Use the ode45 solver for 0<t<100 and initial conditions
    (-1.6,0,1.6).
[t,xb]=ode45(Chua,[0 100],[-1.6,0,1.6],options);
% Plot a time series, y against t. You could also plot x and z
  against t.
plot(t,xb(:,2),'k')
fsize=15;
title('Time series','FontSize',fsize)
```

```
set(gca,'xtick',[0:20:100],'FontSize',fsize)
set(gca,'ytick',[-0.6:0.2:0.6],'FontSize',fsize)
xlabel('time','FontSize',fsize)
ylabel('y','FontSize',fsize)
```

Chua's circuit (see Fig. 1.13) is investigated in some detail in [10] and exhibits many interesting phenomena including *period-doubling* cascades to chaos, *intermittency* routes to chaos, and *quasi-periodic* routes to chaos, for example. For certain parameter values, the solutions lie on a *double-scroll attractor*, as shown in Fig. 1.14. The MATLAB program for the 3D plot is listed below.

```
Chua=inline('[15*(x(2)-x(1)-(-(5/7)*x(1)+(1/2)*(-(8/7)-(-5/7))
      *(abs(x(1)+1)-abs(x(1)-1))));
x(1)-x(2)+x(3);-25.58*x(2)]','t','x');
options = odeset('RelTol',1e-4,'AbsTol',1e-4);
[t,xb]=ode45(Chua,[0 100],[-1.6,0,1.6],options);
% Plot a 3-D figure.
plot3(xb(:,1),xb(:,2),xb(:,3))
title('Chua's Double Scroll Attractor')
fsize=15;
xlabel('x(t)','Fontsize',fsize);
ylabel('y(t)','Fontsize',fsize);
zlabel('z(t)','FontSize',fsize);
```

Click on the **Rotate 3D** icon at the top of the figure plot to rotate the 3-dimensional image. The dynamics are more complicated than those appearing in either the Rössler or Lorenz attractors. Chua's circuit has proved to be a very suitable subject for study since laboratory experiments produce results which match very well with the results of the mathematical model. Recently, the author and Borresen [3] have shown the existence of a bistable cycle for Chua's electric circuit for the first time. Power spectra for Chua's circuit simulations are used to show how the qualitative nature of the solutions depends on the history of the system.

The Simulink toolbox can be used to construct a Chua's circuit simulation. In Fig. 1.15, the three blocks on the left represent the system of three differential equations, the outputs of which are linked to a Mux block which combines its inputs into a single vector output. The Integrator block (denoted by $\frac{1}{s}$) outputs the integral of its input at the current time step. The vector then feeds back to the system of differential equations and is also linked to the Demux block which extracts the components of an input signal and outputs the components as separate signals. The x and y outputs are then linked to the XY Graph block which shows how the solution evolves in time. Note that the three-dimensional attractor is plotted in two dimensions.

Two Chua's circuits can be connected to demonstrate synchronization of chaos which has been well documented in the literature (Fig. 1.16), see for example [2, 7]. Note that the initial conditions for both Chua systems are different (the top circuit has initial conditions $(15,20,30)$ and the lower circuit has initial conditions $(10,20,20)$). If the systems were not coupled they would be behaving in an unsynchronized fashion. By taking the x output from the first Chua's circuit and

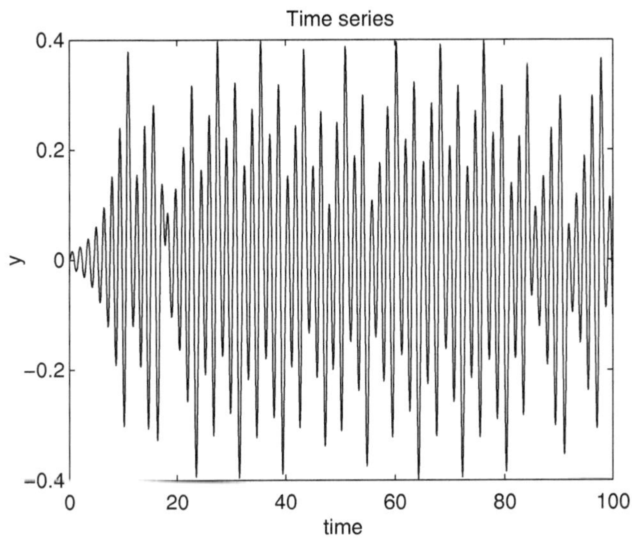

Fig. 1.13 Chua's electronic circuit: Chaotic time series plot for system (1.8) when $a = 15, b = 25.58, c = -5/7$, and $d = -8/7$. The initial conditions are $x(0) = -1.6$, $y(0) = 0$, and $z(0) = 1.6$

Fig. 1.14 Chua's double-scroll attractor: Phase portrait for system (1.8) when $a = 15, b = 25.58, c = -5/7$, and $d = -8/7$. The initial conditions are $x(0) = -1.6$, $y(0) = 0$, and $z(0) = 1.6$

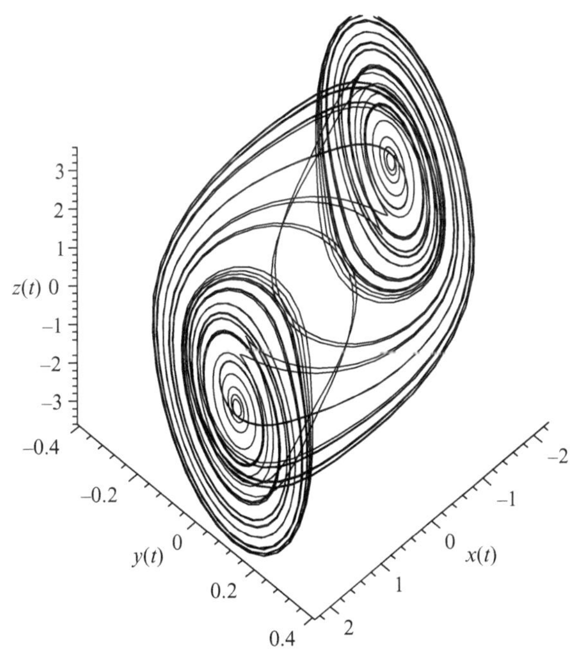

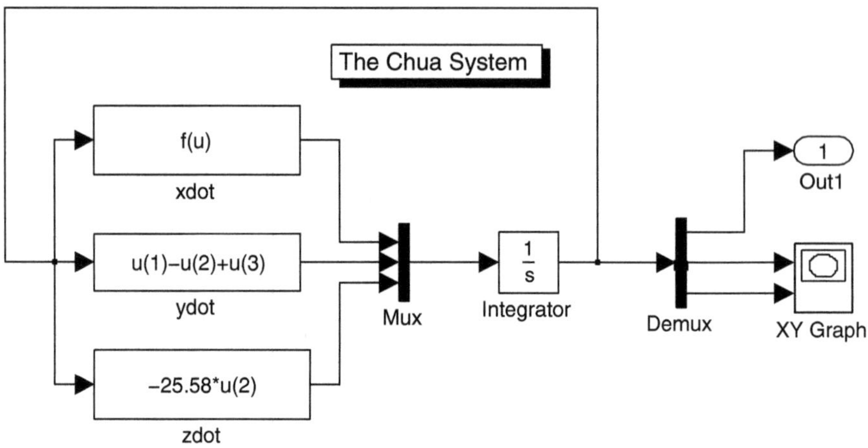

Fig. 1.15 Simulink model of Chua's electronic circuit

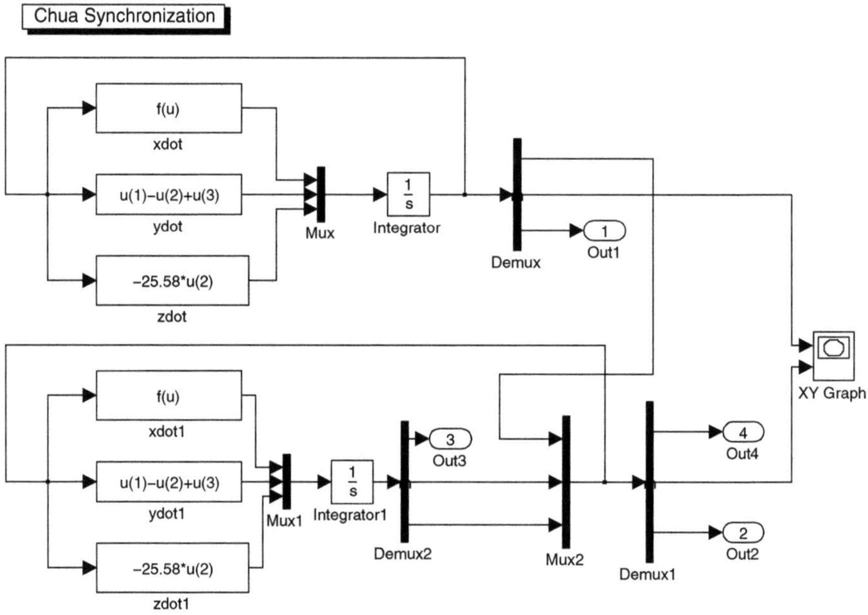

Fig. 1.16 Synchronization between two Chua's electronic circuits

feeding it in to the second Chua's circuit, the circuits can be made to synchronize. If you click on the XY Graph block you get the $y = x$ line indicating that the y outputs from both circuits are identical. Of course, the x and z outputs are also synchronized. For applications of chaos synchronization please see the references listed above.

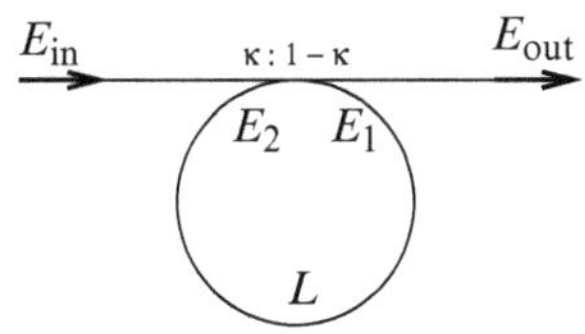

Fig. 1.17 The SFR resonator. The electric field entering the fiber ring is labelled E_1 and the electric field leaving the fiber ring is labelled E_2. The coupler splits the power intensity in the ratio $\kappa : 1-\kappa$

1.5 The Nonlinear Simple Fiber Ring Resonator

This section is concerned with a discrete nonlinear dynamical system. By considering the propagation of light waves through nonlinear optical fiber it is possible to construct a complex iterative map to model the dynamics of a simple fiber ring (SFR) resonator. Consider the all-optical fiber resonator as depicted in Fig. 1.17 and define the slowly varying complex electric fields as indicated.

Note that the power P and intensity I are related to the electric field in the following way:

$$P \propto I \propto |E|^2.$$

If the electric field crosses the coupler, then a phase shift is induced which is represented by a multiplication by j in the equations. Assume that there is no loss at the coupler. Then across the coupler the complex field amplitudes satisfy the following equations:

$$E_1 = \sqrt{\kappa}E_2 + j\sqrt{1-\kappa}E_{in} \tag{1.9}$$

and

$$E_{out} = \sqrt{\kappa}E_{in} + j\sqrt{1-\kappa}E_2, \tag{1.10}$$

where κ is the power-splitting ratio at the coupler. Consider the propagation from E_1 to E_2. Then

$$E_2 = E_1 e^{j\phi}, \tag{1.11}$$

where the total loss in the fiber is negligible (typically about 0.2 dB km^{-1}) and

$$\phi = \phi_L + \phi_{NL}.$$

The linear phase shift is ϕ_L, and the nonlinear phase shift due to propagation is given by

$$\phi_{NL} = \frac{2\pi r_2 L}{\lambda_0 A_{eff}} |E_1|^2,$$

where λ_0 is the wavelength of propagating light in a vacuum, A_{eff} is the effective core area of the fiber, L is the length of the fiber loop, and r_2 is the *nonlinear refractive index coefficient* of the fiber. It is well known that when the optical intensity is large enough, the constant r_2 satisfies the equation

$$r = r_0 + r_2 I = r_0 + \frac{r_2 r_0}{2\eta_0}|E_1|^2 = r_0 + r_2 \frac{P}{A_{\text{eff}}},$$

where r is the refractive index of the fiber, r_0 is the linear value, I is the instantaneous optical intensity, P is the power, and A_{eff} is the effective cross-sectional area of the fiber. If the nonlinearity of the fiber is represented by this equation, then the fiber is said to be of *Kerr type*. In most applications, it is assumed that the response time of the *Kerr effect* is much less than the time taken for light to circulate once in the loop.

Substitute (1.11) into (1.9) and (1.10). Simplify to obtain

$$E_1(t) = j\sqrt{1-\kappa}E_{\text{in}}(t) + \sqrt{\kappa}E_1(t-t_R)e^{j\phi(t-t_R)},$$

where $t_R = \frac{rL}{c}$ is the time taken for the light to complete one loop, r is the refractive index, and c is the velocity of light in a vacuum. Note that this is an iterative formula for the electric field amplitude inside the ring. Take time steps of length equal to t_R. This expression can be written more conveniently as an iterative equation of the form

$$E_{n+1} = A + BE_n \exp\left(j\left(\frac{2\pi r_2 L}{\lambda_0 A_{\text{eff}}}|E_n|^2 + \phi_L\right)\right), \tag{1.12}$$

where $A = j\sqrt{1-\kappa}E_{\text{in}}$, $B = \sqrt{\kappa}$, and E_i is the electric field amplitude at the ith circulation around the fiber loop. Typical fiber parameters chosen for this system are $\lambda_0 = 1.55 \times 10^{-6}$ m, $r_2 = 3.2 \times 10^{-20}\,\text{m}^2W^{-1}$, $A_{\text{eff}} = 30\,\mu\text{m}^2$, and $L = 80$ m.

Equation (1.12) may be scaled without loss of generality to the simplified equation

$$E_{n+1} = A + BE_n \exp\left[j(|E_n|^2 + \phi_L)\right], \tag{1.13}$$

which is sometimes referred to as the *Ikeda map*.

Split (1.13) into its real and imaginary parts by setting $E_n = x_n + jy_n$, and set $\phi_L = 0$. The equivalent real two-dimensional system is given by

$$\begin{aligned} x_{n+1} &= A + B\left(x_n \cos|E_n|^2 - y_n \sin|E_n|^2\right) \\ y_{n+1} &= B\left(x_n \sin|E_n|^2 + y_n \cos|E_n|^2\right), \end{aligned} \tag{1.14}$$

where $|B| < 1$. The Ikeda map can have fixed points of all periods. In this particular case, system (1.14) can have many fixed points of period one depending on the parameter values A and B.

Example 1.4. Determine and classify the fixed points of period one for system (1.14) when $B = 0.15$ and

(i) $A = 1$;
(ii) $A = 2.2$.

Solution. MATLAB is used to solve the equations and to plot the implicit curves, however, you have to be very careful here. The fixed points of period one ($x_{n+1} = x_n = x$, $y_{n+1} = y_n = y$) satisfy the simultaneous equations

$$x = A + Bx\cos(x^2+y^2) - By\sin(x^2+y^2) \quad \text{and}$$
$$y = Bx\sin(x^2+y^2) + By\cos(x^2+y^2).$$

```
>> syms x y.
>> [x,y]=solve('1+0.15*(x*cos(x^2+y^2)-y*sin(x^2+y^2))-x',
   '0.15*(x*sin(x^2+y^2)+y*cos(x^2+y^2))-y')
```

The MATLAB commands above give one solution when $A = 1, B = 0.15$, namely, approximately $(1.048, 0.151)$, which in this case is the only solution. However, if you solve the same set of equations with $A = 2.2, B = 0.15$, then MATLAB again gives one solution, but this is incorrect. To see that there is more than one solution plot the implicit curves defined by (1.14).

```
hold on
axis([0 4 -4 4])
ezplot('2.2+0.15*(x*cos(x^2+y^2)-y*sin(x^2+y^2))-x')
ezplot('0.15*(x*sin(x^2+y^2)+y*cos(x^2+y^2))-y')
hold off
```

(i) When $A = 1$ and $B = 0.15$, there is one solution at $x_{1,1} \approx 1.048$, $y_{1,1} \approx 0.151$. The solution is given graphically in Fig. 1.16a. To classify the critical point $P^* = (x_{1,1}, y_{1,1})$, consider the Jacobian matrix

$$J(P^*) = \left.\begin{pmatrix} \frac{\partial P}{\partial x} & \frac{\partial P}{\partial y} \\ \frac{\partial Q}{\partial x} & \frac{\partial Q}{\partial y} \end{pmatrix}\right|_{P^*}.$$

The eigenvalues of the Jacobian matrix at P^* are easily computed with MATLAB, $\lambda_1 \approx -0.086 + 0.123j$ and $\lambda_2 \approx -0.086 - 0.123j$. Therefore, as the modulus of both eigenvalues is less than one, P^* is a stable fixed point of period one (see Fig. 1.18).

(ii) When $A = 2.2$ and $B = 0.15$, there are three points of period one, as the graphs in Fig. 1.16b indicate. Note that the solve command does not give all three solution here. The fixed points occur approximately at the points $U = (2.562, 0.131)$, $M = (2.134, -0.317)$ and $L = (1.968, -0.185)$. Using the Jacobian matrix, the eigenvalues for U are $\lambda_{1,2} = -0.145 \pm 0.039j$; the eigenvalues for M are $\lambda_1 = 1.360$, $\lambda_2 = 0.017$; and the eigenvalues for L are $\lambda_1 = 0.555$, $\lambda_2 = 0.041$.

Therefore, U and L are stable fixed points of period one, while M is an unstable fixed point of period one. These three points are located within a bistable region of the bifurcation diagram given later in this chapter. The point U lies on the upper

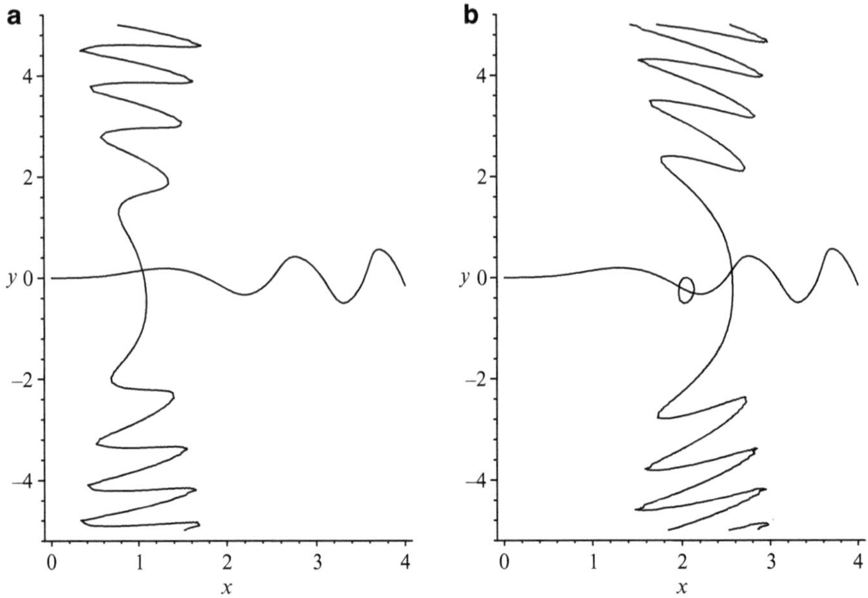

Fig. 1.18 The fixed points of period one are determined by the intersections of the two curves, $x = A + 0.15x\ \cos(x^2 + y^2) - 0.15y\ \sin(x^2 + y^2)$ and $y = 0.15x\ \sin(x^2 + y^2) + 0.15y\ \cos(x^2 + y^2)$; (**a**) $A = 1$ and (**b**) $A = 2.2$. Use the data cursor to locate the fixed points in the figure

Table 1.2 The first two regions of bistability and instability computed for the SFR resonator to three decimal places using a linear stability analysis

B	First bistable region Power/Wm^{-2}	First unstable region Power/Wm^{-2}	Second bistable region Power/Wm^{-2}	Second unstable region Power/Wm^{-2}
0.05	10.970–11.038	12.683–16.272	16.785–17.704	17.878–23.561
0.15	4.389–4.915	5.436–12.007	9.009–12.765	9.554–20.510
0.3	3.046–5.951	1.987–4.704	6.142–16.175	3.633–15.758
0.6	1.004–8.798	1.523–7.930	2.010–24.412	1.461–24.090
0.9	0.063–12.348	1.759–11.335	0.126–34.401	0.603–34.021

branch of the hysteresis loop and the point L lies on the lower branch. Since M is unstable it does not appear in the bifurcation diagram but is located between U and L.

To investigate the stability of the nonlinear SFR resonator, a linear stability analysis will be applied. A first-order perturbative scheme is used to predict the values of a parameter where the stationary solutions become unstable. Briefly, a small perturbation is added to a stable solution and a Taylor series expansion is carried out, the nonlinear terms are ignored, and a linear stability analysis is applied, see [12] for a detailed description. Table 1.2 lists the first two regions of bistability and instability for the SFR for a range of fixed values of the parameter B computed using a linear stability analysis.

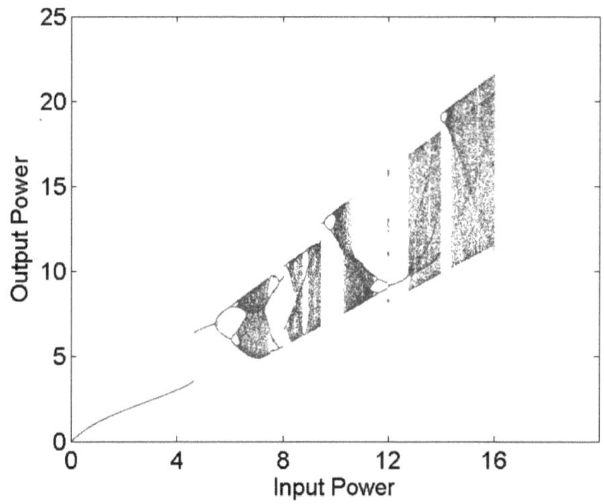

Fig. 1.19 Bifurcation diagram for the SFR when $B = 0.15$ using the first iterative method without feedback

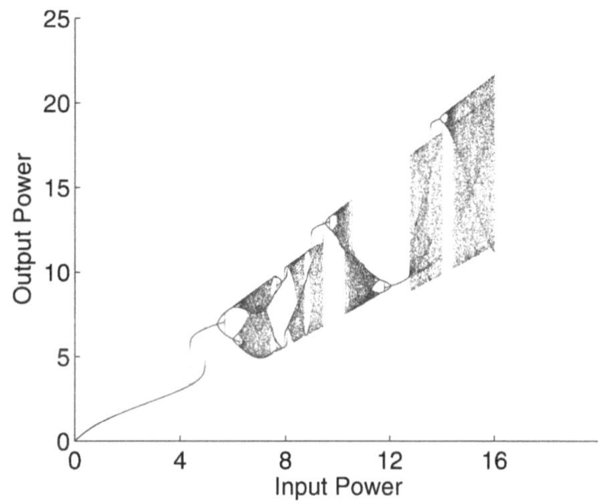

Fig. 1.20 Bifurcation diagram for the SFR when $B = 0.15$ using the second iterative method with feedback. A hysteresis cycle is clearly visible

To end the chapter, bifurcation diagrams will now be plotted for the SFR using the first and second iterative methods given in the second section.

The following programs are used to plot Figs. 1.19 and 1.20 when $B = 0.15$. Figure 1.19 shows the bifurcation diagram using the first iterative method without feedback, note that there is no bistable region in this figure. Figure 1.20 displays the bifurcation diagram with a feedback mechanism incorporated. The first bistable region is clearly visible and makes interesting comparison with the results of Table 1.2. Note that the second bistable region has not materialised due to instabilities. Edit the program below to plot your own bifurcation diagrams for varying values of B and compare the plot with the results in Table 1.2. An animated bifurcation diagram can be downloaded from the MathWorks Central File Exchange.

```
% The first iterative method.
clear
itermax=100;B=0.15;
% Plot the last 30 iterates for each value of Input.
finalits=30;finits=itermax-(finalits-1);
for Input=0:0.01:16
    E=0.5; % Initial value for all Input.
    E0=E;
    for n=2:itermax
        En=sqrt(Input)+B*E0*exp(1i*abs(E0)^2);
        E=[E En];
        E0=En;
    end
    Esqr=abs(E).^2; % The output power.
    p=plot(Input*ones(finalits),
           Esqr(finits:itermax),'.','MarkerSize',1);
    set(p,'Color','black')
    hold on
end
fsize=15;
set(gca,'xtick',[0:4:16],'FontSize',fsize)
set(gca,'ytick',[0:5:25],'FontSize',fsize)
xlabel('Input','FontSize',fsize)
ylabel('Output','FontSize',fsize)
hold off
% End of Program.
```

```
% The second iterative method.
clear
format long;
halfN=19999;N=2*halfN+1;N1=1+halfN;
E(1)=0.4;B=0.15;Pmax=16;

% Ramp the power up
for n=1:halfN
    E(n+1)=sqrt(n*Pmax/N1)+B*E(n)*exp(1i*abs(E(n))^2);
    Esqr(n+1)=abs(E(n+1))^2;
end

% Ramp the power down
for n=N1:N
    E(n+1)=sqrt(2*Pmax-n*Pmax/N1)+B*E(n)*exp(1i*abs(E(n))^2);
    Esqr(n+1)=abs(E(n+1))^2;
end

for n=1:halfN
    Esqr1(n)=Esqr(N+1-n);
    ptsup(n)=n*Pmax/N1;
end

% Plot the bifurcation diagrams
```

```
fsize=15;
hold on
set(gca,'xtick',[0:4:16],'FontSize',fsize)
set(gca,'ytick',[0:5:25],'FontSize',fsize)
plot(ptsup(1:halfN),Esqr(1:halfN),'.','MarkerSize',1);
plot(ptsup(1:halfN),Esqr1(1:halfN),'.','MarkerSize',1);
xlabel('Input Power','FontSize',fsize);
ylabel('Output Power','FontSize',fsize);
hold off
% End of Program.
```

References

1. Attaway, S.: MATLAB: A Practical Introduction to Programming and Problem Solving, Butterworth-Heinemann (2009)
2. Balanov, A., Janson, N., Postnov, D., et al.: Synchronization: From Simple to Complex. Springer, New York (2008)
3. Borreson, J., Lynch, S.: Further investigation of hysteresis in Chua's circuit. Int. J. Bifurc. Chaos **12**(1), 129–134 (2002)
4. Demirkaya, O., Asyali, M.H., Sahoo, P.K.: Image Processing with MATLAB: Applications in Medicine and Biology. CRC Press, (2008)
5. Duffy, D.G.: Advanced Engineering Maths with MATLAB, 3rd edn. CRC Press (2010)
6. Elnashaie, S.S.E.H., Uhlig, F.: Numerical Techniques for Chemical and Biological Engineers Using MATLAB: A Simple Bifurcation Approach. Springer (2006)
7. Femat, R., Solis-Perales, G.: Robust Synchronization of Chaotic Systems via Feedback. Springer, New York (2008)
8. Field, R.J., Gyorgyi, L.: Chaos in Chemistry and Biochemistry. World Scientific, Singapore (1993)
9. Gilat, A.: MATLAB: An Introduction with Applications, 4th edn. Wiley, New York (2010)
10. Kilic, R.: A Practical Guide to Studying Chua's Circuits. World Scientific, Singapore (2010)
11. Kuznetsov, Y.: Elements of Applied Bifurcation Theory, 3rd edn. Springer (2010)
12. Lynch, S.: Dynamical Systems with Applications using MATLAB. Springer, Birkhäuser (2004)
13. Moore, H.: MATLAB for Engineers, 2nd edn. Prentice Hall (2008)
14. Pokorny, P., Schreiber, I., Marek, M.: On the route to strangeness without chaos in the quasiperiodically forced van der Pol oscillator. Chaos, Solitons, Fractals **7**, 409–424 (1996)
15. Pratab, R.: MATLAB: Getting Started with MATLAB: A Quick Introduction for Scientists and Engineers. Oxford University Press, Oxford (2009)
16. Scott, S.K.: Oscillations, Waves, and Chaos in Chemical Kinetics. Oxford University Press, Oxford (1994)
17. Siciliano, A.: Matlab: Data Analysis and Visualization. World Scientific, Singapore (2008)
18. Chien, Y.-S., Liou, C.-T.: Steady-state multiplicity for autocatalytic reactions in a nonideal mixing of CSTR with two unpremixed feeds. Chem. Eng. Sci. **50**(22), 3645–3650 (1995)

Part II
Chaos and Nonlinear Dynamics in Electrical Engineering

Chapter 2
Modeling and Compensation of Flicker in Electrical Networks using Chaos Theory and SVC Systems

Mario Fabiano Alves and Zelia Myriam Assis Peixoto

2.1 Introduction

The number of electric arc furnaces installations in the metallurgic industry has greatly increased in the last decades. Some reasons for this are the necessity of recycling, the profusion of scrap iron and the possibility of producing metallic leagues with high level of quality at relatively low cost.

Arc furnaces are used for melting and refining metals. However, due to its highly nonlinear and dynamic time-varying characteristic, this kind of load has great impact on power quality. This is particularly true in relation to harmonic distortion and flicker generation.

Arc furnace modeling has been extensively investigated. The arc nonlinear, time-varying and unpredictable dynamic behavior represents a great challenge for the researchers [1–3].

In contrast with linear components that usually are represented in the frequency domain, the models for nonlinear elements are, usually, developed in the time domain. Some models are based on the stochastic characteristic [4] presented by the arc and, more recently, on the arc's chaotic nature [1, 3, 5]. Both type of models are capable of a good representation of the process. The stochastic idea is based on the non-periodic and nonlinear arc characteristics using a white noise, which modulates the fundamental component around 0.5%, within a 5 to 10 Hz frequency bandwidth [1, 6, 7].

The sensitive linear interdependence of the system variables in an electric arc furnace is characteristic of a process having a chaotic nature. According to King

M.F. Alves (✉) · Z.M.A. Peixoto
Graduate Program in Electrical Engineering, Pontifical Catholic University of Minas Gerais, Av. Dom Jose Gaspar, 500, Coracao Eucaristico, 30535-901, Belo Horizonte, Minas Gerais, Brazil
e-mail: mfabiano@pucminas.br; assiszmp@pucminas.br

S. Banerjee et al. (eds.), *Applications of Chaos and Nonlinear Dynamics in Engineering – Vol. 1*, Understanding Complex Systems, DOI 10.1007/978-3-642-21922-1_2,

et al. [3], after it has been proved that the arc voltage fluctuations have a chaotic behavior; it becomes natural to use chaos theory for modeling the arc furnace [3].

In this class of models the arc voltage can be modeled by the differential equation governing the voltage-current dynamics of the electric arc furnace, with the arc length as a parameter and the arc current as the forcing function, as proposed by Acha et al. [2]. The model is completed by the introduction of a modulating low frequency chaotic signal generated by Chua's circuit or by Lorenz equations [1–3, 8, 9].

Some researches have compared the performance of the chaotic models with the results obtained from stochastic models and demonstrated that both approaches yield good results when compared to measurements from real arc furnaces [1, 10, 11].

The increasing necessity in dealing with the power systems disturbances, particularly in the last two decades, moved the industry into the development of new power electronics devices and topologies [12, 13]. Some of these were first developed even early, in the seventies. Among them, the Thyristor Controlled Reactor with Fixed Capacitor Bank (TCR-FC) is a classical topology of the Static Var Compensator (SVC) class. The ability of SVC in flicker compensation is limited by delays in reactive power measurements and thyristor ignition, but recent research on the subject indicates that improvements are possible [14, 15]. Thus, in spite of its relatively low efficiency for flicker mitigation, the good cost-benefit relation of the SVC still makes it an attractive solution [4, 15–17].

This chapter presents an integrated solution to predict and simulate the impact of arc furnaces on the electric power system. The arc model uses Chua's oscillator to generate a chaotic signal, which is used to modulate the arc voltage obtained from the differential equations for the arc furnace dynamic deterministic model, thus generating the voltage fluctuations that causes the flicker [5].

A model of the SVC, including a control strategy for flicker mitigation, and an IEC flickermeter model, are also presented. The models are totally implemented in the ATP-Alternative Transients Program environment.

A case study is used to demonstrate the models possibilities, including a methodology to properly adjust the model according to the power system it is connected to, and to the severity level expected for the furnace.

2.2 The Electric Arc Model

The arc presents distinct stages during a typical operation cycle of the furnace, being the melting and refining stages the most relevant when dealing with flicker phenomenon. In the melting stage, when fusion begins, the scrap iron presents a very irregular surface, causing great current fluctuations. In this stage, the arc furnace demands very high active power from the supply system. In the refining stage, a longer and more constant arc and the current fluctuations are lower, when compared to the melting stage.

The electric arc model used in this work was first published in [1] and it is here further detailed in order to demonstrate its integration in the flicker compensated model. The model is developed in three parts.The differential and algebraic equations that represent the dynamic and multi-valued v-i characteristics of the electric arc are solved, yielding a deterministic solution for the problem. A low frequency chaotic signal is used to model the voltage fluctuations imposed by the arc. This chaotic signal is then used to modulate the voltage obtained from the deterministic solution and the result is the multi-valued chaotic voltage-current (v-i) characteristics of the electric arc furnace.

2.2.1 The Arc Furnace v-i Deterministic Characteristic

The electric arc power balance is given by [2],

$$p_1 + p_2 = p_3 \tag{2.1}$$

where p_1, p_2 and p_3 represent the thermical energy transmitted to the external environment, the internal energy of the electric arc and the total energy delivered by the arc and converted to the thermical energy.

The deterministic voltage-current characteristic of the arc furnace can be obtained from a differential equation that represents the dynamic behavior of the arc, based on (2.1) and expressed as [2],

$$k_1 r^n + k_2 r \frac{dr}{dt} = \frac{k_3}{r^{m+2}} i^2 \tag{2.2}$$

where r is the arc radius, and k_1, k_2 and k_3 are constants. The arc voltage v is given by,

$$v = \frac{i}{g} \tag{2.3}$$

where g is defined as the arc conductance and it is given by the following equation:

$$g = \frac{r^{(m+2)}}{k_3} \tag{2.4}$$

Different v-i characteristics may be selected by properly choosing the parameters m (variations of the resistivity with temperature) and n (conditions of cooling). The complete combination of these parameters for the different arc stages can be found in [2]. In this work, these parameters are set as $m = 0$ and $n = 2$, for the melting stage.

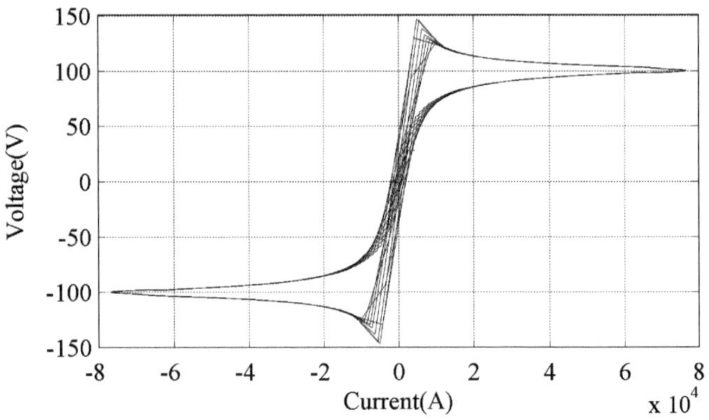

Fig. 2.1 Dynamic *v*-*i* characteristic of electric arc

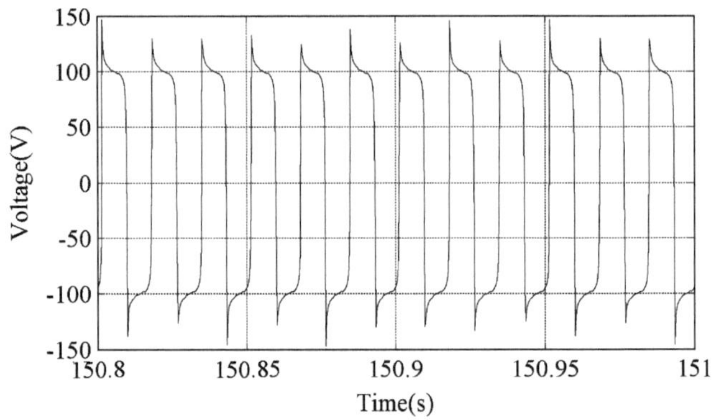

Fig. 2.2 Voltage waveform of electric arc: deterministic solution

Figure 2.1 shows the arc *v*-*i* characteristic resulting by solving (2.2) and (2.3) while Fig. 2.2 shows the arc voltage waveform. Further information on the furnace and power system considered is given in the Appendix.

2.2.2 Arc Furnace's Chaotic Response

As demonstrated in [3], the electric fluctuations in the arc furnace's voltage have a chaotic nature, and chaos theory may be used to model it [1]. In this work, Chua's oscillator circuit, shown in Fig. 2.3, is used to simulate the chaotic characteristic presented by the furnace's voltage fluctuations. The standard Chua's circuit is composed of network passive elements connected to a nonlinear active component called Chua's diode. It is simulated using (2.5) and (2.6) [8, 9]. Chua's circuit is an

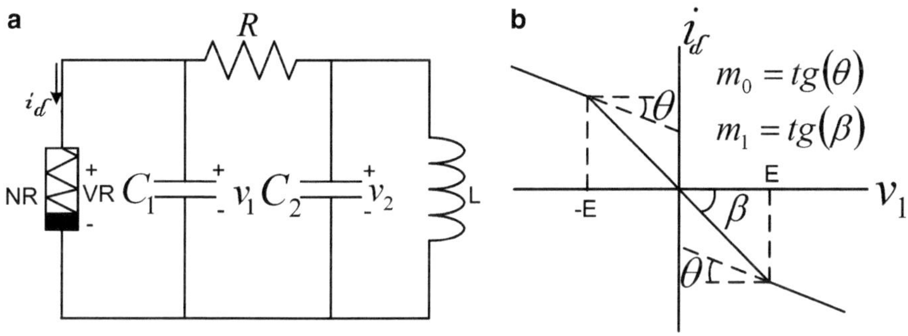

Fig. 2.3 (**a**) Chua's circuit; (**b**) Voltage – current characteristic

oscillator. Starting from some initial conditions ($v_1 = 0$, for instance), the system will proceed for each time step according to (2.5) and (2.6).

$$\begin{aligned} C_1\frac{dv_1}{dt} &= \frac{v_2 - v_1}{R} - i_d(v_1) \\ C_2\frac{dv_2}{dt} &= \frac{v_1 - v_2}{R} - i_L \\ L\frac{di_L}{dt} &= -v_2 \end{aligned} \tag{2.5}$$

Chua's circuit exhibits a broadband spectrum but its major power density is concentrated around the resonant circuit formed by L and C_1. Different circuit dynamics, such as steady fixed points and period of the oscillations, are obtained by varying parameters L, R, C_1 and C_2. More details can be found in [8,9].

L, R, C_1 and C_2 are passive linear elements, and the static linear piecewise curve of Chua's diode is given by (2.6).

$$\begin{aligned} i_d(v_1) &= m_0 v_1 + E(m_0 - m_1), \quad \text{if } v_1 < -E \\ &= m_1 v_1, \quad \text{if } |v_1| \le E \\ &= m_0 v_1 + E(m_1 - m_0), \quad \text{if } v_1 > +E \end{aligned} \tag{2.6}$$

Figure 2.4 presents a block diagram of the electric arc model. The current supplied to the arc is injected as the input to the arc model. The electric arc voltage is obtained from (2.2) to (2.4). The voltage obtained from these equations is modulated by the chaotic signal, producing the final arc furnace voltage output. The model behaves as a controlled source where, at each time step, the current of the arc is the input and the arc voltage is the output.

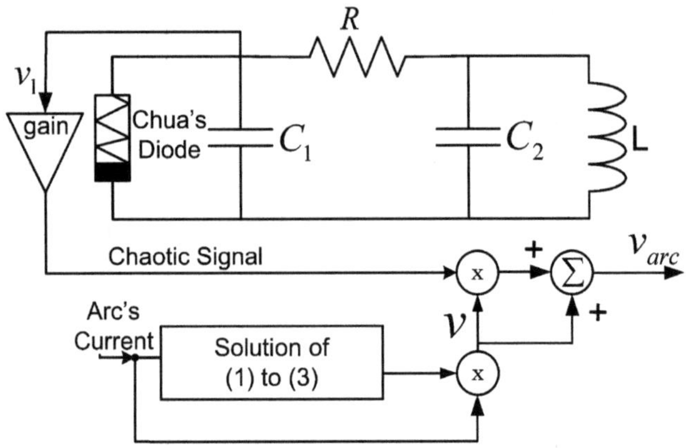

Fig. 2.4 Electric arc model

2.3 Simulation of Arc Furnace Model Under MatLab Environment

Chua's circuit is responsible for the generation of the low frequency, chaotic, modulating signal, that, together with the deterministic voltage defined by (3) and (4), produce the electric arc voltage that causes the flicker [5, 18]. Figure 2.5 shows the Matlab-Simulink block diagram implementation of Chua's circuit.

As written in Sect. 9.2.2, different circuit's dynamics are obtained by varying Chua's circuit parameters. For instance, taking $C_1 = 200\mu$F, $C_2 = 0.2\mu$F, $L =$ 3.6 mH and $G = 5.442e^{-4}$ mho the resulting chaotic signal, given by the voltage across capacitor C_1, presents the spiral attractor shown in Fig. 2.6. The attractor was obtained from 1,722 voltage signals present versus delayed by four sampling intervals observations. As a further investigation, the Lyapunov exponents were calculated for the voltage signal, resulting, for the largest positive exponent, $\lambda =$ 0.04 confirming that the system has a strange or chaotic behavior An attractor for dissipative systems with one or more positive Lyapunov exponents is said to be strange or chaotic [19].

Figure 2.7 shows the MatLab-Simulink block diagram implementation of the electric arc model, for the fusion state of the furnace ($m = 0$ e $n = 2$). The "Gain 4" block in the output of the signal of Chua's circuit is used to adjust the modulation intensity, and, as a consequence, the flicker level. To adjust the frequency of the signal (the flicker dominant frequency), the L and C parameters of Chua's circuit have to be adjusted.

The modulated arc voltage signal is given by (2.7),

$$V_{arc} = v(1 + K_a m(t)) \tag{2.7}$$

where V_{arc} is defined by (2.2), $m(t)$ is the modulating signal delivered by Chua's circuit and K_a is the gain factor referred to above [20].

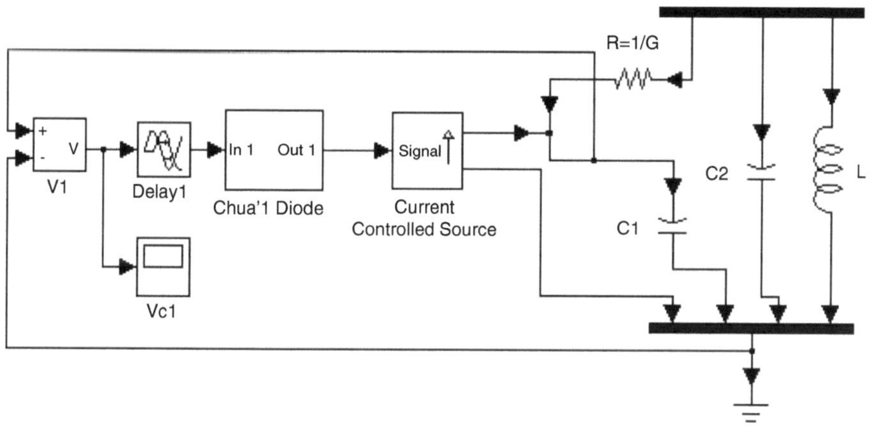

Fig. 2.5 Simulation diagram of Chua's circuit

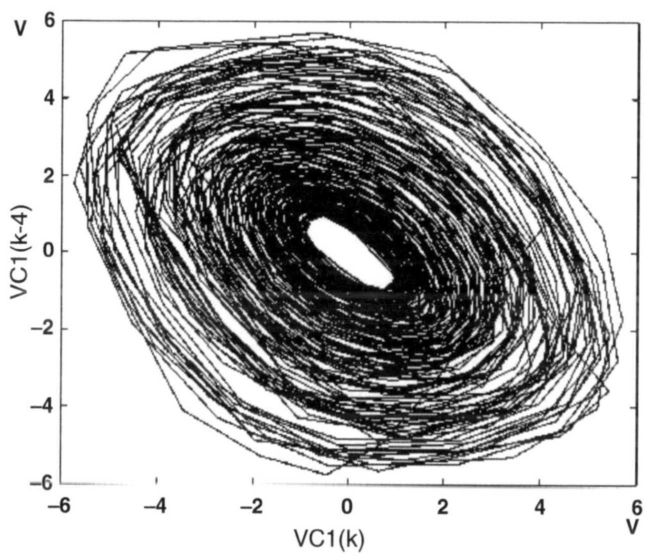

Fig. 2.6 Spiral attractor

The model works as a controlled voltage source, with the electric arc current, measured at the high current bus in the secondary of the furnace's transformer, used as input for the model, as shown in Fig. 2.8.

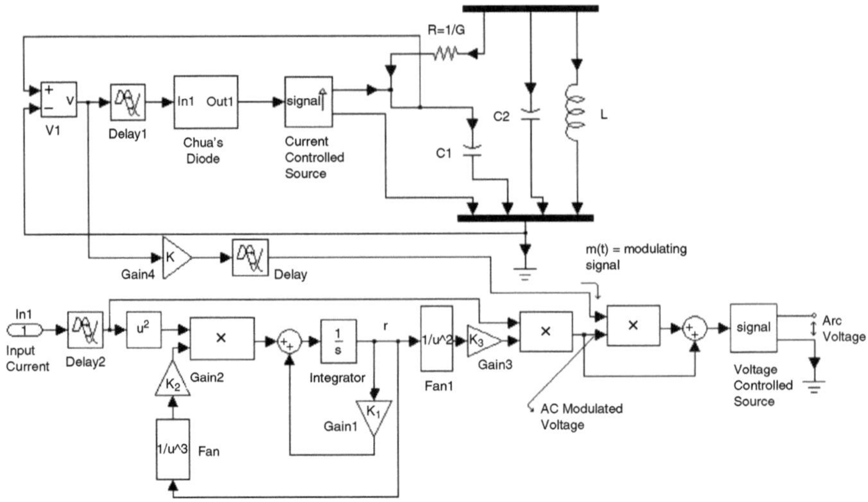

Fig. 2.7 Arc furnace block diagram under Simulink/MatLab environment

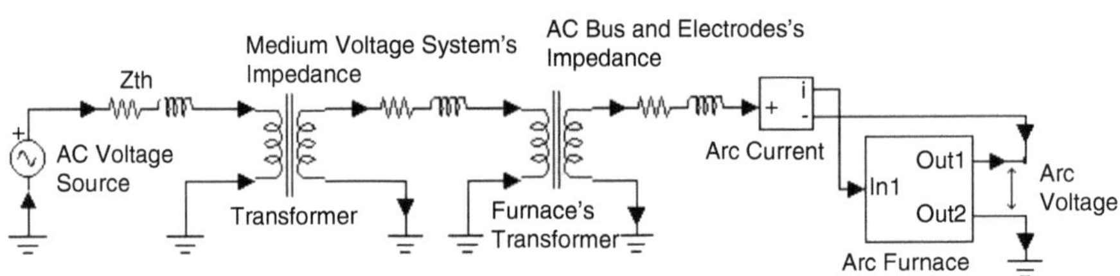

Fig. 2.8 Connection of the arc furnace to electrical network

2.4 Adjustment of the Arc Furnace Model

It is possible to get an estimation of the flicker level caused by a new installation from statistical analysis of the flicker caused by a large number of arc furnaces already in operation. The short term flicker severity, P_{st}, can be estimated by the following expression [21]:

$$P_{st}(99\%) = \frac{k_{st} X_s}{X_s + X_{furnace}} \tag{2.8}$$

$$P_{st}(95\%) = \frac{1}{1.25} P_{st}(99\%) \tag{2.9}$$

k_{st} = Furnace severity factor ranging form 48 to 85;
X_s = Reactance of the supply system;
$X_{furnace}$ = Reactance of the furnace, from the PCC (Point of Common Coupling – see Fig. 2.14) to the electric arc.

The parameters of the arc model can be adjusted to the desired operation conditions, so as to correctly simulate the furnace operation stages (melting or refining stage) and the expected furnace's degree of flicker severity, for each particular arc furnace and power system. Figure 2.9 presents a flow chart showing the sequence of operations needed to adjust the arc furnace model [5, 22].

Estimation of the Pst is achieved by the flickermeter model defined by IEC 61000-4-15 [23]. The IEC flickermeter has five basic blocks as it is shown in Fig. 2.10 [23, 24]. Block 1 is used to adapt the measured voltage to the internal reference level of the flickermeter. The purpose of Block 2 is to square the input signal. Block 3 has three filters in order to emulate the transfer function bulb-to-eye. In Block 4 the signal is squared then filtered to simulate the brain memory. Block's 4 output represents the instantaneous flicker level. Finally, Block 5 performs a statistical evaluation of the instantaneous flicker perception, thus resulting in the short-term perception of the human eye – Pst. Before being used, the flickermeter model was verified by applying the test procedures established in the IEC standard [23]. The weighting filter in Block 3 of the flickermeter, which represents the eye-brain system reaction to flicker effect, was matched to a 60 W, 120 V incandescent light source as the most extensively used in Brazil [24].

2.5 The SVC Model

The variable part of the TCR-FC configuration is made with air core reactors, controlled by thyristors valves, as shown in Fig. 3.11. Fixed capacitor banks are connected in parallel with the TCR, part of which is used for power factor correction and part used to supply the instantaneous reactive power needed to compensate the voltage fluctuations imposed by the electric arc dynamics. The reactive power provided by the capacitor bank, or part of it, constitutes filters, used to reduce to acceptable levels the harmonics in the system.

The TCR-FC acts as a continuously variable equivalent reactance controlled by the firing angle ϕ defined by the control system.

The relation between the reactor's susceptance and the firing angle ϕ, normalized in relation to the reactance of the reactor, is given by [25]:

$$B(\phi)_{pu} = \frac{\sin(2\phi) + 2(\pi - \phi)}{\pi} \tag{2.10}$$

The values for the firing angles ϕ, for each phase, are calculated from (2.10) using the corresponding values for the susceptances B(ϕ) defined by the control system. For each phase, the second thyristor is switched on by symmetry condition. A sawtooth waveform voltage is used to identify the half cycles of the voltage fundamental wave. The intersection of this wave with the firing angle curve defines the switch control signal, as shown in Fig. 3.12.

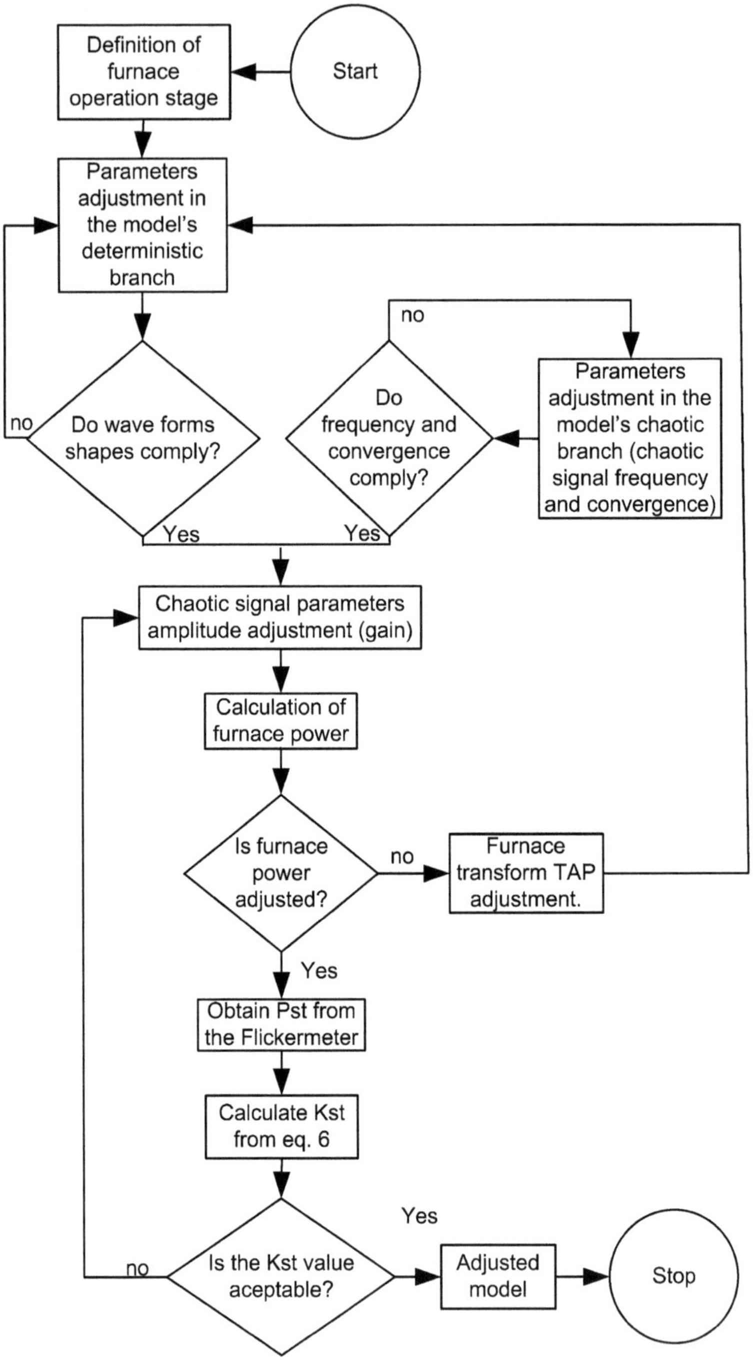

Fig. 2.9 Flow chart for the adjustment of the arc furnace model

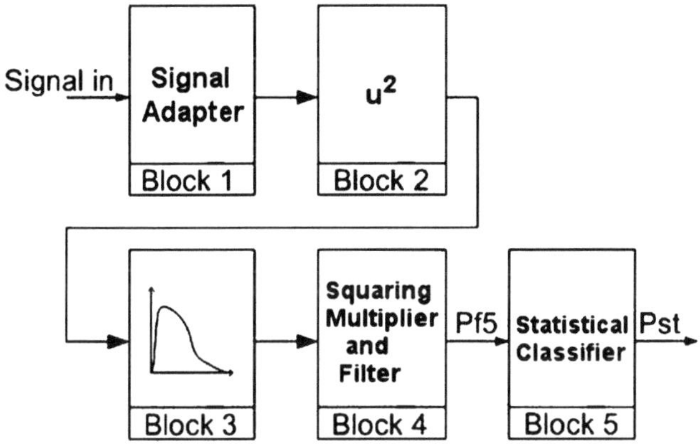

Fig. 2.10 The IEC flickermeter block diagram

Fig. 2.11 One-line diagram for the TCR-FC

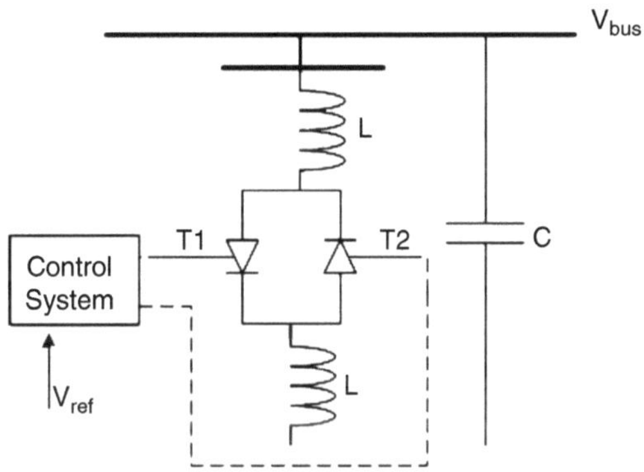

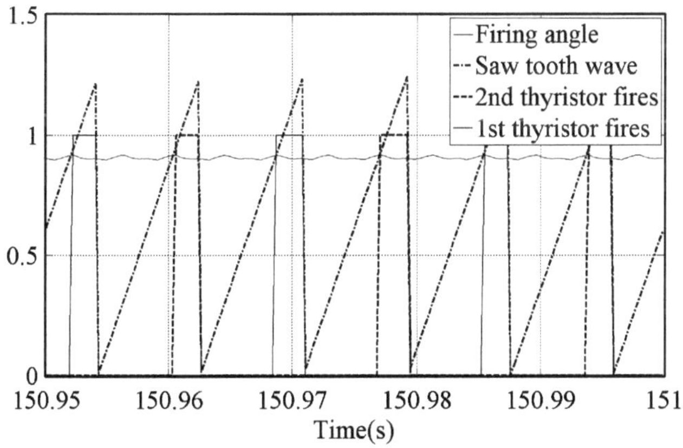

Fig. 2.12 Pulse-width modulator: comparator signals

2.6 The Control System

The flicker mitigation strategy is based on the compensation of all negative sequence currents and the imaginary part of the positive sequence currents drained by the arc furnace, as expressed by (2.11) and (2.12) [25, 26]:

$$Im[\overline{I}_{Eaf,pos}] + Im[\overline{I}_{SVC,pos}] = 0 \tag{2.11}$$

$$\overline{I}_{Eaf,neg} + \overline{I}_{SVC,neg} = 0 \tag{2.12}$$

$\overline{I}_{Eaf,neg}$ and $\overline{I}_{SVC,neg}$ are phasor quantities representing the arc furnace and the compensator currents, respectively.

2.6.1 *Procedure to Determine the SVC's Susceptance*

The desired compensation susceptances, for each phase, are obtained from the admittances of the load, which in turn are obtained from currents and line voltages measurements. The symmetrical components of line currents can be found from a three phase circuit analysis, considering a balanced three phase source with positive phase sequence supplying a unbalanced load.

The line-to-neutral voltage applied to the SVC-Furnace system, in the primary of the furnace's transformer, is given by:

$$\begin{aligned}
\overline{V}_a &= V \\
\overline{V}_b &= \overline{h}^2 V \\
\overline{V}_b &= \overline{h} V
\end{aligned} \tag{2.13}$$

where $\overline{h} = e^{j\frac{2\pi}{3}}$.

The line currents to the furnace are, then, given by:

$$\begin{aligned}
\overline{I}_{Eaf,a} &= \overline{I}_{Eaf,ab} - \overline{I}_{Eaf,ca} \\
&= \left[\overline{Y}_{Eaf,ab}(1-\overline{h}^2) - \overline{Y}_{Eaf,ca}(\overline{h}-1)\right] V \\
\overline{I}_{Eaf,b} &= \overline{I}_{Eaf,bc} - \overline{I}_{Eaf,ab} \\
&= \left[\overline{Y}_{Eaf,bc}(\overline{h}^2-\overline{h}) - \overline{Y}_{Eaf,ab}(1-\overline{h}^2)\right] V \\
\overline{I}_{Eaf,c} &= \overline{I}_{Eaf,ca} - \overline{I}_{Eaf,bc} \\
&= \left[\overline{Y}_{Eaf,ca}(\overline{h}-1) - \overline{Y}_{Eaf,bc}(\overline{h}^2-\overline{h})\right] V
\end{aligned} \tag{2.14}$$

$\overline{Y}_{Eaf,a}$, $\overline{Y}_{Eaf,b}$ and $\overline{Y}_{Eaf,c}$ are the line to line admittances.

Next, using an amplitude invariant transformation, the sequence components of the load current are expressed as a function of the load currents,

$$\begin{aligned}
\overline{I}_{Eaf,zero} &= (\overline{I}_{Eaf,a}+\overline{I}_{Eaf,b}+\overline{I}_{Eaf,c})/3 \\
\overline{I}_{Eaf,pos} &= (\overline{I}_{Eaf,a}+\overline{h}\,\overline{I}_{Eaf,b}+\overline{h}^2\overline{I}_{Eaf,c})/3 \\
\overline{I}_{Eaf,neg} &= (\overline{I}_{Eaf,a}+\overline{h}^2\overline{I}_{Eaf,b}+\overline{h}\,\overline{I}_{Eaf,c})/3
\end{aligned} \tag{2.15}$$

If we now substitute (2.15) in (2.14), the result is

$$\begin{aligned}
\overline{I}_{Eaf,zero} &= 0 \\
\overline{I}_{Eaf,pos} &= (\overline{Y}_{Eaf,ab}+\overline{Y}_{Eaf,bc}+\overline{Y}_{Eaf,ca})V \\
\overline{I}_{Eaf,neg} &= -(\overline{h}^2\,\overline{Y}_{Eaf,ab}+\overline{Y}_{Eaf,bc}+\overline{h}\,\overline{Y}_{Eaf,ca})V
\end{aligned} \tag{2.16}$$

For a delta connected SVC, the symmetrical components of the line currents may be developed in similar fashion, resulting:

$$\begin{aligned}
\overline{I}_{SVC,a} &= \overline{I}_{SVC,ab}-\overline{I}_{SVC,ca} \\
&= \left[jB_{SVC,ab}(1-\overline{h}^2)-jB_{SVC,ca}(\overline{h}-1)\right]V \\
\overline{I}_{SVC,b} &= \overline{I}_{SVC,bc}-\overline{I}_{SVC,ab} \\
&= \left[jB_{SVC,bc}(\overline{h}^2-\overline{h})-jB_{SVC,ab}(1-\overline{h}^2)\right]V \\
\overline{I}_{SVC,c} &= \overline{I}_{SVC,ca}-\overline{I}_{SVC,bc} \\
&= \left[jB_{SVC,ca}(\overline{h}-1)-jB_{SVC,bc}(\overline{h}^2-\overline{h})\right]V
\end{aligned} \tag{2.17}$$

Considering again an amplitude invariant transformation, the positive and negative sequences current phasors for the compensator's line currents can be obtained as

$$\begin{aligned}
\overline{I}_{SVC,zero} &= 0 \\
\overline{I}_{SVC,pos} &= (\overline{I}_{SVC,a}+\overline{h}\,\overline{I}_{SVC,b}+\overline{h}^2\overline{I}_{SVC,c})/3 \\
\overline{I}_{SVC,neg} &= (\overline{I}_{SVC,a}+\overline{h}^2\overline{I}_{SVC,b}+\overline{h}\,\overline{I}_{SVC,c})/3
\end{aligned} \tag{2.18}$$

Substituting the line currents given by (2.17) in (2.18), after some manipulation, we get

$$\begin{aligned}
\overline{I}_{SVC,pos} &= j(B_{SVC,ab}+B_{SVC,bc}+B_{SVC,ca})V \\
\overline{I}_{SVC,neg} &= -j(\overline{h}^2 B_{SVC,ab}+B_{SVC,bc}+\overline{h}B_{SVC,ca})V
\end{aligned} \tag{2.19}$$

Considering (2.16) and (2.19) and using (2.11) and (2.12), it is possible to obtain the phase to phase susceptances equations, $B_{SVC,ab}$, $B_{SVC,bc}$ and $B_{SVC,ca}$, for the ideal compensator, as:

$$\begin{aligned} B_{SVC,ab} &= -\frac{1}{3V}(Im[\overline{I}_{Eaf,pos}] + Im[\overline{I}_{Eaf,neg}] - \sqrt{3}Re[\overline{I}_{Eaf,neg}]) \\ B_{SVC,bc} &= -\frac{1}{3V}(Im[\overline{I}_{Eaf,pos}] - 2Im[\overline{I}_{Eaf,neg}]) \\ B_{SVC,ca} &= -\frac{1}{3V}(Im[\overline{I}_{Eaf,pos}] + Im[\overline{I}_{Eaf,neg}] + \sqrt{3}Re[\overline{I}_{Eaf,neg}]) \end{aligned} \tag{2.20}$$

These equations express the compensating susceptance in terms of the load's line current phasors ($\overline{I}_{Eaf,a}$, $\overline{I}_{Eaf,b}$ and $\overline{I}_{Eaf,c}$), as proposed by [25, 26].

Since the compensation is composed by the combination of the controlled reactor plus the filters (assuming that all the capacitive power is used to constitute the filters), it is possible to obtain the phase to phase susceptances for the TCR, $B_{TCR,ab}$, $B_{TCR,bc}$ and $B_{TCR,ca}$, by subtracting the filters susceptances from the SVC susceptances.

$$\begin{aligned} B_{TCR,ab} &= B_{SVC,ab} - B_{FILTER} \\ B_{TCR,bc} &= B_{SVC,bc} - B_{FILTER} \\ B_{TCR,ca} &= B_{SVC,ca} - B_{FILTER} \end{aligned} \tag{2.21}$$

2.6.2 Determining the Furnace's Current Components

After measuring the furnace line currents, these are transformed from the abc natural reference frame to the $\alpha\beta$ stationary reference frame, with the α axis fixed on the phase a axis, thus decoupling the real and the reactive current components. The current phasor resulting from this transformation can be divided in two parts: the positive and the negative components, with both components referred to a dq coordinate system synchronized with the network frequency, as it is shown next.

The current spacial vector is composed by the line currents, ($i_a(t)$, $i_b(t)$ and $i_c(t)$), given by (2.22),

$$\overline{i}(t) = \frac{2}{3}(i_a(t) + \overline{h}i_b(t) + \overline{h}^* i_c(t)) \tag{2.22}$$

where $\overline{h}^*$ is the complex conjugate of $\overline{h}$.

After measuring the furnace's line currents, the direct axis and the quadrature axis currents in the orthogonal, stationary reference frame is obtained using an amplitude invariant transformation, defined by:

$$\overline{i}^{\,-\alpha\beta}(t) = i_\alpha(t) + j i_\beta(t) \tag{2.23}$$

Next, a dq transformation synchronized with the network frequency, is used to calculate the real and imaginary parts of the furnace line currents, in order to separate the real and reactive components of current.

$$\bar{i}^{dq}(t) = i_d(t) + ji_q(t) \tag{2.24}$$

with

$$\begin{aligned} i_d(t) &= i_\alpha(t)cos(\xi) + i_\beta(t)sin(\xi) \\ i_q(t) &= -i_\alpha(t)sin(\xi) + i_\beta(t)cos(\xi) \end{aligned} \tag{2.25}$$

where $\xi = \omega t$.

In order to get the negative component of the current, one can just subtract the total dq current at a time t from its value at a time delayed by one quarter of a period [26, 27], as indicated by (2.26). This will remove the positive component of the current, since it is a constant in the dq coordinate system.

$$i_{neg}^{-(dq)}(t) = \frac{(i^{-(dq)}(t) - i^{-(dq)}(t - \frac{T}{4}))}{2} \tag{2.26}$$

In order to get the positive component of the current, all that is needed to do is to subtract the negative component of the current from the furnace's total current, as indicated by (2.27).

$$\bar{I}_{Eaf,pos} = \bar{I}_{Eaf} - \bar{I}_{Eaf,neg} \tag{2.27}$$

These components must then be analyzed with respect to their position in relation to the bus voltage so as to identify their real and imaginary parts. A Phase Locked Loop (PLL) is used for this purpose.

The output of the PLL is a unit vector $\bar{e}_{PLL}$ in phase with the supply system's voltage. he transformation of the positive sequence of the arc furnace current into the forward rotating frame, defined by the PLL, gives the real and imaginary parts of the positive sequence load current, $I_{d,pos}$ and $I_{q,pos}$.

$$\begin{aligned} \bar{I}_{Eaf,pos}^{dq} &= i_{d,pos} + ji_{q,pos} \\ &= \bar{I}_{Eaf,pos}\, \bar{e}_{PLL}^{*} \\ &= \bar{I}_{Eaf,pos}\, \bar{e}^{-j\omega t} \\ &= (i_{\alpha,pos} + ji_{\beta,pos})(e_\alpha + je_\beta)^* \\ &= (i_{\alpha,pos}e_\alpha + i_{\beta,pos})e_\beta + j(i_{\beta,pos}e_\alpha - i_{\alpha,pos}e_\beta) \end{aligned} \tag{2.28}$$

with,

$$
\begin{aligned}
i_{d,pos} &= i_{\alpha,pos} e_\alpha + i_{\beta,pos} e_\beta \\
i_{q,pos} &= i_{\beta,pos} e_\alpha - i_{\alpha,pos} e_\beta
\end{aligned} \tag{2.29}
$$

Making the same operation for the negative sequence, but using the backward rotating frame, provides the negative sequence load current, $I_{d,neg}$ and $I_{q,neg}$.

$$
\begin{aligned}
\bar{I}^{dq}_{Eaf,neg} &= i_{d,neg} - j i_{q,neg} \\
&= \bar{I}_{Eaf,neg}\,(\bar{e}^{*}_{PLL})^{*} \\
&= \bar{I}_{Eaf,neg}\,\bar{e}^{j\omega t} \\
&= (i_{\alpha,neg} + j i_{\beta,neg})(e_\alpha - j e_\beta)^{*} \\
&= (i_{\alpha,neg} e_\alpha + i_{\beta,neg}) e_\beta + j(i_{\beta,neg} e_\alpha + i_{\alpha,neg} e_\beta)
\end{aligned} \tag{2.30}
$$

and

$$
\begin{aligned}
i_{d,neg} &= i_{\alpha,neg} e_\alpha - i_{\beta,neg} e_\beta \\
i_{q,neg} &= -(i_{\beta,neg} e_\alpha + i_{\alpha,pos} e_\beta)
\end{aligned} \tag{2.31}
$$

$I_{q,pos}$, $I_{d,neg}$ and $I_{q,neg}$ are the components used as real and imaginary parts for the furnace currents in (2.20).

2.6.3 *Determining the Firing Angle*

After extracting the filters susceptance from (2.20), as indicated in (2.21), it is then possible to obtain the required phase to phase susceptances for the TCR. Finally, using (2.10), the required value for the firing angle ϕ, for each phase, can be calculated. Figure 3.13 shows a schematic flow chart for the control scheme described above.

2.7 Case Study

The corresponding data for the case studied are shown in the Appendix. The installation was in the planning stage when this analysis was taken, and the simulations shown here used design data available from the furnace's design [5,28].

The furnace severity factor K_{st}, which is different for each arc furnace installation, has a typical value between 48 and 85. Using these data, (2.8) and (2.9) give

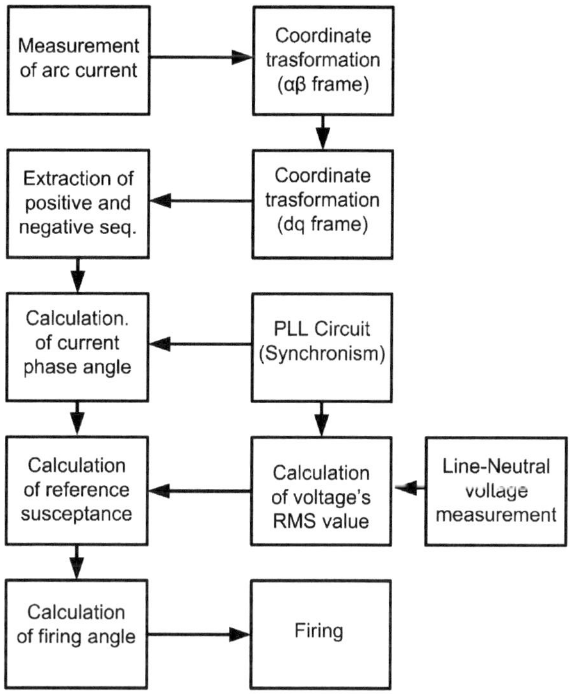

Fig. 2.13 Flow chart for SVC's control system

a estimation for $P_{st}(95\%)$ between 2.62 and 4.64. In the process of defining the set of parameters for the integrated furnace/power grid system,the model is adjusted as explained in Sect. 9.4. Any chosen set of parameters must result in P_{st} levels within these limits. Different P_{st} levels corresponding to a less severe (smaller K_{st}) or a more severe (higher K_{st}) flicker producing furnace.

The compensated arc furnace model, including the arc model and the SVC model, was implemented in the Alternative Transients Program (ATP) environment.

In Fig. 2.14, a single line diagram of the case studied is shown.

2.7.1 System Without Compensation

Waveforms obtained for fusion stage are shown bellow. Figure 2.15 shows the phase to neutral voltage waveforms in the secondary of the furnace transform. They present the typical non-linear characteristics due to the arc load. The modulation frequency calculated for the three phases was between 8 and 10 Hz, as required [24].

Currents waveforms, in the secondary of the furnace transform, are shown in Fig. 2.16, and the resulting v-i characteristic for the electric arc is presented in Fig. 2.1.

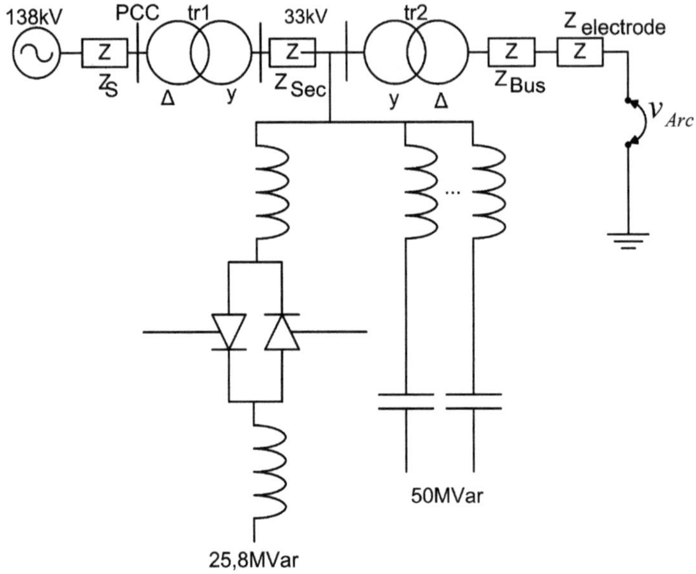

Fig. 2.14 One-line diagram of the case studied

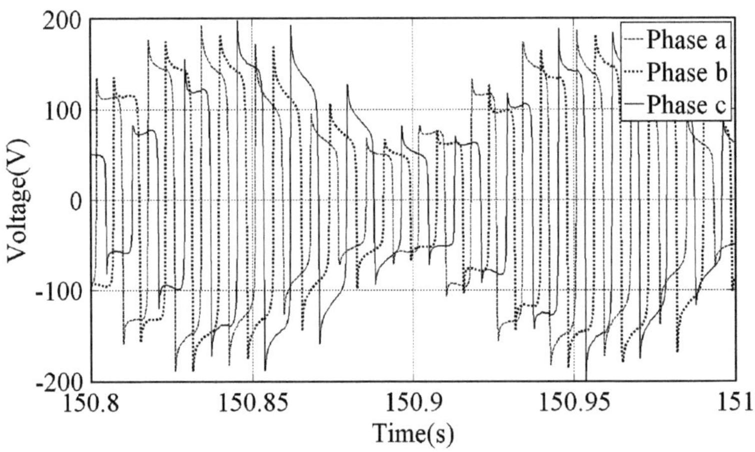

Fig. 2.15 Phase-neutral voltage for the three phases at the secondary of the EAF transformer (tr2), without compensation

Figure 2.17 shows the phase to neutral voltages waveforms at the point of common coupling – PCC, in the high voltage side of the plant's transform. Figure 2.18 shows a detailed view of the phase to neutral voltage in phase a, at the PCC; the voltage fluctuation due to the arc load is clearly seen.

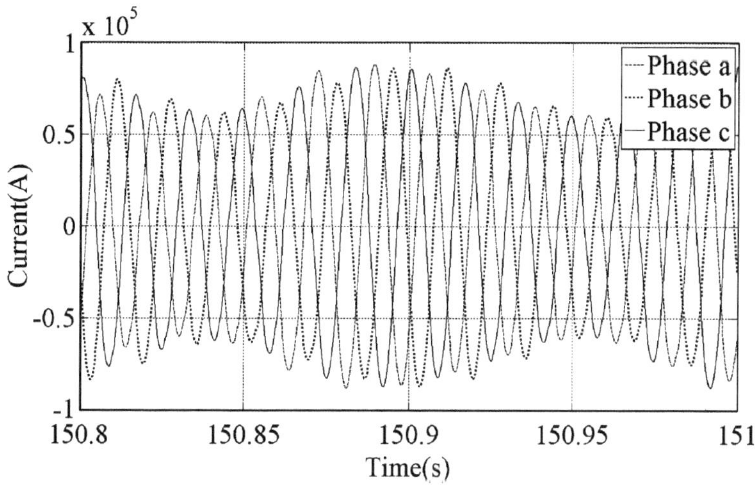

Fig. 2.16 Currents at the secondary of the EAF transformer, without compensation

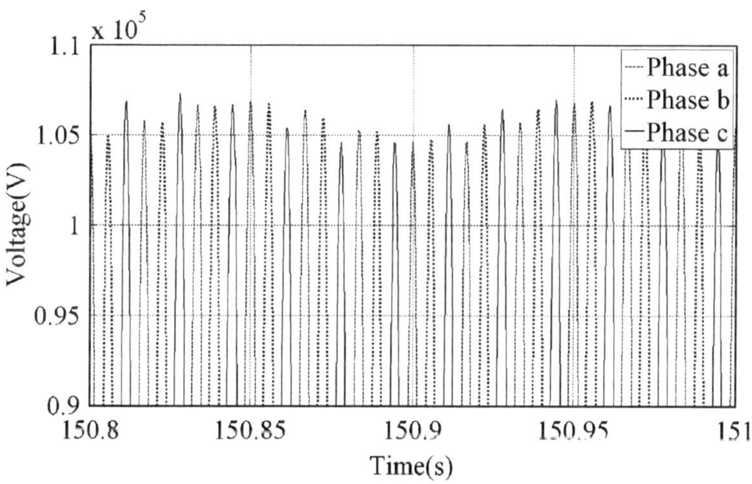

Fig. 2.17 Phase-neutral voltages measured at PCC, without compensation

For the non-compensated system the value indicated by the flickermeter for $P_{st}(95\%)$, at the Point of Common Coupling, is $P_{st}(95\%) = 2.853$. The resulting currents waveforms at the PCC are shown in Fig. 2.19. Figure 2.20 shows the voltage and current in phase a at the PCC.

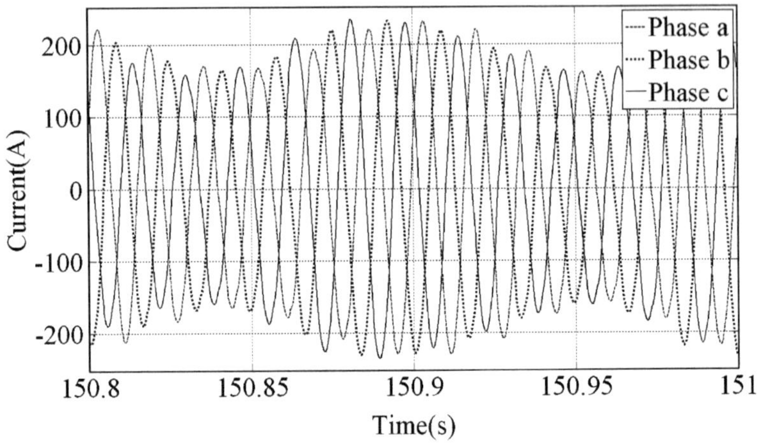

Fig. 2.18 Detail of the phase-neutral voltage at PCC, without compensation

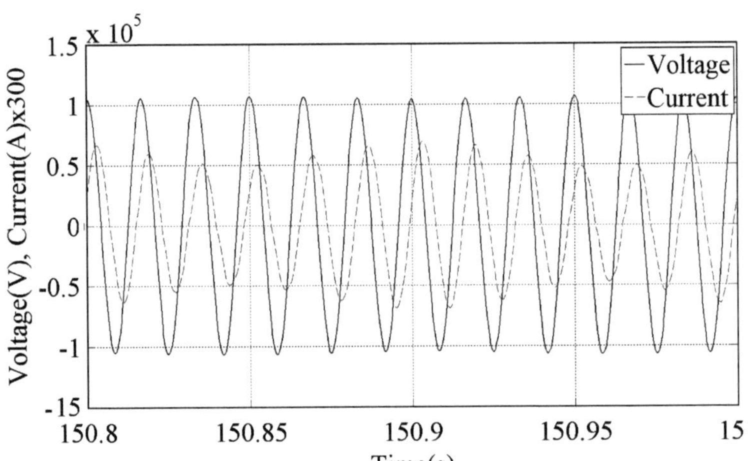

Fig. 2.19 Three phase currents at PCC, without compensation

2.7.2 System with Compensation

The simulation results with compensation are shown in Figs. 2.21–2.23. Figure 2.21 shows the phases to neutral voltages waveforms at the PCC. The effect of the SVC in the voltage waveform is clearly seen in Fig. 2.22, where a detailed view of phase a to neutral voltage is shown. With the compensation, the calculated $P_{st}(95\%)$ reduced

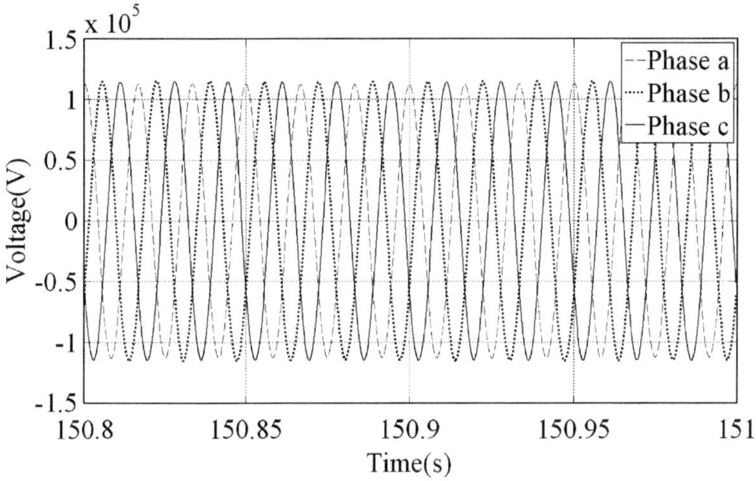

Fig. 2.20 Current and voltage for phase a, at PCC, without compensation

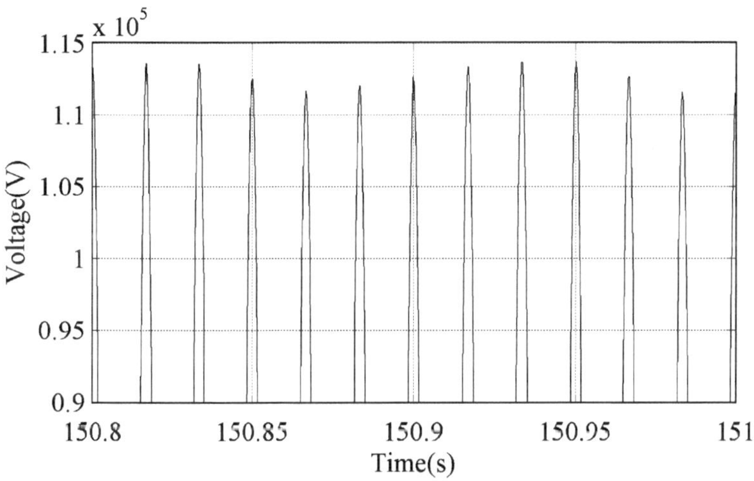

Fig. 2.21 Phase-neutral voltage at PCC, with compensation

to $P_{st}(95\%) = 1.652$, indicating a reduction of 42% from the value calculated for the non-compensated furnace. Finally, in Fig. 2.23, voltage and current in phase a at the PCC are shown. As expected from the control strategy used, a full power factor correction results.

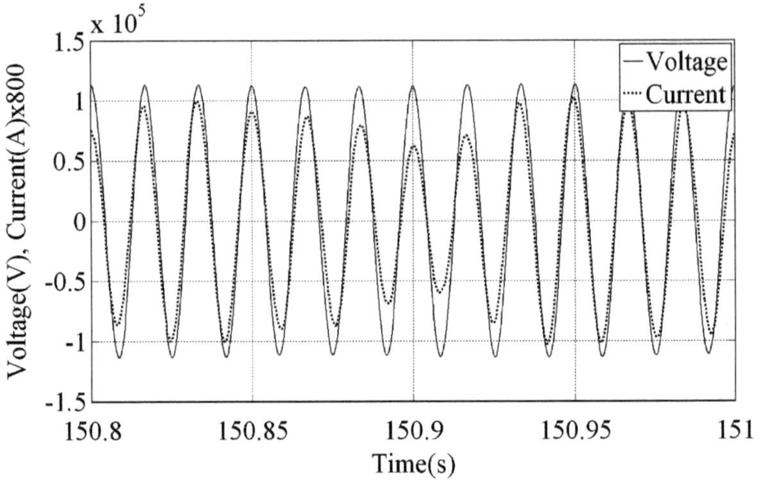

Fig. 2.22 Detail of the phase-neutral voltage (phase a) at PCC, with compensation

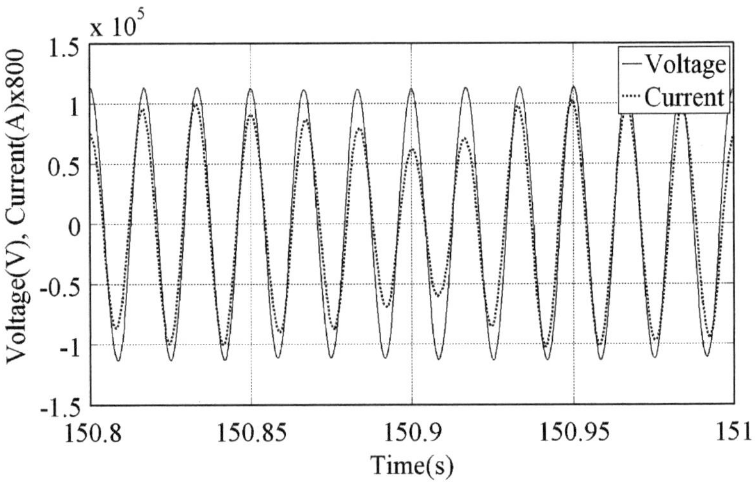

Fig. 2.23 Current and voltage for phase a, at PCC, with compensation

2.8 Conclusions

The electric arc furnace model using the dynamic, multivalued v-i arc characteristics, described by a differential equation, combined with a modulation of chaotic nature, modeled by Chua's circuit, was implemented. A TCR-FC type SVC with a flicker mitigation strategy based on the compensation of all negative sequence currents and the imaginary part of the positive sequence currents drained by the arc furnace, was implemented and the results indicated a potential of up to 40% for flicker reduction. The inherent delay in the SVC's main circuit is its major disadvantage as flicker reduction device. Depending on the control scheme used this delay time will vary from 5 to 10 ms for a 60 Hz system, and flicker level reduction is dependent on that.

A methodology to adjust the combined electric arc furnace – power system model according to expected flicker levels was demonstrated, and the results indicated the validity of the model.

The model was constructed in time domain and can be easily connected to an ATP power system representation as part of an integrated model, allowing for broader system's studies.

Acknowledgements The authors acknowledge the support by CEMIG – Energy Company of Minas Gerais – Brazil, through Research and Development Project 048.

2.9 Case Study Data (See Fig. 2.14)

Source: Ideal sinusoidal AC voltage source; Nominal Voltage = 138 kV (rms);
System Thevenin Impedance = $(1.9 + j37.7)\Omega$.
Transformer 01: Three phase $\triangle$-Y (Grounded); 40 MVA; 138/33 kV;
Reactance = 9%.
Medium Voltage Circuit: $Z_{sec} = j0.098\ \Omega$.
TCR: Three Phase Nominal Power = 25.8 MVA.
Filters: Total Capacitive Power = 50 MVAr;
Second Harmonic Filter: 7.5 MVAr;
Third Harmonic Filter: 15 MVAr;
Fourth Harmonic Filter: 5 MVAr;
Fifth Harmonic Filter: 15 MVAr;
Seventh Harmonic Filter: 7.5 MVAr.
Furnace Transformer: Three Phase Bank Y-$\triangle$; 30 MVA;
33/0.35 kV; Taps: from 0.28 to 0.38 kV; Reactance = 7%.
Arc Furnace: 30 MVA; 0.35 kV; 49.55 kA;
High Current Bus Reactance: $j0.4147\ \Omega$;
Electrode Reactance: $j2.0999654\ \Omega$.
Parameters for Equation (2.2): $k_1 = 2500, k_2 = 1, k_3 = 4, m = 0, n = 2$.
Parameters for Chua's Circuit: $C_1 = 15.05\mu F, C_2 = 150\mu F, L = 2H,$
$R = 482.4655\Omega, m_0 = -0.00037, m_1 = -0.00068, E = 1.1$ V.

References

1. Ozgun, O., Abur, A.: Flicker study using a novel arc furnace model. IEEE Trans. Power. Deliv. **17**(4), 1158–1163 (2002)
2. Acha, E., Semlyen, A., Rajakoviė, N.: A harmonic domain computational package for nonlinear problems and its application to electric arcs. IEEE Trans. Power. Deliv. **5**(3), 1390–1397 (1990)
3. King, P.E., Ochs, T.L., Hartman, A.D.: Chaotic responses in electric arc furnaces. J. Appl. Phys. **76**(4), 2059–2065 (1994)
4. Chen, C.S., Chaung, H.J., Hsu, C.T., Tseng, S.M.: Stochastic voltage flicker analysis and its mitigation for steel industrial power systems. IEEE Porto Power Tech Conference 10th–13th. Porto, Portugal (2001)
5. Alves, M.F., Peixoto, Z.M.A., Garcia, C.P., Gomes, D.G.: An integrated model for study of flicker compensation in electrical networks. Elec. Power Syst. Res., vol. 80, no. 10, pp. 1299–1305. Elsevier (2010)
6. Montanari, G.C., Logginil, M., Cavallinil, A. et al.: Arc-furnace model for the study of flicker compensation in electrical networks. IEEE Trans. Power. Deliv. **9**(4), 2026–2036 (1994)
7. Varadan, S., Makram, E.B., Girgis, A.A.: A new time domain voltage sourge model for an arc furnace using EMTP. IEEE Trans. Power. Deliv. **11**(3), 1685–1691 (1996)
8. Kennedy, M.P.: Three steps to chaos – part I: evolution, fundamental theory and applications. IEEE Trans. Circ. Syst. **40**(10), 640–656 (1993)
9. Kennedy, M.P.: Three steps to chaos – part II: a Chua's circuit primer. IEEE Trans. Circ. Syst. **40**(10), 657–674
10. O'Neill-Carrillo, E., Heydt, G.T., Kostelich, E.J.: Nonlinear deterministic modeling of highly varying loads. IEEE Trans. Power. Deliv. **14**(2), 537–542 (1999)
11. Carpinelli, G., Iacovone, F., Russo, A., Varilone, P.: Chaos-based modeling of DC arc furnaces for power quality issues. IEEE Trans. Power. Deliv. **19**(4), 1869–1876 (2004)
12. Grunbaum, R.: SVC light: a powerful means for dynamic voltage and power quality controle in industry and distribution. Power Electronics and Variable Speed Drives Conference Publication No. 475 0 IEE 2000 (2000)
13. Doleial, J., Castillo, A.G., Valouch, V.: Topologies and control of active filters for flicker compensation. ISIE'2000. Cholula, Puebla, Mexico (2000)
14. Samet, H., Parniani, M.: Predictive method for improving SVC speed in electric arc furnace sompensation. IEEE Trans. Power. Deliv. **22**(1), 732–734 (2007)
15. Samet, H., Golshan, M.E.H.: Employing stochastic models for prediction of arc furnace reactive power to improve compensator performance. IET Generation, Transmission and Distribution, **2**(4), 1751–8687 (2008)
16. Sharmeela, C., Uma, G., Mohan, M.R., Karthikeyan, K.: Voltage flicker analysis and mitigation – case study in AC electric arc furnace using PSCAD/EMTDC. International Conference on Power System Technology – POWERCON. Singapore (2004)
17. Poudel, S., Watson, N.R.: Assessment of light flicker mitigation using shunt compensators. International Conference on Power System Technology – POWERCON. Singapore (2004)
18. Machado Neto, J.P.: Aplicacao das Tecnicas de Identificacao de Sistemas Nao-Lineares a Modelagem de Fornos Eletricos a Arco. MSc Dissertation, Pontifical Catholic University of Minas Gerais Belo Horizonte Brazil (2005)
19. Wolf, A., Swift, J.B., Swinney, H.L., Vastano, J.A.: Determining Lyapunov exponents from a time series. Physica, **16D**, pp. 285–317, Amsterdam, North-Holland (1985)
20. Oppenheim, A.V., Schafer, R.W.: Discrete-Time Signal Processing. Prentice Hall, New Jersey (1999)
21. Robert, A., Couvreur, M.: Recent experience of connection of big arc furnaces with reference to flicker level. CIGRE Paper 36–305 (1994)
22. Alves, M.F., Peixoto, Z.M.A., Garcia, C.P., Gomes, D.G., Machado Neto, J.P.: An electric arc furnace model for flicker estimation. WSEAS International Conferences on Power Engineering Systems – ICOPES'05. Rio de Janeiro, Brazil (2005)

23. IEC Flickermeter – Functional and Design Specifications. In: IEC 61000-4-15 International Standard, Electromagnetic Compatibility (EMC) 1st edn. Part 4: Testing and Measurement Techniques Section 15 (1997)
24. UIE Part 5 Flicker and Voltage Flutuation. Qualit de l'alimentation "Power Quality" Working Group WG 2. Prepared by de travail GT (1999)
25. Miller, T.J.E.: Reactive Power Control in Electric Systems. John Wiley, New York (1982)
26. Larsson, T.: Voltage Source Converters for Mitigation of Flicker Caused by Arc Furnaces. Ph.D. Thesis, Department of Electric Power Engineering Division of High Power Electronics, Royal Institute of Technology, Stockholm, Sweden (1998)
27. Le, T.N.: Kompensation schnell vernderlicher Blindstrme eines Drehstromverbrauchers. etzArchiv, Bd. 11, H.8, pp. 249–253 German (1989)
28. Gomes, D.G.: Estrategia de Controle Para Mitigacao de Cintilacao Luminosa Causada por Fornos Eletricos a Arco Utilizando Compensador Estatico de Reativos. MSc Dissertation, Pontifical Catholic University of Minas Gerais Belo Horizonte Brazil (2005)

Chapter 3
Nonlinear Optical Fibre Resonators with Applications in Electrical Engineering and Computing

S. Lynch and A.L. Steele

3.1 Introduction

In 1969, Szöke et al. reported their first experiments on optical bistability [39] and Seidel filed his patent [29] on a bistable optical circuit. Both reported optical hysteresis and suggested applications. The two essential ingredients for bistability are nonlinearity and feedback. Bistability is a phenomenon where a given optical device can have two possible output powers for a given input power. In 1976, Gibbs [10] demonstrated bistability using a sodium-filled Fabry-Perot interferometer. Bistability is a hysteresis effect that is dependent upon the history of the system. There are two stable steady-states within a bistable region and the path followed is dependent upon whether the input power is increasing or decreasing. Consider Fig. 3.1, the power is increased from zero and the output power follows the route A, B, C. The input power is then decreased back down to zero and the output power follows the route C, D, A, giving a counterclockwise bistable region. For an input power, x, say, there are two possible outputs depending on whether the input power is increasing or decreasing. This bistable behavior can be introduced through an absorption process or a dispersion process. With the first process a saturable absorber could be contained within a cavity [29]. At relatively weak optical powers the light in the cavity is absorbed, but as the power is increased the saturation point of the absorber could be reached and the absorber is bleached allowing the light to pass around with little or no absorption. Reducing the power again whilst the absorber is in this bleached state will produce an optical output that can be higher

S. Lynch (✉)
School of Computing, Mathematics and Digital Technology, Manchester Metropolitan University, Manchester M1 5GD, UK
e-mail: s.lynch@mmu.ac.uk

A.L. Steele
Department of Electronics, Carleton University, Ottawa, Ontario, K1S 5B6, Canada
e-mail: asteele@doe.carleton.ca

S. Banerjee et al. (eds.), *Applications of Chaos and Nonlinear Dynamics in Engineering – Vol. 1*, Understanding Complex Systems, DOI 10.1007/978-3-642-21922-1_3,

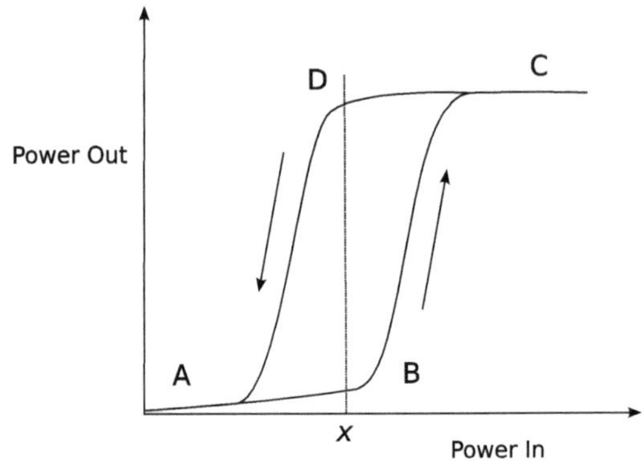

Fig. 3.1 Transfer function showing optical bistability

than the earlier power when the light was being increased, for the same input power. This process follows the path shown by the arrows in Fig. 3.1. The second process, the dispersive process, relies on a change in the refractive index of a material due to the optical intensity within an optical cavity. This was first achieved using sodium vapour in a Fabry-Perot interferometer [10]. The dispersive process is perhaps less intuitive than the absorptive one, but if one considers that the light circulating in an optical cavity can have a phase shift induced because of the induced refractive index change then one can perhaps see how variations in the phase at different points in the cavity can cause light to mix in and out of phase within the cavity. In this chapter we will focus only on dispersive systems and so a deeper understanding will come with the analysis in the later sections. We should also make one further categorization of optical bistable devices, which is a device can be active, for example a laser, or passive (no optical amplification). Again, in this chapter we will restrict analysis to only passive devices.

In 1979, Ikeda [16] showed that a simple optical ring system with a two-level absorber can display not only bistability, but multistability and instabilities, or optical chaos, if there is sufficient input power. This is an important finding as it shows that it is possible to have an optically bistable device that can become unstable, but it also shows that it is possible to have a device that can generate optical chaos. The observation of optical chaos was made by Gibbs et al. in 1981 [12] using a hybrid optical system composed of an optical ring with an electrically introduced delay line. The first all-optical system was reported in 1983 by Nakatsuka et al. [25]. This all-optical system was formed from a single-mode optical fibre with the input light coming from a Q-switched YAG laser that produced 140 ps pulses. They reported the observation of period doubling and chaos in the output, with the optical fibre introducing a nonlinearity through a nonlinear refractive index.

Optical fibre produces a more compact ring cavity than one composed of bulk optical elements such a mirrors and cells containing the nonlinear material. There are also issues related to the alignment of optical beams. The optical waveguiding and the nonlinear characteristics of optical fibres make them good candidates for investigation of the nonlinear dynamics of optical resonators. Indeed, with the inclusion of optical amplification in a optical fibre cavity, usually with erbium

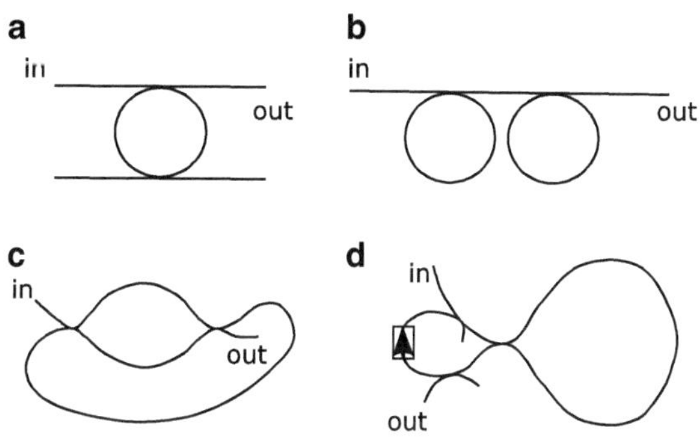

Fig. 3.2 Some fibre resonator configurations. (**a**) Two couplers [11]; (**b**) double-ring resonator [17]; (**c**) Mach-Zehnder with feedback [2], and (**d**) NOLM with feedback, [30, 34]

doped optical fibre [7], one can produce fibre lasers [9]. Although those device are interesting to study, in this chapter we will restrict our examination of nonlinear dynamics to passive, non-amplifying, optical fibre cavity structures.

Since the early work in the 1970s and 1980s there has been a range of optical fibre resonators, with a sample shown in Fig. 3.2. These include ring resonators with two couplers, double-ring resonators, a Mach-Zehnder interferometer with a return delay-line and resonators based on nonlinear optical loop mirrors (NOLMs) [8]. These latter devices are themselves interesting as they are Sagnac interferometers that use the nonlinear refractive index to induce phase differences with the counter-propagating beams in the Sagnac loop which ultimately can lead to input light being passed between output ports. This switching between two output ports can lead to a nonlinear behaviour that can lead to interesting dynamics if the output from one port is reinjected back into the input port.

The nonlinear simple fibre ring resonator was investigated in the MATLAB chapter of this book. Section 3.2 of this chapter provides an investigation of the NOLM using a linear stability analysis and bifurcation theory. For completeness we now discuss the applications of the NOLM in engineering. The NOLM, a fibre Sagnac interferometer, was first studied by Doran and Wood [8] in 1988. The NOLM is widely acknowledged to be an indespensible tool in a diverse range of optical signal processing applications. Due to its flexibility and self-switching ability the NOLM can be applied to the fields of mode-locking [31], optical switching [6], phase conjugation [19], pedestal suppression and pulse compression [18], stabilizing soliton pulses [24], wavelength demultiplexing [33], and more recently in the telecommunications industry the NOLM has been validated as an operative 2R regenerative component [4].

Section 3.3 provides an investigation of a microfibre ring resonator. As already mentioned, optical fibre offers a more compact and stable environment than a bulk optical system, however, the length of conventional fibre needed can still be an issue, preventing minaturization. In recent years, there has been the development of microfibres [40], nanofibres [5] and resonators [35] composed of these types of fibre. Optical fibre microwires and nanowires are beneficial in that they have enhanced evanescent fields [40], and they are flexible, configurable, robust and compact. Sumetsky has recently published a paper on the theory of microfibre loop resonators

[36]. Now the fibre diameter has been reduced down to the nanoscale and resonator ring lengths are of the order of millimetres. Because of the narrowness of the fibre diameter, a significant proportion of the guided electric field can extend beyond the optical fibre core [41], known as the evanescent field, which makes them of interest in optical sensing applications. Changes in the evanescent field can cause changes in the refractive index of the fibre and hence the power of light propagating in the fibre. The changes in the evanescent field can be caused by temperature variation, radiation variation, small changes in the concentration of chemical or biological species, and the presence of microparticles, for example. Examples of applications can be found in the papers [5, 35, 37, 38]. As well as the sensing applications there have also been studies of the microfibre ring exhibiting optical bistability [42, 43].

In 1984, Smith [32] published an article in *Nature* with the enthralling title "Towards the Optical Computer", and in 1985 Gibbs published the book entitled "Optical Bistability: Controlling Light with Light" [13], unfortunately over 25 years later all-optical computers have failed to materialize. The transistor remains a fundamental building block of the modern computer and computers remain comprised of electronic components which operate relatively slowly and suffer from interference from magnetic and electric fields. Quantum computers, biomolecular computers, spin computers and chemical computers are amongst the candidates for the future of computing. The future aim is to produce very fast low power computers and this goal may be achieved with all-optical computers. In 2004, Ibrahim et al. [15] constructed an all-optical NOR logic gate based on symmetric GaAs-AlGaAs microring resonators. In [44], Zhou et al. present a silicon microring resonator-based reconfigurable optical lattice filter for on-chip signal processing. In 2009, Hwang et al. [14] successfully created an optical transistor from a single molecule which they claim has brought them one step closer to an optical computer. Intel have also recently announced a breakthrough in optical connection technology which could lead to novel computer design.

More recently, one of the authors Lynch and a colleague Jon Borresen [3] of the Manchester Metropolitan University Novel Computation Group, have filed a UK patent application for a new type of silicon chip technology which offers an alternative to those listed above. Potentially, this technology could lead to low power, high performance computing which links in with current hardware architecture. The authors of the patent are currently seeking industrial partners in order to fabricate a prototype.

3.2 Analysis of a Nonlinear Optical Loop Mirror with Feedback

This section presents an investigation of the NOLM using complex iterative equations. A schematic of the NOLM is given in Fig. 3.3 and a mathematical description of the device was provided by Shi [30] in 1994 and is included here for completeness.

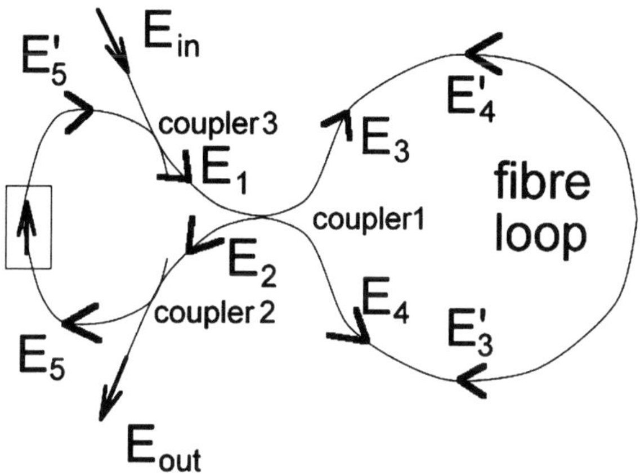

Fig. 3.3 A schematic of the NOLM with feedback. The electric field at position, say, p, is E_p and the fibre coupling at coupler, say, c, is given by $\kappa_c : (1-\kappa_c)$

The electric field E_1 can be expressed as

$$E_1 = \sqrt{\kappa_3} E_{\text{in}} + j\sqrt{1-\kappa_3} E_5', \tag{3.1}$$

where E_5' is the unidirectional feedback field incident on coupler 3. In the main fibre loop, the electric fields are expressed as

$$\begin{aligned} E_3 &= \sqrt{\kappa_1} E_1, \\ E_4 &= j\sqrt{1-\kappa_1} E_1. \end{aligned} \tag{3.2}$$

Since the light circulating in the NOLM is a continuous wave then both the counter-propagating fields will contribute to the phase shifts. The nonlinear phase shifts induced in the clockwise and counter-clockwise directions of Fig. 3.3 can be expressed, respectively, by

$$\begin{aligned} \phi_C &= \frac{2\pi n_2 L}{\lambda_0 \text{A}_{\text{eff}}} (2-\kappa_1) P_1, \\ \phi_{CC} &= \frac{2\pi n_2 L}{\lambda_0 \text{A}_{\text{eff}}} (1+\kappa_1) P_1, \end{aligned} \tag{3.3}$$

where n_2 is the nonlinear refractive index coefficient, L is the length of the fibre, A_{eff} is the effective core area of the fibre, λ_0 is the wavelength of the propagating light and $P_1 = |E_1|^2$ is the optical power entering coupler 1 from position 1. The electric fields circulating clockwise and counterclockwise around the main loop will have suffered respective phase shifts of:

$$\begin{aligned} E_3' &= E_3 e^{-j(\phi_L + \phi_C)}, \\ E_4' &= E_4 e^{-j(\phi_L + \phi_{CC})}, \end{aligned} \tag{3.4}$$

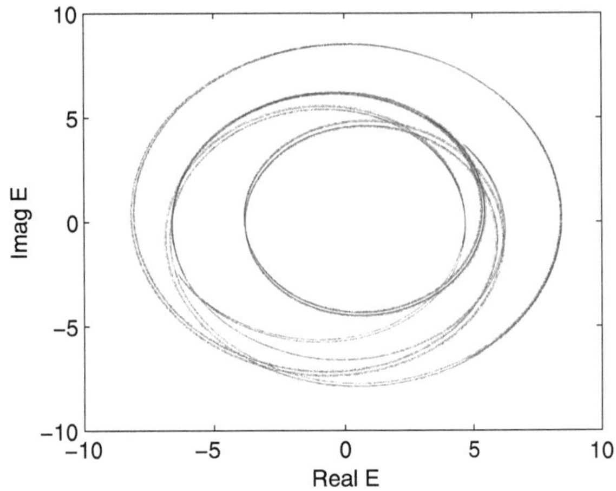

Fig. 3.4 Complex iterations, the output electric field is split into its real and imaginary components. Chaotic attractor for the NOLM when $\lambda_0 = 1.55 \times 10^{-6}, n_2 = 3.2 \times 10^{-20}, A_{\text{eff}} = 30 \times 10^{-6}\,\text{m}^2, L = 80\,\text{m}$, $\kappa_1 = 0.25$, $\kappa_2 = \kappa_3 = 0.8$ and $P = 100\,\text{W}$

where ϕ_L is a linear phase shift due to propagation. In the unidirectional feedback section there are two couplers and a length of optical fibre, which has an optical isolator so constraining light to propagate through the fibre in one direction only. Assume that the feedback fibre's length is short enough to ignore nonlinear phase shifts, and so a linear phase shift, say, ϕ_f, need only be considered. The feedback field incident on coupler 3 is then given by

$$E_5' = E_5 e^{j\phi_f}. \tag{3.5}$$

where

$$\begin{aligned} E_2 &= \sqrt{\kappa_1} E_3' + j\sqrt{1-\kappa_1} E_4' \\ E_5 &= j\sqrt{1-\kappa_2} E_2. \end{aligned} \tag{3.6}$$

The output electric field is given by

$$E_{\text{out}} = \sqrt{\kappa_2} E_5. \tag{3.7}$$

Equations (3.1) to (3.7) can be used to plot a chaotic attractor for the NOLM. Figure 3.4 shows a chaotic attractor for the parameter values

$$\lambda_0 = 1.55 \times 10^{-6},\ n_2 = 3.2 \times 10^{-20},\ A_{\text{eff}} = 30 \times 10^{-6}\text{m}^2 \text{ and} L = 80\,\text{m}, \tag{3.8}$$

when $\kappa_1 = 0.25$, $\kappa_2 = \kappa_3 = 0.8$ and the input power is $P = 100\,\text{W}$.

Substituting equations (3.2) to (3.6) into equation (3.1), gives an equation for the electric field at position 1 that is iterative with respect to the cavity period, say, t_R. The iterative equation becomes

$$E_1(t) = \sqrt{\kappa_3} E_{\text{in}} - K E_1(t - t_R) \left(\kappa_1 e^{-j(\phi + \phi_C)} - (1-\kappa_1) e^{-j(\phi + \phi_{CC})} \right), \tag{3.9}$$

where $K = \sqrt{(1-\kappa_2)(1-\kappa_3)}$, and $\phi = \phi_L - \phi_f$.

Suppose that E_S is a steady-state solution for equation (3.9), then

$$\kappa_3 |E_{\text{in}}|^2 = |E_S|^2 + 2K|E_S|^2 [\kappa_1 \cos(\phi_C + \phi) - (1-\kappa_1)\cos(\phi_{CC} + \phi)] \\ + K^2 \left[\kappa_1^2 |E_S|^2 + (1-\kappa_1)^2 |E_S|^2 - 2(1-\kappa_1)\kappa_1 |E_S|^2 \cos(\phi_C - \phi_{CC})\right], \tag{3.10}$$

where $E_S = E_1(t) = E_1(t - t_R)$. This gives a bistable relationship equivalent to the graphical method, which is well documented in the literature and was adopted by Shi [30], unfortunately this method gives no information on instabilities. Steele et al. [34] considered four different methods of analysis including the graphical method, the first and second iterative methods to produce bifurcation diagrams and a linear stability analysis. The linear stability analysis will now be described and bifurcation diagrams will be plotted as a means of checking the results.

Consider a slight perturbation from the stable situation in a cavity period, such that

$$E_n = E_S + \epsilon_n, \quad |\epsilon_n| \ll |E_n|, \tag{3.11}$$

where n indicates the number of cavity periods the light has travelled. To simplify the equations, let

$$A = \sqrt{\kappa_3} E_{\text{in}}, \quad B_1 = \frac{2\pi n_2 L(2-\kappa_1)}{\lambda A_{\text{eff}}}, \quad B_2 = \frac{2\pi n_2 L(1-\kappa_1)}{\lambda A_{\text{eff}}},$$

equation (3.9) can then be written as a complex iterative equation of the form

$$E_S + \epsilon_{n+1} = A - K(E_S + \epsilon_n) \\ \left[\kappa_1 e^{-j\phi} e^{-jB_1(E_S+\epsilon_n)(E_S^*+\epsilon_n^*)} - (1-\kappa_1) e^{-j\phi} e^{-jB_2(E_S+\epsilon_n)(E_S^*+\epsilon_n^*)}\right]. \tag{3.12}$$

Take a Taylor series expansion and ignore powers of ϵ_n greater than one to get

$$\epsilon_{n+1} = -K\left[\kappa_1 e^{-j(B_1|E_S|^2+\phi)}\left(\epsilon_n - jB_1\left(E_S^2\epsilon_n^* + |E_S|^2\epsilon_n\right)\right) \\ -(1-\kappa_1)e^{-j(B_2|E_S|^2+\phi)}\left(\epsilon_n - jB_2\left(E_S^2\epsilon_n^* + |E_S|^2\epsilon_n\right)\right)\right]. \tag{3.13}$$

Since ϵ is real, it may be split into its positive and negative frequency parts as follows:

$$\epsilon_n = E_+ e^{\lambda t} + E_- e^{\lambda t} \quad \text{and} \quad \epsilon_{n+1} = E_+ e^{\lambda(t+t_R)} + E_- e^{\lambda(t+t_R)}, \tag{3.14}$$

where $|E_+|$ and $|E_-|$ are much smaller than $|E_S|$, and λ is the amplification rate of a small fluctuation added to a stable solution. Substitute equation (3.14) into equation (3.13), then the validity of (3.13) at all times t requires that

$$\begin{pmatrix} M_{11}-e^{\lambda t_R} & M_{12} \\ M_{21} & M_{22}-e^{\lambda t_R} \end{pmatrix} \begin{pmatrix} E_+ \\ E_-^* \end{pmatrix} = \begin{pmatrix} 0 \\ 0 \end{pmatrix}, \tag{3.15}$$

where

$$\begin{aligned} M_{11} &= U - jB_1U\left|E_S\right|^2 - V + jB_2V\left|E_S\right|^2, \\ M_{12} &= -jB_1UE_S^2 + jB_2VE_S^2, \\ M_{21} &= jB_1U^*E_S^{2*} - jB_2V^*E_S^{2*}, \\ M_{22} &= U^* + jB_1U^*\left|E_S\right|^2 - V^* - jB_2V^*\left|E_S\right|^2, \end{aligned}$$

and

$$\begin{aligned} U &= K\kappa_1 e^{-j\left(B_1\left|E_S\right|^2+\phi\right)}, \\ V &= K\left(1-\kappa_1\right)e^{-j\left(B_2\left|E_S\right|^2+\phi\right)}. \end{aligned}$$

Using the non-trivial solution of (3.15), the stability edges of the system are found when $e^{\lambda t_R} = \pm 1$, which yields the stability conditions

$$\begin{aligned} D_{+1} &= 0, \\ D_{-1} &= 4K\Big[\kappa_1\cos\left(B_1\left|E_S\right|^2+\phi\right) - \left|E_S\right|^2 B_1\kappa_1\sin\left(B_1\left|E_S\right|^2+\phi\right) \\ &\quad - \left(1-\kappa_1\right)\cos\left(B_2\left|E_S\right|^2+\phi\right) + \left|E_S\right|^2 B_2\left(1-\kappa_1\right)\sin\left(B_2\left|E_S\right|^2+\phi\right)\Big]. \end{aligned}$$

Thus, the system is stable as long as

$$0 < D < D_{-1}, \tag{3.16}$$

where

$$D = \frac{d\left|A\right|^2}{d\left|E_S\right|^2}.$$

The condition $D_{+1} = 0$ marks the boundaries between the branches of positive and negative slope on the graph of $\left|E_S\right|^2$ versus $\left|A\right|^2$, and hence defines the region where the system is bistable. These results match those given by the graphical method as employed by Shi [30]. The system becomes unstable at the boundary where $D = D_{-1}$. Take the parameter values as in equation (3.8) as taken by Shi [30] and

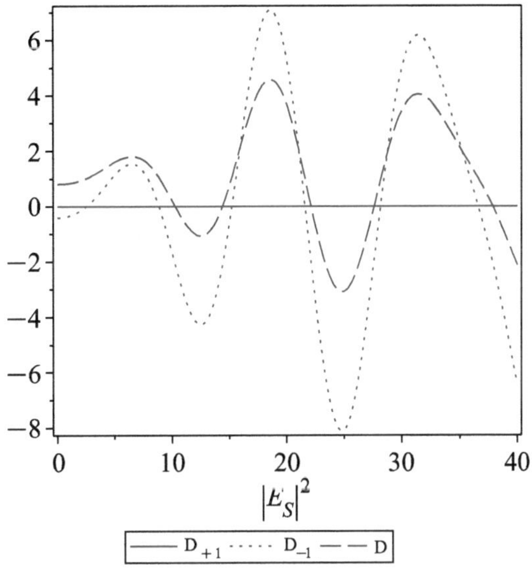

Fig. 3.5 The stability curves D_{+1}, D_{-1}, and $D = \frac{d|A|^2}{d|E_S|^2}$ for the NOLM under conditions (3.8) when $\kappa_1 = 0.25$, $\kappa_2 = 0.8$, $\kappa_3 = 0.8$, and $\phi = 0$

Steele et al. [34]. Figure 3.5 shows the stability curves for the parameter values listed in the figure caption. The intersection points where $D = 0$, determine the boundary conditions for bistable behaviour. Equation (3.10) is used to determine $|A|^2$ which in turn can be used to compute $|E_{in}|^2$. Similarly, the boundary conditions for unstable behaviour are determined by solving the equation $D = D_{-1}$; the intersection points of the curves D and D_{-1} can be computed with a suitable mathematics package and again equation (3.10) can be used to determine the regions of instability in terms of the power in, $P_{in} = |E_{in}|^2$.

A bifurcation diagram is plotted in Fig. 3.6 for the parameter values $\kappa_1 = 0.25$ and $\kappa_2 = \kappa_3 = 0.8$. Referring to Fig. 3.6, the upper figure shows the output power against the number of iterations along with the Gaussian input power curve, which can be defined by the function

$$P_{in}(x) = P_{max} \exp\left(-0.02\left(x - \frac{P_{max}}{2}\right)\right),$$

where P_{max} is the maximum power of the input pulse. It is clear from the figure that the bistable cycle is encroached by a period doubling instability. Note that Fig. 3.6 also shows that the maximum output power for a given input power is limited, this is due to the behaviour of the Sagnac interferometer section of the NOLM that causes part of the bifurcation diagram to fold over on itself. The results make interesting comparison when considering the results from the linear stability analysis as highlighted in Table 3.1. When $\phi = 0$, the linear stability analysis predicts a first bistable region in the range $11.906 < P_{in} < 15.512$ and a first unstable region for $13.856 < P_{in} < 37.413$. As ϕ is increased from zero to 1.75π, the

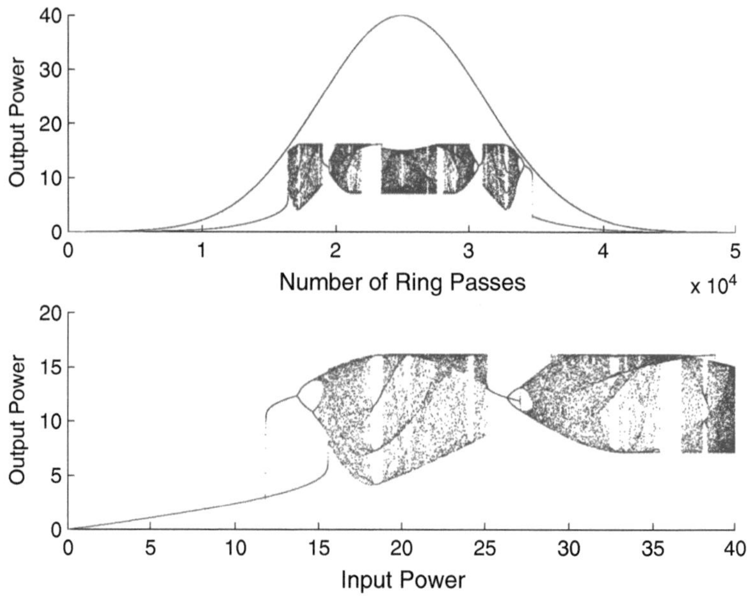

Fig. 3.6 The upper figure shows the number of ring passes (50 000 iterations) against the output power (the Gaussian input power curve is also included) and the lower figure shows the input power versus the output power under conditions (3.8) when $\phi = 0$, $\kappa_1 = 0.25$ and $\kappa_2 = \kappa_3 = 0.8$

Table 3.1 First regions of bistability and instability computed using a linear stability analysis

ϕ	Bistable, P_{in} (W)	Unstable, P_{in} (W)
0.0	11.906 – 15.512	13.856 – 37.413
0.25π	10.724 – 12.773	12.984 – 34.729
0.5π	9.509 – 10.277	12.194 – 31.817
0.75π	8.150 – 8.159	11.481 – 28.701
π	17.241 – 27.545	10.857 – 25.397
1.25π	15.741 – 24.491	10.369 – 21.897
1.5π	14.375 – 21.431	10.163 – 18.103
1.75π	13.108 – 18.419	11.284 – 13.015

bistable region detaches from the unstable region and gets smaller and smaller until it disappears for low input powers and a large unstable region, which shows period doublings and undoublings to and from chaos, shifts to lower input powers. A series of figures is displayed below which shows the bifurcation diagrams for some of the values of ϕ appearing in Table 3.1.

The bifurcation diagrams plotted in Fig. 3.7 make interesting comparison with the results listed in Table 3.1. The bistable region disappears from the bifurcation diagrams when $\phi = \pi$ as seen in Fig. 3.7e. According to the tabulated results, the first bistable region should occur for the input power range 17.241 – 27.454 W, however, the first unstable region has already been entered and prevents bistability from occurring.

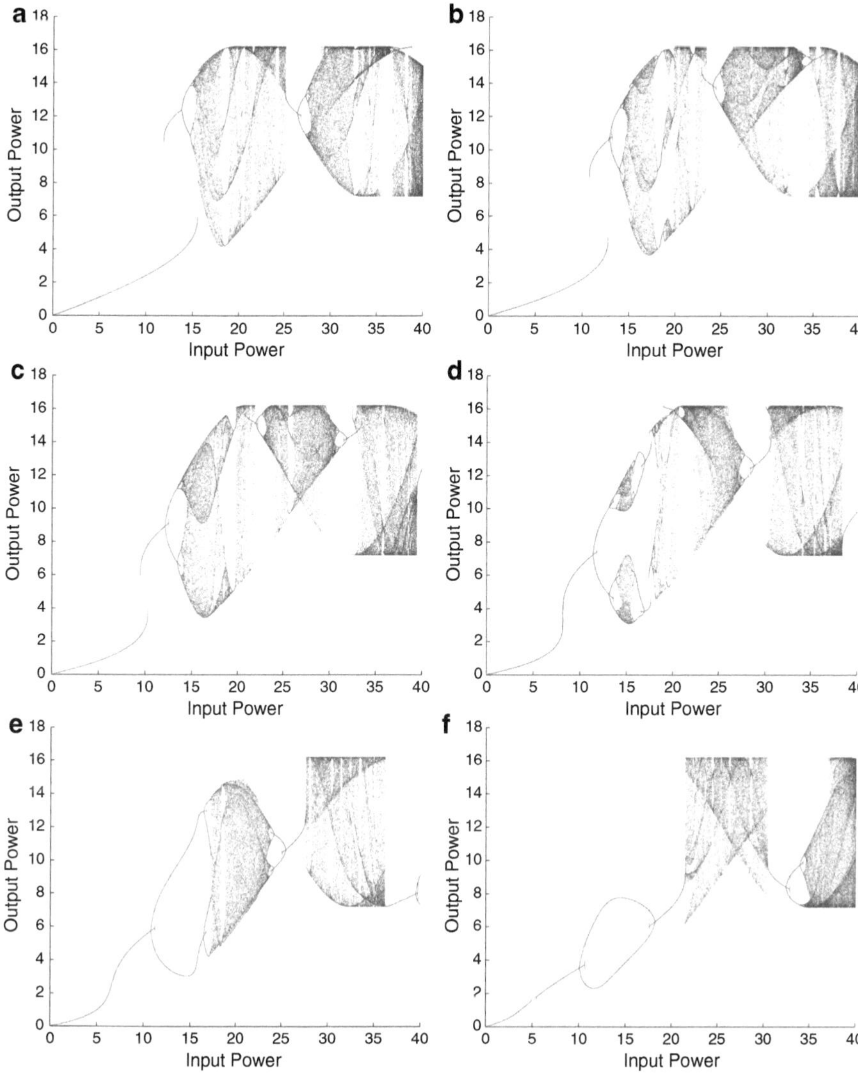

Fig. 3.7 A gallery of bifurcation diagrams for the NOLM under conditions (3.8) when $\kappa_1 = 0.25, \kappa_2 = 0.8, \kappa_3 = 0.8$ and (**a**) $\phi = 0$, (**b**) $\phi = 0.25\pi$, (**c**) $\phi = 0.5\pi$, (**d**) $\phi = 0.75\pi$, (**e**) $\phi = \pi$ and (**f**) $\phi = 1.5\pi$

The gallery of bifurcation diagrams displayed in Fig. 3.7 show how the bistable region of the NOLM is affected by a changing linear phase shift ϕ. It is also interesting to see how the dynamics of the NOLM vary as other parameters in the system vary. The use of mathematical packages such as Maple™ [20], Mathematica® [21], and MATLAB® [22] can really help with this problem. For example, Lynch has posted an animated bifurcation diagram on the Wolfram

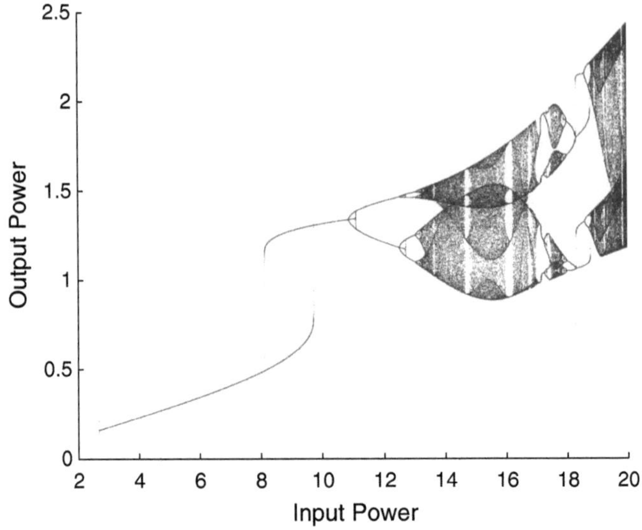

Fig. 3.8 Bifurcation diagram for the NOLM under conditions (3.8) when $\phi = 1.5\pi$, $\kappa_1 = 0.95$, $\kappa_2 = 0.1$, and $\kappa_3 = 0.95$. There is an isolated bistable region at a relatively low input power

Demonstrations web site where readers can watch a preview and download working Mathematica notebooks. It is recommended that the reader watch the animated movies to see how the bifurcation diagrams are altered as one or more parameters vary. Two interesting cases are considered below.

In Fig. 3.8, there is a large bistable region at a relatively low power input for the parameter values $\phi = 1.5\pi$, $\kappa_1 = 0.95, \kappa_2 = 0.1$, and $\kappa_3 = 0.95$. The linear stability analysis predicts the first bistable region for $8.151 < P_{in} < 9.727$, and a first unstable region for $10.905 < P_{in} < 31.200$. The results from the linear stability analysis match well with those shown on the bifurcation diagram.

For the parameter values $\phi = 1.8\pi$, $\kappa_1 = 0.1, \kappa_2 = 0.95$, and $\kappa_3 = 0.95$, Fig. 3.9 shows two isolated bistable regions separated by an unstable region. The linear stability analysis predicts a first bistable region in the range $32.095 < P_{in} < 32.546$ and a second bistable region in the range $48.829 < P_{in} < 50.126$. The linear stability analysis predicts a first unstable region in the range $35.948 < P_{in} < 45.141$.

3.3 Microfibre Resonators

As mentioned in the introduction of this chapter there has been recent interest in the area of optical microfibres. The diameter of these fibres is reduced compared to conventional singlemode fibres, indeed the diameters can be of the order of the wavelength of light that is being guided or less, Tong et al. [40] reported diameters down to 50nm. With diameters of this size, overlapping fibres can structurally align

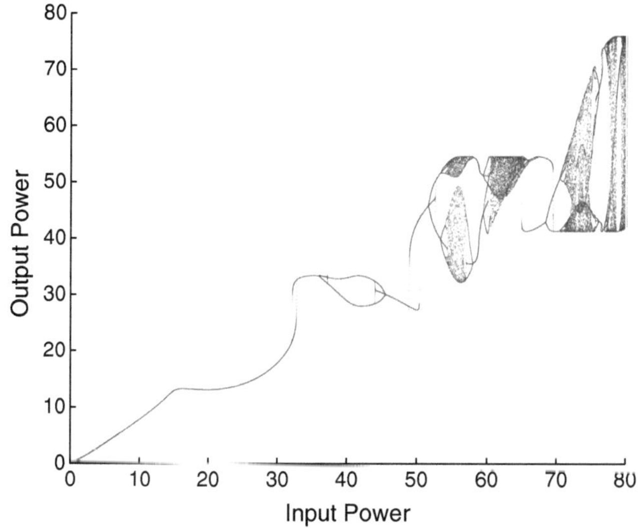

Fig. 3.9 Bifurcation diagram for the NOLM under conditions (3.8) when $\phi = 1.8\pi$, $\kappa_1 = 0.1$, $\kappa_2 = 0.95$, and $\kappa_3 = 0.95$. There are two isolated bistable regions with an unstable region between them

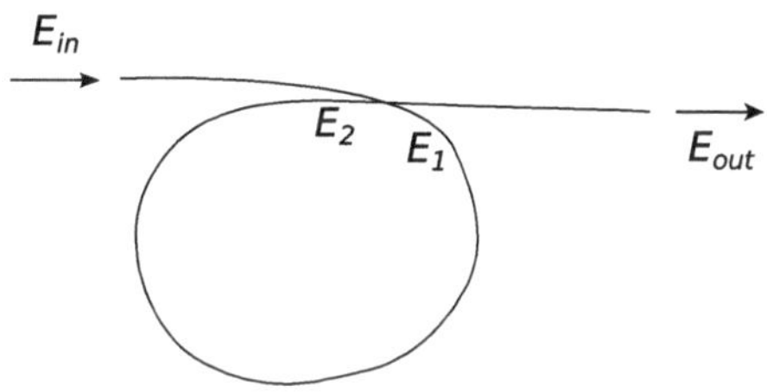

Fig. 3.10 Schematic of the microfibre ring, with electric fields marked

themselves through Van der Waals and electrostatic forces and introduce coupling of the evanescent fields from one fibre to another [27, 35]. Therefore, ring resonators have been fabricated with a loop of fibre [27, 35, 36] to diameters of only a few millimetres.

With light guided in very small cross-sections of fibre there is the potential for the intensity of the light to induce the Kerr nonlinearities and in a loop structure there is the potential for nonlinear dynamics, such as bistability and instabilities. This was investigated theoretically by Vienne et al. [42] and bistability was reported. In this section we shall investigate similar structures as reported in that work, but we shall adopt a slightly different approach with the simulations.

The structure to be considered is shown in Fig. 3.10. This is a standard fibre ring structure composed of a four port coupler with two of the ports, on opposite sides of the coupler, being connected with a length of optical fibre. The key difference here from the SFR device simulated in the MATLAB chapter of this book is that in this case the loop is formed of a microfibre. So, although we will use similar equations for conventional optical fibre the parameters will differ.

First let us consider the coupler. This is described by the equations

$$E_{\text{out}} = \sqrt{1-\rho}\left(i\sqrt{\kappa}E_{\text{in}} + \sqrt{1-\kappa}E_2\right). \tag{3.17}$$

$$E_1 = \sqrt{1-\rho}\left(\sqrt{1-\kappa}E_{\text{in}} + i\sqrt{\kappa}E_2\right). \tag{3.18}$$

The equations are similar to those that have been used elsewhere in this chapter, and the variable names have been chosen to be the same as those in Ref. [42]. This allows the reader to easily compare the analysis and results. The electric field at the different stages of the coupler are denoted in Fig. 3.10, κ is the coupling coefficient and ρ is the fractional power loss. The moving of the light through the fibre of length L, from port 1 to 2 of the coupler, is described by

$$E_2 = E_1 \exp\left(-\frac{\alpha L}{2} + i\varphi\right), \tag{3.19}$$

where α is the power loss through the fibre and phase shift φ is given by

$$\varphi = \beta L + \phi_{nl}. \tag{3.20}$$

Since the cavity length is of the order of millimetres and the loss coefficient, α, is approximately 0.002dB mm^{-1} [42] we shall assume the losses around the fibre are negligible.

Regarding the phase shift, the two terms in equation (3.20) account for the linear phase shift due to the propagation constant, β, and a nonlinear phase shift, ϕ_{nl}. The nonlinearity we will consider here will be only due to the nonlinear optical Kerr effect [1], although it should be noted that bistability due to thermal effects have also been reported in microfibre ring systems [43]. The thermal effects are being neglected due to their slow response (around 6 ms [43]), whereas the nonlinear optical Kerr effect is assumed to be instantaneous [42]. Therefore the nonlinear phase shift, ϕ_{nl}, is given as

$$\phi_{nl} = \gamma_{\text{eff}} LP, \tag{3.21}$$

where γ_{eff} is the effective nonlinear coefficient and P is the optical power at the start of the passage around the loop (within the cavity), at coupler point 2. If L is the length of the fibre forming the circular loop, then assuming the coupler has a negligible length, the diameter of the circular loop is given by $D = \frac{L}{\pi}$.

If we consider the ring to be close to resonance, with just a small low power detuning of ϕ_0, such that $\phi_0 \ll 1$ then for the phase shift around the fibre ring

$$\phi = \phi_0 + \gamma_{\text{eff}} LP. \tag{3.22}$$

However, we need to consider the phase shift contribution of the coupler too. Consideration of equation (3.17) shows the input and the output electric field coupling is proportional to $i\sqrt{\kappa}$, so there is a further phase difference of $-\frac{\pi}{2}$ between the two fields, which would need to be accounted for at resonance in simulations, see [28, 42].

Using the above equations we can simulate the electric fields circulating around the cavity to see the evolution of the field within and emerging from the microring resonator. In the next section, we reproduce some work from Ref. [42]. In that work the authors establish equations for the input power, the intracavity power and the output power. To identify the threshold they use the conditions

$$\frac{d^2P_{\text{in}}}{dP^2} = \frac{dP_{\text{in}}}{dP} = 0, \tag{3.23}$$

where P_{in} is the power entering the microring input and P is the intracavity power. This condition enables us to determine critical values for the power within the cavity P^{cr} and the cavity detuning, ϕ_0^{cr}. These are given [42] as

$$P^{cr} = \frac{2}{\sqrt{3Bx}} \tag{3.24}$$

and

$$\phi_0^{cr} = -\sqrt{\frac{3}{B}}. \tag{3.25}$$

For brevity we shall state two more results from [42], which are the relationships between the input power, P_{in}, and the intracavity power, P, as well as the output power, P_{out}. The equations are derived from (3.17) to (3.23) above. (The interested reader can examine Ref. [42] for details, see equations (16) and (17) in that paper). The equations are

$$\frac{P}{P_{\text{in}}} = \frac{\kappa}{(1-\kappa)}\frac{1}{1+B\phi^2} \tag{3.26}$$

and

$$\frac{P_{\text{out}}}{P_{\text{in}}} = \frac{\kappa B\phi^2}{1+B\phi^2}, \tag{3.27}$$

where it is assumed the losses due to propagation are negligible, $B = \kappa/(1-\kappa)^2$, $\phi = \phi_0 + \phi_{nl}$ and ϕ_0 is a small detuning phase shift, with the condition $\phi \ll 1$.

By using (3.26) and (3.27) it is possible to show that there will be bistability. Figure 3.11 shows a clear region of optical bistability in the region bounded by an input power of 25W to 40W. The parameters used to generate this graph were for silica optical fibre, so $\gamma_{\text{eff}} = 0.09\ \text{W}^{-1}\text{m}^{-1}$, a coupler with $\kappa = 0.95$ and $\rho = 0.05$, a ring diameter of 1mm and a detuning phase $\phi_0 = 1.6\,\phi_0^{cr}$. This figure matches that shown in Fig. 7 of [42]. To get these results we applied a numerical approach similar to the graphical method of Marberger and Felber [23].

Having established the bistable behaviour from equations (3.26) and (3.27) let us now look at the dynamical behaviour of the microfibre ring resonator and use a simulation approach not used in [42]. To do this we shall use equations (3.17) to (3.22). To carry out dynamic simulations an electric field E_{in} is applied

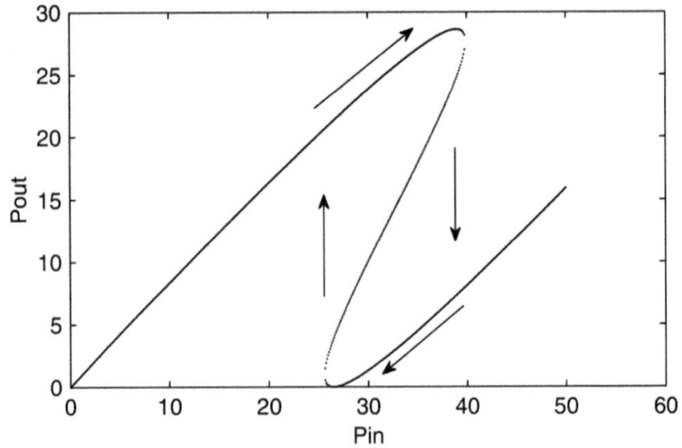

Fig. 3.11 Optical bistability for $\gamma_{\text{eff}} = 0.09\ \text{W}^{-1}\text{m}^{-1}$, $\kappa = 0.95$, $\phi_0 = 1.6\,\phi_0^{cr}$ and a ring diameter of 1mm. The arrows illustrate the bistable region and the path followed for P_{out} depending on whether P_{in} is increasing or decreasing. This is a clockwise bistable cycle

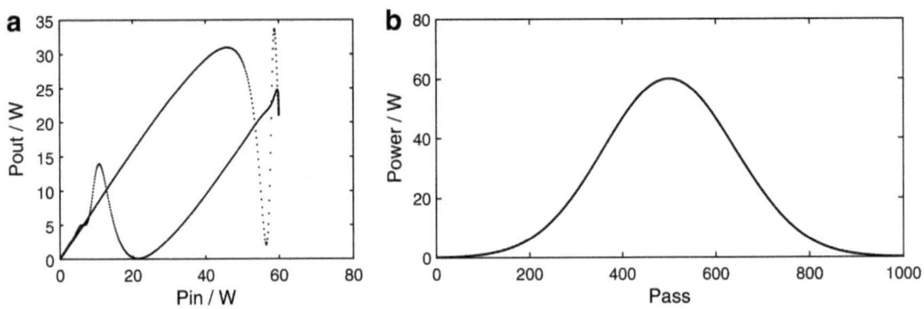

Fig. 3.12 (**a**) Dynamic bistability for the same parameters as used in Fig. 3.11, note the ringing at both ends of the bistable region. (**b**) The Gaussian input pulse

that rises and then falls over time (following a Gaussian pulse) and the evolving electric field around the ring is calculated each pass, similar to the second iterative approach described in [42]. The output field E_{out} can be calculated and then a plot of corresponding power out as a function of power in can be made. One should be reminded that in such a simulation that there needs to be a $-\frac{\pi}{2}$ phase shift included in the calculation (as mentioned earlier) to account for the near resonance conditions.

Figure 3.12 illustrates the dynamical behaviour for similar conditions used to obtain the results in Fig. 3.11. We can see from Fig. 3.12 that the bistable region is wider in input power compared to the bistable region in Fig. 3.11. There is also some oscillation or ringing at the bistable transition point. This type of output response has been reported by Ogusu [26] who examined conventional singlemode fibre ring resonators. This is not surprising as the structures are similar, albeit the microring is smaller in both ring and fibre diameters. Ogusu also showed that it is possible to

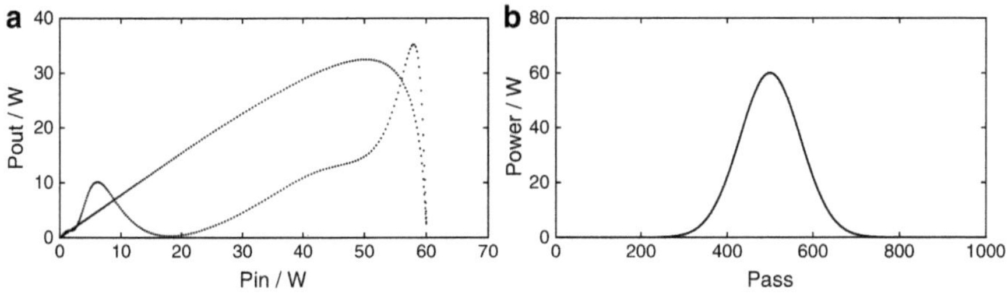

Fig. 3.13 (**a**) Dynamic bistability for the same parameters as used in Fig. 3.11. (**b**) A narrower Gaussian input pulse

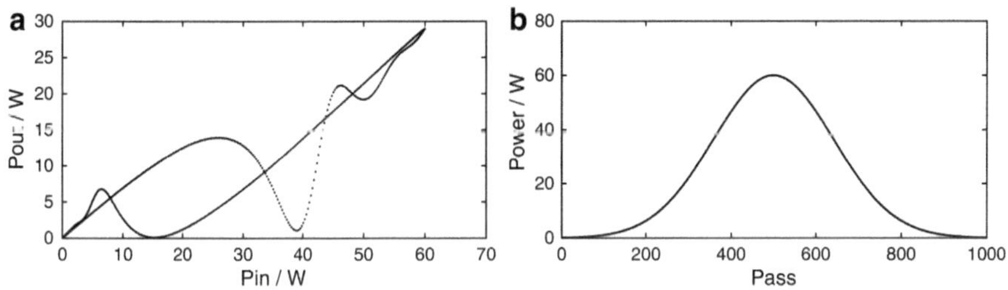

Fig. 3.14 (**a**) Dynamic bistability for the same parameters as used in Fig. 3.11. (**b**) Gaussian input and reducing the detuning phase to $1.2\,\phi_0$

change the nature of the bistable regions by varying the input pulse width and the detuning parameter. We too can see similar variations as shown in Figs. 3.13 and 3.14, where the pulsewidth and the detuning phase are changed from the parameters used to generate the bistable region in Fig. 3.11.

The effect of the detuning phase was also examined in [42] and they too showed the effect on the bistable region (see Fig. 8 in [42]), although they did not use the numerical approach taken which shows the dynamic behaviour.

The results in this section show that microring resonators can not only be small but can offer bistable behaviour within reasonable power levels. The numerical approach taken of following the electric fields for a pulsed input not only shows the bistable behaviour but also ringing at the bistable transition points, something that a graphical method approach does not reveal. Similarly, the variation in the pulse width affects the bistable region which is not revealed with the graphical method. Within the power regions considered here there is no evidence of instabilities. So, given the small size, the lack of instabilities and the power levels involved, microfibre resonators may be a good candidate for optical bistable devices. It should also be noted that to reduce the power level of the bistable region even further microfibres could be made from different glasses, such as those made from heavy metal oxide glasses or chalcogenide glasses. Reference [42] does discuss these glasses and gives the values of $\gamma_{\text{eff}} \approx 5.2\,\text{W}^{-1}\,\text{m}^{-1}$ for a tellurite (a heavy metal glass) fibre and $\gamma_{\text{eff}} \approx 47\,\text{W}^{-1}\,\text{m}^{-1}$ for a $Ge_{20}Ga_5Sb_{10}S_{65}$ (chalcogenide) fibre. The interested reader may want to conduct their own simulations, based on the above approaches, to examine the impact these effective nonlinearities will have on the power levels of the bistable regions.

3.4 Conclusion

In this chapter we have investigated two different kinds of optical resonator, namely the NOLM with feedback and the microring, and examined the nonlinear dynamics of these devices. The resonators considered are quite different in structure, but both types rely on the nonlinear refractive index to induce the nonlinear behaviour. The output from both types of resonators show bistable behaviour and other nonlinear effects, such as instabilities for the NOLM with feedback and overshoot and undershoot with the ring system. Different numerical techniques have been used to examine the nonlinear behaviour. We have shown that although the graphical technique can help identify the regions of bistability, to examine the dynamical behaviour we have used an iterative technique that follows the electric field as it passes around the resonator.

To use these resonators for all-optical switching it would be preferable to have a low-power and stable device. We can see then that the recently developed microfibre resonators could be the preferred option, ignoring any manufacturing difficulties, as they offer low switching power and no instabilities apart from the ringing that can occur at the transition points. It would be interesting to investigate how this ringing can be reduced for these devices.

In order to enable the reader to reproduce the bifurcation diagrams appearing in this chapter, MATLAB function and script files will be made available to download from the MATLAB Central File Exchange.

References

1. Agrawal, G.P.: Nonlinear Fiber Optics, 4th edn. Academic (2006)
2. Babkina, T.K., Bass, F.G., Bulgakov, S.A., Grigor'yants, V.V., Konotop, V.V.: On multistability on nonlinear fiber interferometer with recirculating delay line. Optics Comm. **78**, 398–402 (1990)
3. Borresen, J., Lynch, S.: Binary Half Adder and other Logic Circuits. UK patent pending application number 1011110.2 (2010)
4. Boscolo, S., Turitsyn, S.K., Blow, K.J.: Nonlinear loop mirror-based all-optical signal processing in fiber-optic communications. Opt. Fiber Tech. **14**, 299–316 (2008)
5. Brambilla, G.: Optical fibre nanowires and microwires: a review. J. Optic. **12**, 043001 (2010)
6. Cao, W., Wai, P.K.A.: Comparison of fiber-based Sagnac interferometers for self-switching of optical pulses. Opt. Comm. **245**, 177–186 (2005)
7. Desurvire, E.: Erbium-doped fiber amplifiers: principles and applications. Wiley, New York (1994)
8. Doran, N.J., and Wood, D.: Nonlinear-optical loop mirror. Optics Lett. **13**, 56–58 (1988)

9. Duling, I.N. III: All-fiber ring soliton laser mode locked with a nonlinear mirror. Optics Lett. **16**, 539–541 (1991)
10. Gibbs, H.M., McCall, S.L., Venkatesan, T.N.C.: Differential gain and bistability using a sodium-filled Fabry-Perot interferometer. Phys. Rev. Lett. **36**, 1135–1138 (1976)
11. Fraile-Pelaez, F.J., Capmany, J., Muriel, M.A.: Transmission bistability in a double-coupler fiber ring resonator, Optics Lett. **16**, 907–909 (1991)
12. Gibbs, H.M., Hopf, F.A., Kaplan D.L., et al.: Observation of chaos in optical bistability. Phys. Rev. Lett. **46**, 474–477 (1981)
13. Gibbs, H.M.: Optical Bistability: Controlling Light with Light. Academic (1985)
14. Hwang, J., Pototschnig, M., Lettow, R., et al.: A single-molecule optical transistor. Nature **460**, 76–80 (2009)
15. Ibrahim, T.A., Amarnath, K., Kuo, L.C., et al.: Photonic logic NOR gate based on two symmetric microring resonators. Optics Lett. **29**, 2779–2781 (2004)
16. Ikeda, K.: Multiple-valued stationary state and its instability of the transmitted light by a ring cavity system. Optics Comm. **30**, 257–261 (1979)
17. Ja, Y.H.: Multiple bistability in an optical-fiber double-ring resonator utilizing the Kerr effect. IEEE J. Quant. Electron. **30**, 329–333 (1994)
18. Lee, J.H., Kogure, T., Richardson, D.J.: Wavelength tunable 10-GHz 3-ps pulse source using a dispersion decreasing fiber-based nonlinear optical loop mirror. IEEE J. Quant. Electron. **10**, 181–185 (2004)
19. Lim, H.C., Futami, F., Kikuchi, K.: Polarization independent wavelength-shift-free optical phase conjugator using a nonlinear fiber Sagnac interferometer. IEEE Photo. Technol. Lett. **11**, 578–580 (1999)
20. Lynch, S.: Dynamical Systems with Applications using Maple, 2nd edn. Springer, Birkhäuser (2010)
21. Lynch, S.: Dynamical Systems with Applications using Mathematica. Springer, Birkhäuser (2007)
22. Lynch, S.: Dynamical Systems with Applications using MATLAB., Springer, Birkhäuser (2004)
23. Marburger, J.H., Felber, F.S.: Theory of a lossless nonlinear Fabry-Perot interferometer. Phy. Rev. A **17**, 335–342 (1978)
24. Matsumoto, M., Hasegawa, A.: Adiabatic amplification of solitons by means of nonlinear amplifying loop mirrors. Optics Lett. **19**, 1019–1021 (1994)
25. Nakatsuka, H., Asaka, S., Itoh, H., et al.: Observation of bifurcation to chaos in an all-optical bistable system. Phys. Rev. Lett. **50**, 109–112 (1983)
26. Ogusu, K.: Dynamic behavior of reflection optical bistability in a nonlinear fiber ring resonator. IEEE J. Quant. Electron. **32**, 1537–1543 (1996)
27. Pal, P., Knox W.H.: Fabrication and characterization of fused microfiber resonators. IEEE Photonics Tech. Lett. **21**, 766–768 (2009)
28. Schwelb, O.: Transmission, group delay, and dispersion in single-ring optical resonators and add/drop filters – a tutorial overview. J. Lightwave Tech. **22**, 1380–1394 (2004)
29. Seidel, H.: Bistable optical circuit using saturable absorber within a resonant cavity. U.S. Patent 3610731 October (1971)
30. Shi, C.-X.: Nonlinear fiber loop mirror with optical feedback. Optics Comm. **107**, 276–280 (1994)
31. Simova, E., Golub, I., Picard, M.J.: Ring resonator in a Sagnac loop. J. Opt. Soc. Am. B **22**, 1723–1730 (2005)
32. Smith, S.D.: Towards the optical computer. Nature **307**, 315–316 (1984)
33. Sotobayashi, H., Sawaguchi, C., Koyamada, Y., et al.: Ultrafast walk-off-free nonlinear optical loop mirror by a simplified configuration for 320-Gbit/s time-division multiplexing signal demultiplexing. Optics Lett. **27**, 1555–1557 (2002)
34. Steele, A.L., Lynch, S., Hoad, J.E.: Analysis of optical instabilities and bistability in a nonlinear optical fibre loop mirror with feedback. Optics Comm. **137**, 136–142 (1997)

35. Sumetsky, M., Dulashko, Y., Fini, J.M., et al.: Optical microfiber loop resonator. Appl. Phys. Lett. **86**, 161108-1-3 (2005)
36. Sumetsky, M., Dulashko, Y., Fini, J.M., et al.: The microfiber loop resonator: theory, experiment, and application. J. Lightwave Tech. **24**, 242–250 (2006)
37. Sumetsky, M.: Optical micro- and nanofibers for sensing applications. Proc. SPIE **6556**, 65560J (2007)
38. Sumetsky, M.: Recent progress in the theory and applications of optical microfibers. Proc. SPIE **6593**, 659302 (2007)
39. Szöke, A., Daneu, V., Goldhar, J., et al.: Bistable optical element and its applications. Appl. Phys. Lett. **15**, 376–379 (1969)
40. Tong, L., Gattass, R.R., Ashcom, J.B., et al.: Subwavelength-diameter silica wires for low-loss optical wave guiding. Nature **426**, 816–819 (2003)
41. Tong, L., Lou, J., Mazur E.: Single-mode guiding properties of subwavelength-diameter silica and silicon wire waveguides. Optic. Express **12**, 1025–1035 (2004)
42. Vienne, G., Grelu, P., Pan, X., et al.: Theoretical study of microfiber resonator devices exploiting a phase shift. J. Optics A: Pure Appl. Optics **10**, 025303 (2008)
43. Vienne, G., Yuhang, Li, Limin, Tong, et al.: Observation of a nonlinear microfiber resonator. Optics Lett. **33**, 1500–1502 (2008)
44. Linjie, Zhou, Djordjevic, S.S., Fontaine N.K., et al.: Silicon microring resonator-based reconfigurable optical lattice filter for on-chip optical signal processing. LEOS Annual Meeting Conference Proc., 2009. LEOS '09. IEEE, pp. 501–502 (2009)

Part III
Chaos and Nonlinear Dynamics in Building Mechanism and Fluid Dynamics

Chapter 4
Application of Chaos Control Techniques to Fluid Turbulence

Sridhar Muddada and B.S.V. Patnaik

4.1 Introduction

In non-linear dynamics, sensitive dependence on initial conditions is known as chaos, which is pervasive in fluids. All fluid mechanical systems in nature and technology are prone to chaos. The popular butterfly effect is often stated as follows: *if a butterfly flutters in Brazil, it can cause torandoes in Texas* [1]. Understanding chaos in fluids is the chief outstanding problem in non-linear science and is often regarded with reverence. To keep it straight and simple, chaos in the context of fluids is known as turbulence and its various facets are well investigated by researchers. However, the outstanding nature of the turbulent flows, continues to attract the attention and fascination of physicists, engineers and Mathematicians alike. The non-linear partial differential equations, given by Navier-Stokes exactly describe turbulent flows. Ideally, it is possible to employ a fine grid resolution and simulate the entire spectrum of eddy sizes that are present in the flow. Thus, temporal and spatial behaviour of fluid motion is described by partial differential equations. By approximating the spatial co-ordinates with a finite grid, a set of algebraic equations can be derived and time marching strategies can be devised. The level of complication increases as fluid dynamics involves multi-degree of freedom systems [2].

Turbulent flows and the associated mass, momentum, thermal and species transport is a key element in a number of engineering systems that involve energy and environment, efficient production processes, securing high quality products etc. Progress in turbulence would impact future technological developments and in turn the quality of human life. In terms of traditional engineering goals, these

S. Muddada · B.S.V. Patnaik (✉)
Department of Applied Mechanics, Fluid Mechanics Laboratory, Indian Institute of Technology Madras, Chennai 600 036, India
e-mail: sridhar_fmlab@yahoo.co.in; bsvp@iitm.ac.in

S. Banerjee et al. (eds.), *Applications of Chaos and Nonlinear Dynamics in Engineering – Vol. 1*, Understanding Complex Systems, DOI 10.1007/978-3-642-21922-1_4,

would translate into turbulent drag reduction, noise suppression, enhancement of heat transfer, better mixing, faster chemical reactions etc. [3]. The major aim of turbulence research has been: (a) to understand highly nonlinear nature of its mechanics, (b) to develop predictive methods for turbulent flow phenomena and (c) to devise suitable control schemes [4].

Good mixing can be promoted by exploiting turbulence since mixing by diffusion is a very slow process. Control of turbulent mixing is even more important as restrictions on NOx emissions and Unburnt Hydro Carbons (UHC) require focus on smart sensing and control strategies. In traditional control methods such as twisted tapes, blowing and suction, acoustics, polymer additives etc., feed back component is missing. However, under smart control of turbulence, sophisticated interactive control schemes can be devised by exploiting micro-machining technology and a feedback loop. The complex spatio-temporal features associated with turbulence and its exploitation can be better understood by studying control strategies in other non-linear dynamical systems which exhibit chaos. Therefore, simulation, accurate prediction and control of chaotic systems is central to the understanding of fluid turbulence.

Ott, Grebogi, and Yorke [5], were the first to demonstrate the control of chaos on a Hénon map. Chaos control techniques have emerged for a wide range of chaotic systems such as mechanical, electronic, and chemical, as well as solid-state lasers, heart tissues, pattern dynamics, and turbulence in the Ginzburg-Landau equation were studied [6]. Turbulence is of primary concern in a wide diversity of applications ranging from nuclear reactors to molecular chaos. A major goal of turbulence study is to predict and control turbulence so that the phenomenon can be exploited (either suppressed or enhanced at will) as circumstances dictate in a specific application [7]. Although a wide spectrum of techniques are available in the literature, only a few methods will be discussed here. Special attention will be focused on the control of shear flows such as, wakes as it makes a good precursor to turbulence control. Flow past a circular cylinder is chosen for a thorough analysis. Bluff body wakes are complex as they involve interaction with three type of shear layers, i.e., a boundary layer, a separating free shear layer, and a wake [8]. Dictated by the Reynolds number, the flow behind a cylinder exhibits a variety of complicated phenomena, such as the transition from a steady pair of eddies to an unsteady vortex shedding in the wake, and the scenario of Kármán vortex street formation, transition, and turbulence. The shear layer instability and the wake at low Reynolds numbers, indeed forms a supercritical bifurcation of Hopf type and can be described by the Ginzburg-Landau equation. As the Reynolds number increases, the Kármán vortex street develops due to self-excited and self-sustained oscillations [9]. To contain these oscillations, a coupled synchronization scheme is constructed, based on a reactive feedback control developed through sensors and actuators [10]. The control of alternating lift forces is achieved by annihilating the wake structures through a novel momentum injection strategy [11, 12]. The basis behind all these wake and turbulence control strategies will be analyzed. With the above background, the present review focuses on the following objectives:

- To summarize and review how chaos is manifested in non-linear dynamics and exemplified in fluid turbulence,
- To review popular chaos control strategies and verify their applicability to fluid turbulence,
- To investigate a particular instance of fluid turbulence viz., the bluff body wakes and control of these vortex structures.

4.2 Relevance of Chaos to Fluid Turbulence

Random, erratic fluctuations which are present in a spatio–temporal signal is popularly called turbulence in fluids. This is the subject of discussion not just in fluids but in a variety of other fields ranging from financial markets to human physiology. Understanding and controlling these signals of either price movements or neuronal origin is of societal benefit. In the context of developing new technologies, engineers have toiled millions of hours to understand turbulence and its effect on device designs. It is also the subject of discussion in scientific meetings, where it is seen with reverence and respect, and often treated as the chief outstand difficulty of our subject. The British physicist, Horace Lamb, told a meeting of the British association for the advancement of Science: "I am an old man now, and when I die and go to heaven there are two matters on which I hope for enlightenment. One is quantum electrodynamics, and the other is the turbulent motion of fluids. And about the former I am rather optimistic".

Understanding chaos is vital to understanding turbulence. Turbulence is manifested with a broad spectrum of scales which range from integral (as big as the size of largest length scale in that system) to dissipative scales. The latter is popularly known as Kolmogorov's scale eddies. Typically, this entire range of eddy scales are present in a turbulent flow at sufficiently high Reynolds number.

The interlink that exists between chaos and turbulence is often not apparent. In fluids, the geometric nonlinearity that exists in the convective terms exemplify and generate turbulence. However in fluids the main objectives are [13]:

1. To seek order within apparent chaos,
2. To "explain" important "mechanisms" in the flow in terms of its elemental structures, and
3. To identify how these "important" structures can be manipulated to achieve engineering goals such as drag reduction, enhanced heat transfer, reduced noise, improved mixing etc.

Spatio-temporal chaos manifests itself in the random-like occurrence of certain local spatial structures, and might possess an underlying low-dimensional chaotic attractor [2]. By projecting the dynamics onto rapidly expanding directions, Egolf et al. [14] have identified dynamical processes of relevance to large scale systems (see Fig. 4.1). These structures are related to novel phenomena, such as nonlinear waves, spikes, fronts, boundary effects, coherent behaviour. The presence of

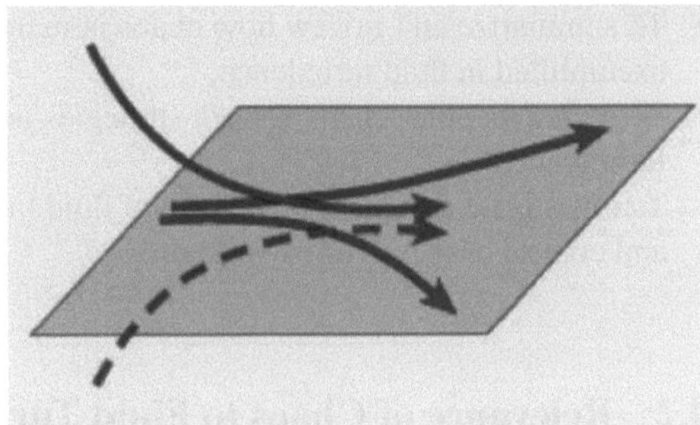

Fig. 4.1 Manifestation of spatiotemporal is exemplified in this figure [14]. In a chaotic system, nearby trajectories separate exponentially along some directions in phase space, while they converge along other directions (*blue arrows*)

spatio-temporal chaos can be shown through measurement and/or simulation. But, it cannot be detected by just looking at the equations and parameters. The presence of spatio–temporal chaos (not the pure temporal chaos) was detected in problems such as, epidemics and population dynamics.

Many phenomena that look random in first glance are indeed examples of deterministic chaos which offers a better way to understand complex systems [15]. Phenomena that have been shown to be chaotic include the transition to turbulence in fluids [16], many mechanical vibrations [17], irregular oscillations in chemical reactions [18], the rise and fall of epidemics [19]. Several studies have argued that certain cardiac arrhythmias are instances of chaos [20]. This is important because the identification of a phenomenon as chaotic may make new therapeutic strategies possible.

4.2.1 Convection Cells and Chaos

The Lorenz Attractor is a crude model for the atmospheric flows that mimics the global weather patterns for forecasting through the simulations in a rotating annulus. Several drastic approximations were applied to simplify the equations for flow in a differentially heated annulus. The system of equations that Lorenz [21] proposed are:

$$\frac{dx}{dt} = -\sigma(y-x); \quad \frac{dy}{dt} = Rx - y - xz; \quad \frac{dz}{dt} = -Bz + xy. \tag{4.1}$$

Here, $x(t)$ is proportional to the intensity of convective motion, while $y(t)$ and $z(t)$ reflects some broad features of the temperature field in the fluid. The Parameters σ, B and R associated with the system are,

$$\sigma = 10,\ B = \frac{8}{3},\ R = 28 \tag{4.2}$$

The mathematical model for this fluid motion was originally developed by Lord Rayleigh in 1916. The Lorenz attractor is shown in Fig. 4.2. The Lorenz system represents a model of thermal convection which describes the motion of some

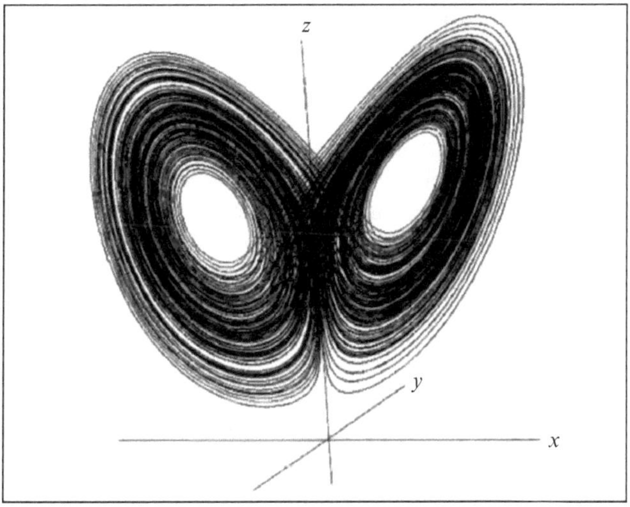

Fig. 4.2 The Lorenz attractor – some trajectories from the lorenz attractor

viscous fluid or atmosphere and information about distribution of heat, and the driving force for thermal convection. There are two main sheets in which trajectories spiral outwards. When the distance from the center of such a spiral becomes larger than some particular threshold, the solution is ejected from the spiral and attracted by the other spiral, where it again begins to spiral out and the process is repeated.

Earth's surface is like a hot plate and when cold air comes in contact, it gets heated up and the warm air rises. This is an important factor in the atmospheric weather patterns. Convective air currents may accumulate and give rise to cells of several types and, when forced more vigorously, may produce turbulent motion in the atmosphere. Examples of convection cells are cylindrical rolls and structures, which are called Bénard cells, resembling a honeycomb from above. In these hexagonal cells the warm portions of the fluid rise in the center, get colder near the top and sinks back to the surface around the cell boundaries. The Lorenz system is related more to the cylindrical roll type of fluid motion in which one of the dimensions can be disregarded pretending that these rolls extend to infinity. It assumes that all convection happens in a rectangular region whose bottom is heated such that the temperature difference between bottom and top remains constant. With certain parameter configurations in this model, Lorenz obtained a system of differential equations for the time dependent amplitudes in the Rayleigh model.

Turbulence is abundant and omnipresent in nature [7]. The fascinating complexity of flow turbulence has attracted the attention of philosophers, poets and scientists alike for centuries. Turbulence is advantageous in many circumstances, such as dispersion of pollutants, fuel mixing in a combustion chamber and heat dispersion in atmosphere etc. However, it is undesirable if it poses danger to aircraft stability due to wake turbulence, flutter etc. apart from causing additional drag.

The control of turbulence is thus of great particular importance. Although the engineering perspective of turbulence control has been studied over decades in terms of passive and active flow control [3], theoretical understanding from the point of view of nonlinear dynamical systems and controlling chaos has been rarely pursued. In the context of turbulence, all the degrees of freedom take part in the entire dynamics of spatio–temporal process. A number of similarities do exist between a variety of signals such as, cardiac, neuronal, price fluctuations etc. and a turbulent flow signal. Although the former appears chaotic, the presence of noise and its elimination can bring order to the flow.

4.3 Turbulent Shear Flows

There is wide spread interest in the simulation of Navier-Stokes equations as it impacts device and process designs. The millennium prize by the Clay foundation [22] on proving the global smoothness and regularity of the flows described by the Navier-Stokes equations is a testimony to this interest among mathematicians as well. The mass, momentum conservation equations govern the fluid motion starting from Newton second law of motion, we can derive the governing equation for a viscous, incompressible, Newtonian fluid as,

$$\frac{\partial u_i}{\partial t} + u_j \frac{\partial u_i}{\partial x_j} = -\frac{1}{\rho}\frac{\partial p}{\partial x_i} + \nu \frac{\partial^2 u_i}{\partial x_j \partial x_j} \tag{4.3}$$

mass conservation equation is given as,

$$\frac{\partial u_i}{\partial x_i} = 0 \tag{4.4}$$

The equations described above are sufficient to simulate the entire range of eddy scales that are present in a turbulent flow. However, such a calculation should have a grid resolution as small as the dissipative scales involved in fluid turbulence, which is in the realm of direct numerical simulations (DNS). It should be mentioned that, at the moment DNS, is beyond the scope and requirements of engineering problem solving. For turbulent flows, u_i and p are dependent on both space and time and appears highly random, chaotic and unpredictable. However, its statistical properties such as $\bar{u}_i$ and $\bar{u_i^2}$ are predictable and repeatable. To that end, statistical averaging route to Navier-Stokes equations was propounded Reynolds [23]. The Reynolds averaging involves splitting the instantaneous variable into a mean component plus a fluctuating component as follows:

$$u_i = \bar{u}_i + u_i'; \qquad p = \bar{p} + p' \tag{4.5}$$

Substituting (4.5) into (4.3) and (4.4) and apply rules of averaging yields the following equations:

$$\frac{\partial \bar{u}_i}{\partial x_i} = 0 \tag{4.6}$$

$$\frac{\partial \bar{u}_i}{\partial t} + \bar{u}_j \frac{\partial \bar{u}_i}{\partial x_j} = -\frac{1}{\bar{\rho}} \frac{\partial p}{\partial x_i} + \nu \frac{\partial^2 \bar{u}_i}{\partial x_j \partial x_j} - \frac{\partial (\overline{u'_i u'_j})}{\partial x_j} \tag{4.7}$$

The last term on RHS of (4.7) is known as Reynolds stress tensor term and represents influence of velocity fluctuations on the mean flow. It should be noted that six additional equations will be required to achieve closure. A variety of modeling approximations can be designed by employing zero, one or two equation based models [13].

Precise definition of turbulence itself is not easy. Therefore, researchers prefer approximate description, which is far from holistic. Turbulent flow is a state of continuous instabilities [24]. A tangle of interacting vortex tubes [25] are some apt definitions. Furthermore, understanding of same basic mechanisms in free and wall bounded turbulence has increased substantially in the last few years. This understanding suggests that the taming of turbulence is to eliminate some of its deleterious effects, while enhancing its useful traits [26]. In an authoritative compilation on transition and turbulence control Gad-el-Hak [26] is of the view that research in flow control is more hotly pursued than any other subject of fluid mechanics.

Fluid turbulence has been differentiated by Osbore Reynolds [23] through a subtle, simple and classic experiment in pipe flow [23]. Beyond a critical parameter, the fluid flow was found to exhibit sinuous oscillations. This is now popularly called Reynolds number, which signifies the ratio of inertial to viscous forces and is given by, $Re = \frac{U_{ref} L}{\nu}$, where L and U_{ref} are characteristic length scale and velocity scale respectively. and ν represents the kinematic viscosity of the fluid. In Fig. 4.3, typical schematic for a general class of shear flows is presented. Here, characteristic length scale refers to pipe diameter (D) for the internal flows problem, (or) cylinder diameter (D) for the external flow configuration.

4.3.1 Modeling Turbulent Fluid Motion

In 1922, Richardson [25] summed up his observations of atmospheric weather patterns in the form of a famous stanza as follows:

Big whirls have little whirls
that feed on their velocity and
little whirls have lesser whirls
and soon to viscosity

He hypothesized energy cascade as shown in Fig. 4.4, in which energy is transferred from large eddies to small eddies, small eddies to smaller eddies and so

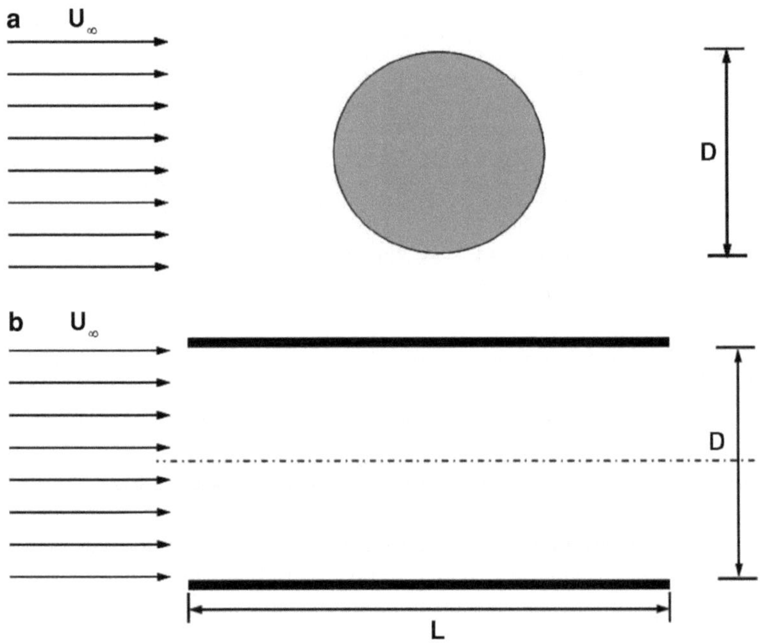

Fig. 4.3 Typical shear flow of problems in nature and technology can be classified as: (**a**) External flow past a body. (**b**) Internal flow through a pipe. Internal and external flow mentioned above are popularly known as closed and open flows

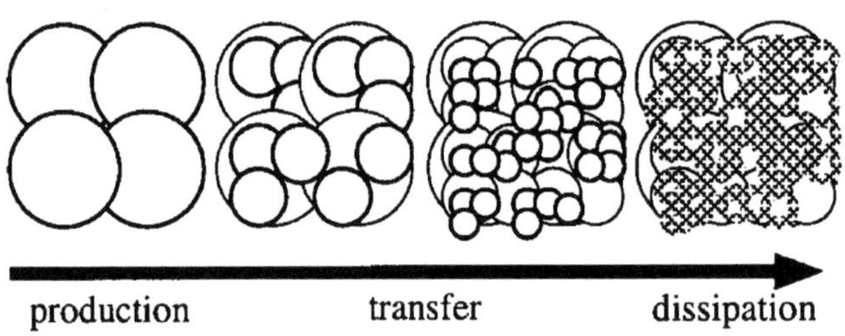

Fig. 4.4 Illustration of the Kolmogorov cascade from large scales to small scales where fluctuating eddies are blurred by viscous dissipation [101]

on... this energy transfer process comes to a halt, when Reynolds number based on small scale eddies is of order unity. This idealized model of energy transfer assumes no loss of energy in the transfer process and complete dissipation happens only at the level of Kalmogorov's scale eddies. His remarkable contribution includes the notion of local isotropy, where the small scales which are smaller than the integral scale eddies and much larger than the dissipative scale eddies display universal features. In the range $\eta \ll r \ll l$, popularly known as inertial subrange, they are completely oblivious of the way they have been generated. [110] postulated that, as

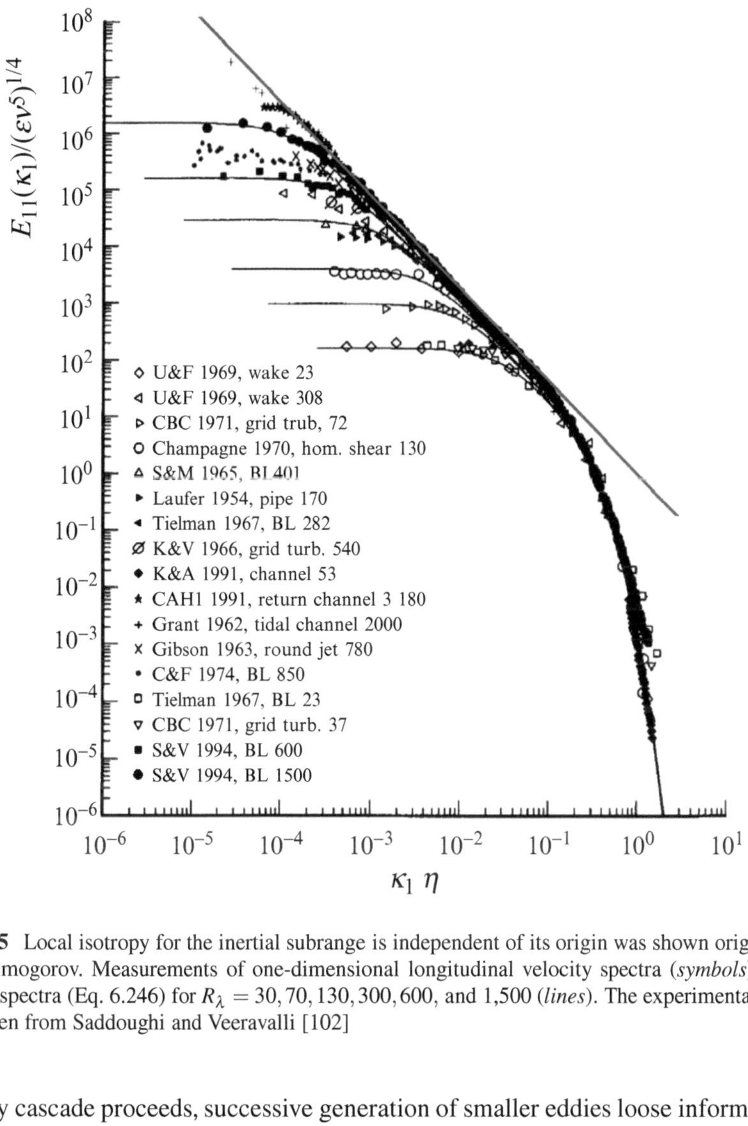

Fig. 4.5 Local isotropy for the inertial subrange is independent of its origin was shown originally by Kolmogorov. Measurements of one-dimensional longitudinal velocity spectra (*symbols*), and model spectra (Eq. 6.246) for $R_\lambda = 30, 70, 130, 300, 600$, and 1,500 (*lines*). The experimental data are taken from Saddoughi and Veeravalli [102]

energy cascade proceeds, successive generation of smaller eddies loose information of the large scale structure of the flow and depict a power law relation as shown in Fig. 4.5. The spatio temporal scales of the streamwise vortices involved in a wide range of applications is shown in Fig. 4.6. It can be noticed that higher the Reynolds number, the finer will be the small scale coherent structures.

Unsteady Reynolds averaged Navier-Stokes (URANS) and Direct numerical simulations (DNS) based calculations are indeed at two ends of the spectrum, both in terms of computational requirements and their ability to resolve the fine scale structure of turbulent flow. Although DNS can simulate all the turbulent eddy scales accurately ranging from Kolmogorov's dissipative scale to the integral scale,

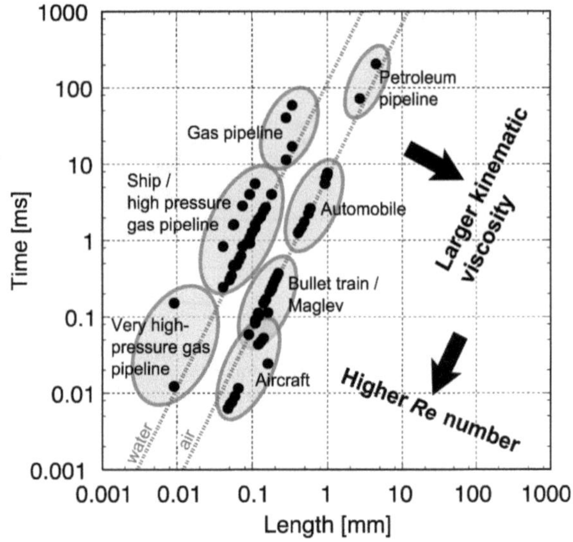

Fig. 4.6 Spatio-temporal scales of coherent structure in real applications [26]

enabling fine grid resolution is very expensive. This computational expenditure shoots up approximately as Re^3 [27]. URANS can only approximate the influence of the entire range of turbulent eddies. Under URANS based turbulence models, the Reynolds stress tensor needs an accurate approximation. A wide range of turbulence models are built depending on the premise used in this approximation. At the lower end, simple zero equation models indulge in a linear approximation to the mean velocity gradient without involving any additional partial differential equations. At the other end, an additional equation for each component of the Reynolds stress tensor and an equation for the rate of dissipation of turbulent kinetic energy (ϵ) can be introduced. A total of seven additional equations come under Reynolds Stress Modeling (RSM), over and above the Reynolds averaged mass, momentum conservation equations. Solving additionally derived equations could again be very expensive both in terms of CPU and RAM space requirements. To this end, a simple alternative would be to stick to the standard two-equation models of turbulence which are popular in industrial practice and their non-linear variants. However, all these models rely on turbulent eddy viscosity hypothesis which assumes a linear approximation between Reynolds stress tensor and the strain rate tensor.

The Reynolds stresses $\overline{u_i'u_j'}$ appearing in the statistically-averaged Navier-Stokes equations are approximated by solving an evolution equation for TKE and its rate of dissipation. The standard $k-\epsilon$ model of Launder and Spalding [28] uses the eddy viscosity hypothesis of Boussinesq:

$$\overline{u_i'u_j'} = \nu_t \left(\frac{\partial \bar{u}_i}{\partial x_j} + \frac{\partial \bar{u}_j}{\partial x_i} \right) - \frac{2}{3} k \delta_{ij}. \qquad (4.8)$$

This concept works on the assumption that the Reynolds stresses are linearly related to the mean velocity gradients through a proportionality constant, given by the turbulent eddy viscosity (ν_t). In the $k-\epsilon$ model, the turbulent eddy viscosity is related to the turbulent kinetic energy k and to its rate of its dissipation ϵ as

$$\nu_t = c_\mu \frac{k^2}{\epsilon}. \tag{4.9}$$

4.3.2 Wall Bounded and Free Shear Flows

Majority of the flows, which are encountered in engineering practice are either wall bounded or free shear flows. When the velocity gradient is predominantly in the direction normal to the flow, such flows are called shear flows. Jets, wakes, and mixing layers, etc. are characterized as free shear flows, while flow through channels, pipes, flow over flat or curved surfaces are known as wall bounded shear flows. Viscosity-dominated, wall-bounded shear flows are generally more difficult to control due to favourable pressure gradients, which make these flows stable. Free-shear flows and separated boundary layers, on the other hand, are intrinsically unstable and therefore readily amenable to manipulation. Typical wake vortices which are reflective of free shear flows in nature and technology are presented in Fig. 4.7. Flow past obstacles, at sufficiently high Reynolds numbers results in the formation of von Kármán vortex street. In the case of flow past Gran Canaria islands, satellite images of cloud pattern past the island are visualized in Fig. 4.7c. The traces of unstable manifolds with eye of the vortex indicated by A, B, C, D etc. is shown in Fig. 4.7a. Corresponding numerical visualization [11] past a circular cylinder in presented in Fig. 4.7b.

4.3.3 Coherent Structures

Coherent structures do not reflect a well defined behavior in turbulent flows. However they are persistent structures without a precise characterization. This makes it difficult for the quantification tools. A variety of internal flow problems such as flow through pipes, channels and other contrivances and passages come under the broad classification of shear flows. Turbulent structures or quasi-coherent structures in this class of shear flows can be identified either by visualization or by computations. Extraction of these features typically involves conditional sampling techniques, eduction methodologies etc. [13]. The idea is that they are regions of space and time in which the flow field has a characteristic coherent pattern which are significantly larger than the smallest flow or turbulent scales. Different instances of the structure occur at different positions and times, and their flow fields certainly differ in its detail. These coherent structures can explain the behavior of near-wall turbulent flows. There are many structures within the

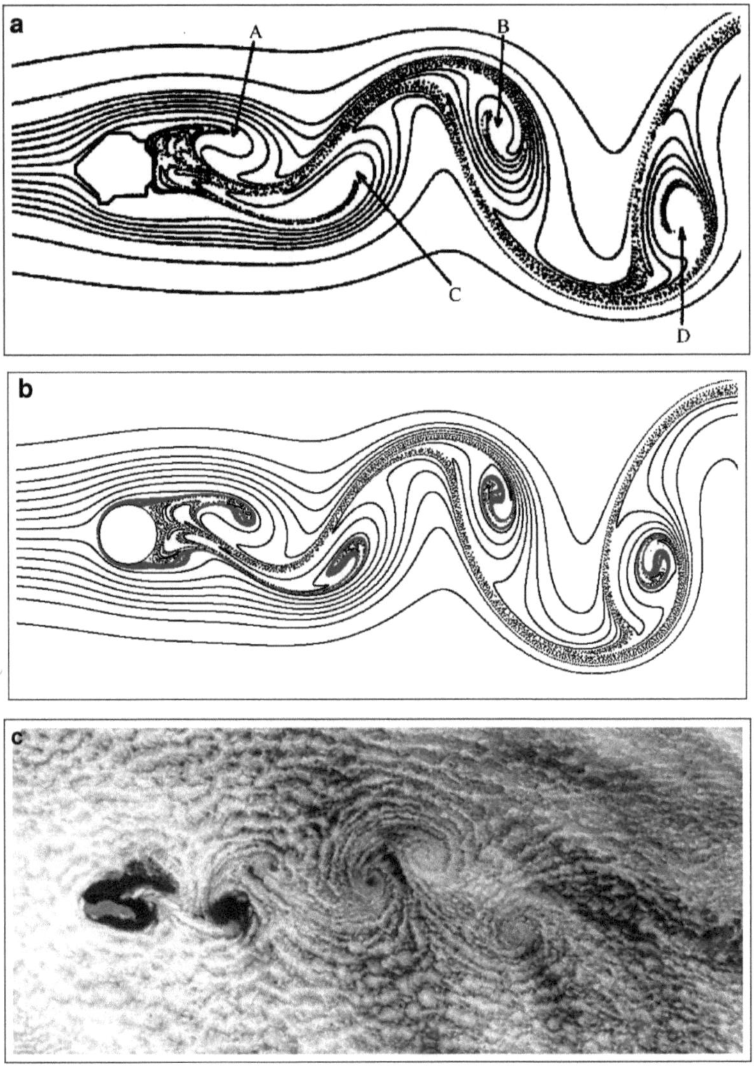

Fig. 4.7 Traces of continuously injected dye for sea current around Gran Canaria island. Streakline plots are shown for (**a**) behind the island [103]; (**b**) behind a circular cylinder; (**c**) satellite image of cloud pattern around Guadalupe Island, August 20, 1999 (NASA SEAWIFS image) [92]

random background, and the deterministic and stochastic interactions are difficult to interpret. Quasi-coherent structures in channel flow and boundary layers have been categorized by Kline and Robinson [29] and Robinson [30] into categories such as, sweep, ejection, vortical structures, near wall pockets, etc. Reviews of experimental work and discussion on these structures are available in Sreenivasan [31], Cantwell [32], and Gad-el-Hak and Bandyopadhyay [33], among others.

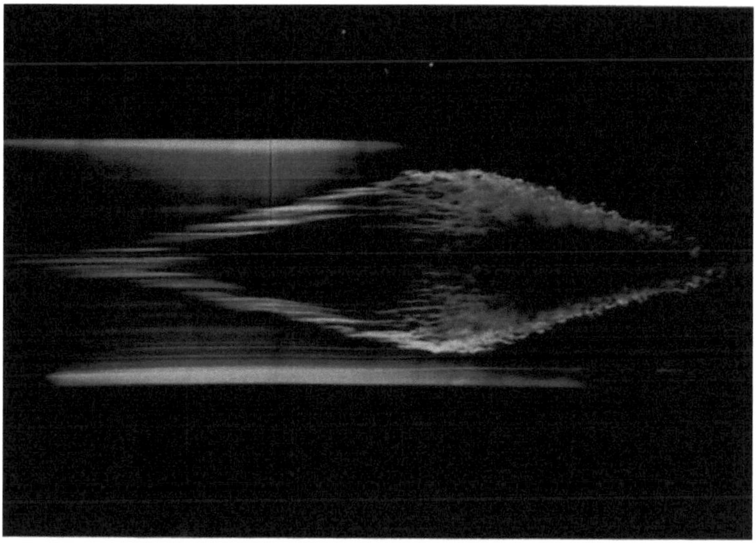

Fig. 4.8 Top view of an artificially generated turbulent spot evolving in a laminar boundary layer. The displacement thickness Reynolds number at the spot's initiation point is $Re_\theta = 625$ [45]

Figure 4.8 depicts the top view of an artificially generated turbulent spot evolving in a laminar boundary layer [34]. The displacement thickness based Reynolds number at the point of spot initiation is 625, which is well above the critical *Re* for linear instability. Laser-induced fluorescence imaging was used to visualize different cuts through the growing turbulent structure. In this figure, the laser sheet is parallel to and very near the flat wall. The dynamics within the spot appear to be controlled by many individual eddies similar to those within a turbulent boundary layer. The spot grows in the spanwise direction by an efficient mechanism, which Gad-el-Hak et al. [34] have termed the growth by destabilization process. Near the edges of the spot, the dye lines are sharp, which is indicative of the initial breakdown into chaotic motion. Toward the middle, on the other hand, the dye becomes more diffused because the turbulence there is older and more mixing has taken place.

A typical coherent structure obtained a shadowgraph image is shown in Fig. 4.9, for a mixing layer. Here, one can notice transient but "organized" patterns of motion or density, lasting a long time and of relatively large scale in comparison with small diffusive scale eddies. They are usually detected in natural flows by correlation based techniques or conditional averaging, but may also be visible in satellite images of the sea surface or in photographs of foam created by breaking waves and may be indications of processes underlying turbulent transfers [35]. They often appear similar in form to the dominant mode of instability in a laminar flow that has a distribution of velocity and density similar to the averages of those of the turbulent flow.

The turbulent boundary-layer close to the wall is complicated due to near wall physics that influences the skin friction drag. There are several types of flow

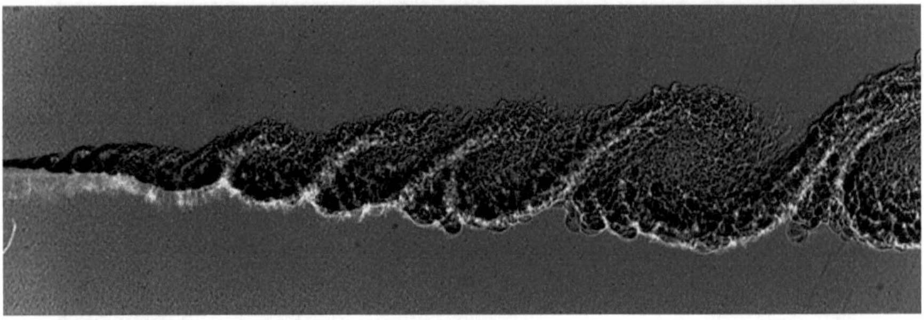

Fig. 4.9 Coherent structures made visible by shadowgraphy for the mixing layer between two streams of different gases [104]. Two gases of different velocities, which are initially separated by means of a splitter plate move from left to right. Eddies begin to grow, as the gases come into contact downstream of the splitter plate. The large eddies co-exist with a smaller-scale incoherent field of non-homogeneous turbulent motion

structures in the near-wall region, which is usually defined as consisting of the viscous sublayer, the transitional layer and the inner part of the log-law region. Historically, hair pin vortices are among the first type of flow structures to be identified by Theodorsen [36]. Later, Kline et al. [37] discovered the sublayer streaks. These are regions that are narrow in the spanwise and wall-normal directions but elongated in the streamwise direction, having streamwise velocities that alternate in the spanwise direction between being higher or lower than the mean velocity. Kline et al. [37] postulated that they were created by a vertical "lift-up" mechanism produced by the induced velocity due to the legs of the hairpin vortices as shown in Fig. 4.10. However, Moin and Mahesh [38] indicate that the streamwise vortex cells located near the wall do not have to always appear in pairs. In fact, even a randomized field of wall-normal velocity perturbations can create streaks [39]. However, in turbulent boundary layers, streamwise vortices in some form are usually considered to be responsible.

4.3.3.1 Convective and Absolute Instabilities

In addition to grouping the different kinds of hydrodynamic instabilities as inviscid or viscous, the other way to classify them is based on convective or absolute instability (Huerre and Monkewitz 1990). This can be studied by looking at system response to an imposed perturbation. A flow is said to be convectively unstable if disturbance grows in the downstream, leaving the source unaffected. Convective instabilities occur when there is no mechanism for upstream propagation of the disturbance. If the disturbance is removed, then perturbation propagates downstream, and the flow relaxes to an undisturbed state. Effective suppression of convective instabilities can be achieved by applying a perturbation near the point, where it originates. If a disturbance grows at the spatial position where it originates,

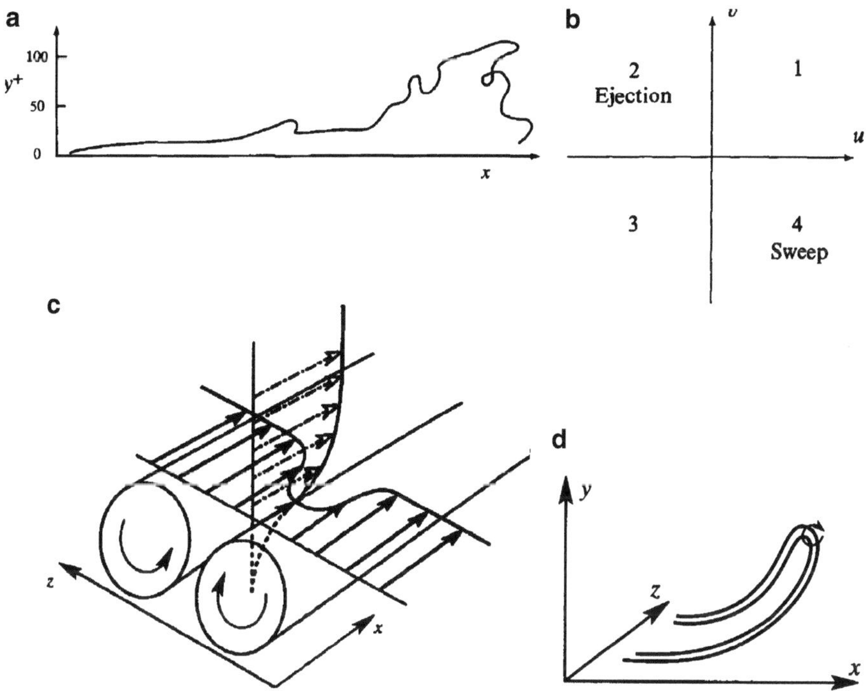

Fig. 4.10 (**a**) A dye streak in a turbulent boundary layer showing the ejection of low-speed near-wall fluid. (**b**) The u-v sample space showing the numbering of the four quadrants, and the quadrants corresponding to ejections and sweeps. (**c**) A sketch of counter-rotating rolls in the near-wall region. (**d**) The hairpin vortex suggested by [111]

where mechanism exists for propagation of disturbance upstream. For flow over a backward-facing step, flow recirculation provides such a mechanism. In this case, some of the growing disturbances can travel back upstream and continually disrupt the flow even after the initial disturbance is neutralized. Therefore, absolute instabilities are much more difficult to control. The two-dimensional flow behind a bluff body has a zone of absolutely and convectively unstable region. Identifying and controlling these zones requires a number of strategies. The upstream addition of acoustic or electric feedback can change a convectively unstable flow to an absolutely unstable one, and self-excited flow oscillations can thus be generated. In any case, identifying the character of flow instability facilitates in identifying an effective amplification or suppression strategy for the perturbations.

For many nonparallel flows such as spatially developing mixing layers, flat–plate boundary layers, and homogeneous jets (see Fig. 4.11a), the mean flow is locally convectively unstable everywhere with respect to vortical fluctuations. In such systems, any initial disturbance is advected by the flow as it is amplified. The medium is extremely sensitive to external coherence forcing and the flow can be thought of as a collection of spatially evolving vortical instability waves of different

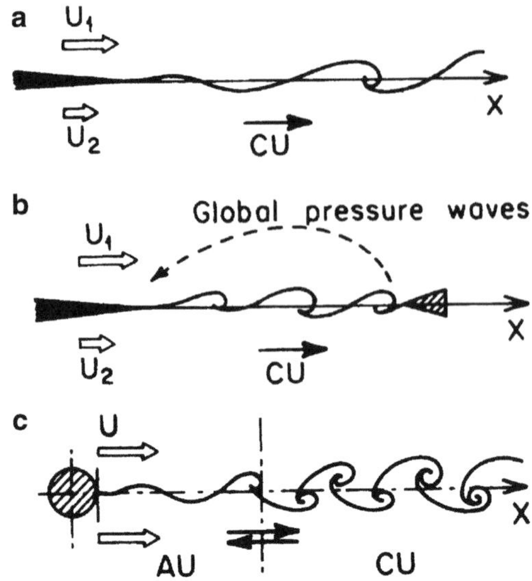

Fig. 4.11 Classification of spatially developing flows [93]. (**a**) Extrinsic flows: no resonances. (**b**) Intrinsic flows: hydroacoustic resonances. (**c**) Intrinsic flows: with hydrodynamic resonances

frequencies traveling in the downstream direction. Measured frequency spectra are generally broadband. Self-sustained oscillations do not seem to be possible in this class of flows although significant feedback effects could be induced by global pressure fluctuations present in far-field region as sketched in Fig. 4.11a.

In contrast, strong self-sustained oscillations can be obtained when, in the same developing free shear flow, one introduces a second streamlined or blunt body at a finite distance downstream (see Fig. 4.11b). For particular values of the distance between the two objects, a resonance can be triggered which gives rise to monochromatic acoustic radiation known as wake tones, jet tones or edge tones, depending on the specific shear-layer-solid-body configuration. The flow is still locally convectively unstable from the point of view of vorticity fluctuations but the dynamics of the flow is dominated by a feedback loop (Fig. 4.11b): The downstream branch consists of rotational instability waves rolling up into vortices. The interaction between the vortical structures and the downstream body then generates global irrotational pressure disturbances which travel along the upstream branch of the loop. Resonance occurs when vortical shedding at the trailing edge of the upstream body is in phase with the vertical velocity induced by the global pressure fluctuations.

There is now increasing experimental and theoretical evidence that in wakes behind bluff bodies and in inhomogeneous jets, self-sustained oscillations can also be produced by purely hydrodynamic means, without the need for a second downstream body (Fig. 4.11c). In this third class of spatially developing flows, the nature of the instability changes from locally absolute to locally convective at a particular downstream station X_t. This transition allows global oscillations of the separated flow to develop. It is conjectured that the feedback loop is made up of temporally growing vorticity waves propagating in both flow directions. The observed flow is then the finite-amplitude saturated state associated with these waves. It is found experimentally that such flows are relatively insensitive to infinitesimal external perturbations: Discrete frequency spectra are obtained with one or several fundamental components and their harmonics. The archetype of such a flow regime is the von Kármán vortex street behind a circular cylinder at low Reynolds numbers.

4.4 Chaos Control Methodologies

In the theory of dynamical systems, the so-called butterfly effect denotes sensitive dependence of initial conditions on the dynamics of evolutions described by the differential equations. Phase–space solutions which are initially very close and together, separate exponentially. The solution of nonlinear dynamical systems of three or more degrees of freedom may be in the form of a strange attractor whose intrinsic structure contains a well-defined mechanism to produce a chaotic behavior without requiring random forcing. Chaotic behavior is complex, aperiodic and, though deterministic, appears to be random.

Controlling chaos is a process wherein an unstable periodic orbit (UPO) embedded in a chaotic attractor is stabilized by means of tiny perturbations given to the system. These perturbations lead to goal oriented feedback techniques which act either on the state variables of the system or on the control parameters. Arecchi et al. [40] review some theoretical schemes and experimental implementation for the control of chaos. Controlling chaos consists of perturbing a chaotic system in order to stabilize a given UPO embedded in the chaotic attractor (CA). The UPO's constitute the skeleton of chaotic dynamics, which, indeed, can be seen as a continuous irregular jumping process in the neighborhood of different periodic behaviors [41]. Thus, control of chaos implies extracting a desired periodic motion out of a chaotic one, by the application of judiciously chosen small perturbations. The process enables the exploitation of a single dynamical system for the production of a large number of different periodic behaviors, with an extreme ability in switching from one to another. This phenomenon is exemplified in Fig. 4.12, which demarcates phase-lock in between natural and excitation frequencies [9]. Ignoring initial transients, under lock-in, the excitation frequency (f_{ex}) is able to influence the shedding frequency (f_s). It should be noted that a single periodic orbit exists when the shedding frequency was driven into the excitation frequency, while UPO's persist when there is no control.

4.4.1 OGY Control

The OGY algorithm has been widely applied for nonlinear dynamical systems. In several cases, successful experimental chaos control methodologies have been implemented for electronic circuits, chemical reactions, lasers and biological systems

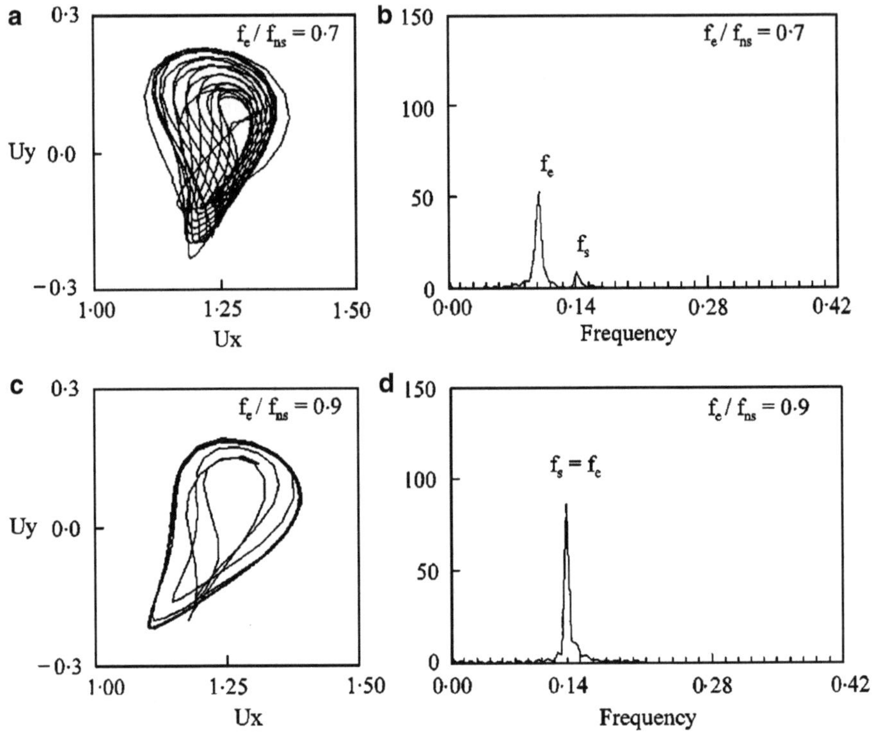

Fig. 4.12 Phase plane and FFTs for a vibrating circular cylinder at $Re = 80$ [9]. (**a**) No-lock-in. (**b**) Lock-in and its corresponding FFT's in (**c**) and (**d**) respectively. The exploitation one of the stable periodic orbits under lick-in can be noticed

etc. The OGY theory was first applied experimentally to controlling the chaotic vibrations of a magnetoelastic ribbon [42] and subsequently to a diode resonator circuit [43] and to the chaotic output of lasers [44]. Granfinkel [15] have found that it is possible to control a chaotic cardiac arrhythmia using the same basic properties of chaotic systems that were exploited by OGY but with some modifications.

Chaos control strategy by Ott-Grebogi-Yorke (OGY) hinges on the fact that, behind every chaotic and unpredictable system, there exists a back bone in the form of an ordered structure. The dynamical evolution passes through series of periodic patterns one after another, which creates the impression of randomness over time. However, an appropriately controlled system, gets locked into one particular type of repeating motion. In such a reactive control, the dynamical system reaches a stable behavior. The OGY method can be best illustrated by means of a schematic in Fig. 4.13 [45]. The state of the system is represented as the intersection of a stable manifold and an unstable one. The control is applied intermittently whenever the system departs from the stable manifold by a prescribed tolerance; otherwise, the control is turned off. The control attempts to put the system back to the stable manifold so that the dynamical state of the system converges toward the

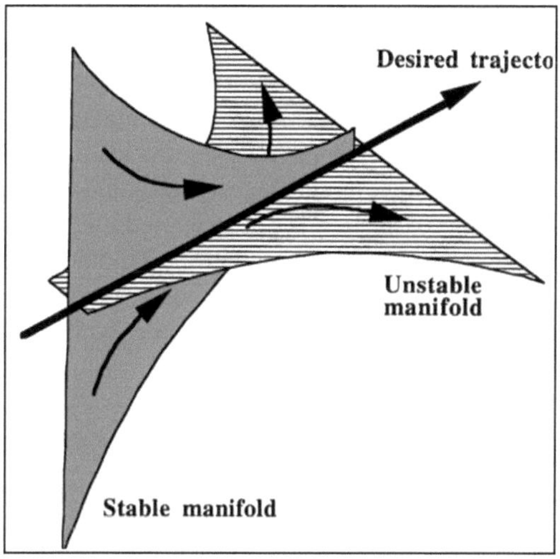

Fig. 4.13 The OGY method for controlling chaos [45]

desired trajectory. Noise may be induced due to unmodeled components in the system and a tendency for the state to wander off in the unstable direction. The intermittent control prevents that, and the desired trajectory is achieved.

OGY algorithm [5] is based on the following ideas:

1. That chaotic sets (attractors, saddles) contain an infinite number of hyperbolic periodic orbits,
2. Furthermore, due to the existence of a natural distribution, the state of the system visits arbitrarily small neighbourhoods of any periodic orbit sooner or later.

An important feature of the algorithm is that, it only applies small controlling effects. Such a control is economical, since tiny external perturbations are sufficient to allow the system to reach the desired behaviour [2].

The first step of the OGY algorithm [5] is to choose a desired periodic orbit on the chaotic set, with the aim of directing the system's dynamics towards the dynamics of this orbit. Next, a small neighbourhood of the orbit, for example a circle around one of the cycle points, is taken, and control is applied only if the state point enters this neighbourhood. Within this region, the linearised dynamics must be known, in particular the eigenvalues and the stable and unstable manifolds of the hyperbolic cycle should be determined. The control itself is maintained by changing one of the system parameters from the closest point of the pre-selected cycle. Instead of the original map $r_{n+1} = M(r_n, \mu)$, a modified map, $r_{n+1} = M(r_n, \mu + \delta\mu_n) \equiv M(r_n, \mu)$, was applied with a parameter perturbation $\delta\mu_n$ proportional to the distance r_n from the cycle point. Note that the originally autonomous system is thus converted into a non-autonomous one,

$$M \rightarrow M_n, \tag{4.10}$$

whose form also depends on the time instant. A feedback has been created between the instantaneous state of the system and its parameter. The parameter perturbation is chosen so that the image under the action of M_n falls on the stable manifold of the pre-selected cycle in the linearised dynamics. Knowing the map M and its linearised version, the required $\delta\mu_n$ can be explicitly determined. In a piece-wise linear system, control would be maintained by such a single control step since the iterated point then approaches the cycle along the straight line of the stable manifold. In general, however, the algorithm should be kept active in later steps also, since the linear approximation is not exact and therefore the image point falls somewhat off the exact manifold. The controlled dynamics converges towards the pre-selected cycle quite quickly, and remains in its vicinity over a long time. By means of a parameter perturbation, we can thus convert an unstable cycle into an attracting limit cycle.

4.5 Application of Chaos Control Techniques to Flow Control

A variety of techniques have been proposed for controlling complex dynamical systems since the first work by OGY [5]. The basic idea of OGY is to take advantage of the sensitivity to small disturbances of chaotic systems to stabilize the system in a desirable unstable periodic orbit naturally embedded in the chaotic attractor. While detailed system equations are not required for the control, a learning process is necessary to obtain the essential information required for the control, such as the location and the eigenvalues of the desirable unstable periodic orbit in the phase space. Nevertheless, it is generally difficult to apply the OGY idea to high-dimensional systems, such as the turbulence. However, a more practical method proposed by Pyragas [46] utilizes a time delayed feedback to some dynamical variables of the system. A periodic orbit embedded in the chaotic set can be stabilized when its period matches the delay time. Control of spatiotemporal chaos is investigated by means of complex Ginzburg-Landau equation (CGLE) [47–49]. Often, in the literature, the popular Ginzburg-Landau (G-L) equation is used to model the cylinder wake as it is straightforward to numerically integrate the amplitude equation, which enables rapid prototyping of control strategies [50, 51]. Suppression of spurious oscillations in spatiotemporal systems was studied and an angular momentum injection technique was proposed for taming the wake turbulence behind a bluff body [10, 52]. However, the viability and effectiveness of using this techniques developed for low-dimensional chaotic systems over the past 10 years for controlling true flow turbulence described by the incompressible Navier-Stokes equations have not been fully exploited. Though the control of chaos, especially spatiotemporal chaos, Guan et al. [79] have investigated the control of flow turbulence by using chaos control strategies. It is emphasized that true flow turbulence governed by the incompressible Navier-Stokes equations differs fundamentally from the “amplitude/defect turbulence” and the “phase turbulence” within the spatio temporal chaos regime governed by the complex Ginzburg-Landau equation. First of all, the complex Ginzburg-Landau equation is an amplitude equation, and it describes the dynamics near the Hopf bifurcation in spatially extended systems. In contrast, the incompressible Navier-Stokes equations have a vector equation and a coupled scalar equation and they describe real hydrodynamic flows. Second, the complex Ginzburg-Landau equation is a prototype of spatio temporal chaos, which might be regarded as a population of chaotic oscillators locally coupled by diffusion. Its transition from stationary states to unsteady states has been studied in terms of Benjamin-Feir instability and the Eckhaus instability, with detailed phase diagram having been identified. Nevertheless, for the Navier-Stokes equations, the transition from laminar flow to turbulent flow remains a long-lasting open question for many real-world systems.

Encountering vibrations due to flow induced oscillations is a common experience in industrial practice. A variety of passive flow control techniques are often available for the suppression of vortex induced oscillations such as, splitter plates, axial slats, fairings, helical strakes etc. [53]. As opposed to passive techniques, active flow control techniques enable dynamic adaptability. However, these techniques require additional power input and can be applied without any modifications to the geometrical configuration. On the cylinder surface, blowing and suction is used as a means of control [54,55], for the purpose of wake stabilization. A systematic control method was developed by Min and Choi [56] using a suboptimal feedback control scheme. Oscillatory motion of the cylinder [57, 58] and external forcing through acoustic excitation [59, 60] are also popular in the literature. With the advent of MEMS based technologies and development of smart materials, which involves intelligent sensing and actuator techniques, methods such as, electromagnetic forcing [61, 62] and piezo-electric actuation [63] are becoming popular. In a combustion application [64] electromagnetic flap actuators are used to manipulate the flame characteristics. Lifted flames could be controlled to achieve combustion stability and reduced toxic emission.

Development of low-order models is useful to analyze high dimensional m-dof problems. Gerhard et al. [65] have presented a low-dimensional Galerkin model for the suppression of vortex shedding behind a circular cylinder. These lower order models are typically constructed from Karhunen-Loève decomposition in the from of Proper orthogonal decomposition (POD) tools. Lumley and Blossey [4] have laid out a broad framework for the control of turbulence, which can be generalized for the purpose of understanding generic flow control mechanisms. The essential ingredients of this approach include a large number of sensors (that report the state of the flow), a number of actuators (to enable actions to be performed and in what proportion and at what instant) and a control algorithm (to decide and parametrize the extent of these interactions).

4.5.1 Advantages of Passive and Active Flow Control

The need for flow control directly translates into fuel savings. For example, the main sources of drag on an aircraft are due to: (a) skin friction, (b) the formation of shock waves, (c) drag due to flow separation leading to form drag [66]. For modern airliners, such as the Airbus A300-600, skin-friction drag accounts for approximately 50% of the total drag at cruise conditions. Most of the other resistance is divided between induced and wave drag, with form drag making only a small contribution at cruise conditions. Induced drag is inevitable, as it is an essential by-product of lift generation for a conventional aircraft [67]. Small drag reductions have been achieved by fitting winglets to the wingtips, in conventional aircrafts. Induced drag is proportional to the square of the aircraft's weight, the largest reductions are likely to come from reducing the total drag, thereby reducing the weight of fuel required. Wave drag reduction was achieved by supercritical wings [68]. To achieve large drag reduction, one should target skin-friction drag.

Broadly, there are two main strategies for reducing skin-friction drag can be followed: (a) maintain laminar flow, since the level of skin friction drag less for laminar flow than for turbulent flow. (b) Try to reduce the skin friction drag of the fully turbulent flow. It is conceivable that by careful design, natural laminar flow can be maintained over a substantial part of the wing. However, to go further than this using current proven technology would necessitate the use of laminar-flow control in the form of boundary-layer suction [45,69], compliant walls [45,70] etc., which offer a passive method for laminar-flow control but are not practical for use in aircraft [71].

Among passive flow control strategies, riblets have been found to be phenomenally successful with the commercial airlines [72]. Hydrodynamics of flow past shark surfaces have been extensively investigated [73] to gain this understanding. Riblets take the form of minute streamwise ridges and valleys on the aircraft's surface. A polymeric riblet film with triangular-shaped ridges and valleys is produced by the 3M company and used by Cathay Pacific on their Airbus A300-600 aircraft. To achieve a reduction in skin-friction drag, the spacing between adjacent ridges has to be in the range of 25–75 μm for flight conditions. Approximately 6% reduction in skin-friction drag was achieved with the application of such riblets to the surface. Better drag reduction is possible with the more advanced riblet designs developed by Bechert et al. [73]. Different roughness treatments such as, randomized, chevron-shaped, "roughness" elements of Sirovich and Karlsson [74] (see also [75]), approximately 10% drag reduction was achieved. The right choice of roughness element and distribution can delay the onset of laminar-turbulent transition was demonstrated in Ref. [76]. For all these various types of surface treatment, 10–12% is probably the ultimate reduction potential that can be achieved through skin-friction drag for commercial airliners.

4.5.2 Model Control

Abergel and Temam [77] have laid out the basic framework for minimizing turbulence in an evolutionary Navier-Stokes flow from the view point of optimal controls. This would encompass minimization of the cost functional, which is based on turbulence measured by the L_2 norm of the curl vector. Typically the state of the flow with a known physical domain and initial configuration is chosen, and we can act upon the fluid through various devices (body forces, boundary values, temperature, etc.), then the issue of determining the optimal action would depend on the system in order to minimize turbulence within the flow. Often control with feedback is required for two reasons: (a) To change the dynamics of the system–to make the response stable and sufficiently fast. (b) To reduce uncertainty and improve system sensitivity. Designing suitable controllers and analyzing the resulting controlled system is therefore plausible through such a feedback, to see if the specifications are satisfied and further controller tuning is required. Medjo et al. [78] have recently reviewed a mathematical and numerical framework, with emphasis on contemporary methods and optimization algorithms for flow control models.

4.5.3 Turbulence Control: The Grand Problem in Non-Linear Dynamics

Since the pioneer work of Ott, Grebogi, and Yorke in 1990, controlling chaos has been extensively investigated. A variety of approaches, such as the OGY scheme, openloop strategy, feedback technique and adaptive method, have been developed for the purpose of chaos control. Chaos control have been implemented for spatially extended dynamical systems, such as coupled map lattices and partial differential equations. These works are motivated by potential applications in laser and plasma physics, chemical reactions, electric circuits, neuronal networks as well as secure communication. One of the most complicated spatiotemporal systems is the fluid turbulence. Guan et al. [79] have shown that, flow turbulence governed by the Navier-Stokes equations can be effectively controlled by global and local pinning methods. A variety of synchronization based coupling methods are available in nonlinear dynamics community. Their utility to turbulence control is yet to be proven.

4.5.4 Control Goals and Means

The potential benefits of turbulence control in fluid flow systems are numerous. It ranges from saving billions of dollars in fuel costs for land, air, and sea vehicles to achieving economically competitive and environmentally friendly systems and devices for industrial practice. Choosing a particular type of control strategy depends on the kind of flow control goal. Flow-control goals are strongly, often adversely, interrelated, and there lies the challenge of making the tough compromises. There are several different ways for classifying control strategies to achieve a desired effect. Presence or lack of walls, Reynolds and Mach numbers, and the character of the flow instability are all important considerations for the type of control to be applied [3].

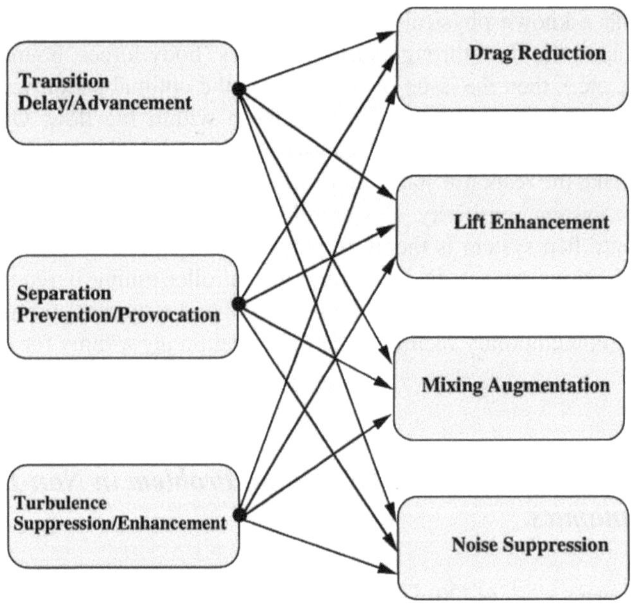

Fig. 4.14 Engineering goals and corresponding flow changes [45]

An engineer, typically aims to achieve: (a) drag reduction; (b) lift enhancement; (c) augmenting mixing of mass, momentum, or energy; (d) suppression of flow induced noise etc. To achieve any of these useful end results for either free-shear or wall-bounded flows, transition from laminar to turbulent flow may have to be delayed or advanced, flow separation may have to be prevented or provoked, and finally turbulence levels may have to be suppressed or enhanced [3]. All these engineering goals and the corresponding flow changes intended to effect them are schematically depicted in Fig. 4.14. Although some of these goals are easy to achieve, some of them are mutually exclusive. Therefore, developing a simple device that is inexpensive to build as well as to operate would require optimization. For external wall-bounded flow, such as that developing on the exterior surfaces of an aircraft or a submarine, can be manipulated to achieve transition delay, separation postponement, lift increase, skin-friction and pressure drag (Pressure drag includes contributions from flow separation, displacement effects, induced drag, wave drag, etc.) reduction, turbulence augmentation, heat-transfer enhancement, or noise suppression [3]. These objectives are not necessarily mutually exclusive. For flow over an aircraft wing if the boundary layer becomes turbulent, its resistance to separation is enhanced, and more lift could be obtained at higher angles of attack. On the other hand, the skin-friction drag for a laminar boundary layer can be as much as an order of magnitude less than that for a turbulent case. If transition is delayed, lower skin friction as well as lower flow-induced noise is possible.

In the context of flow past an aerofoil the laminar boundary layer separates, for moderate Reynolds numbers. This would result in transition to turbulence

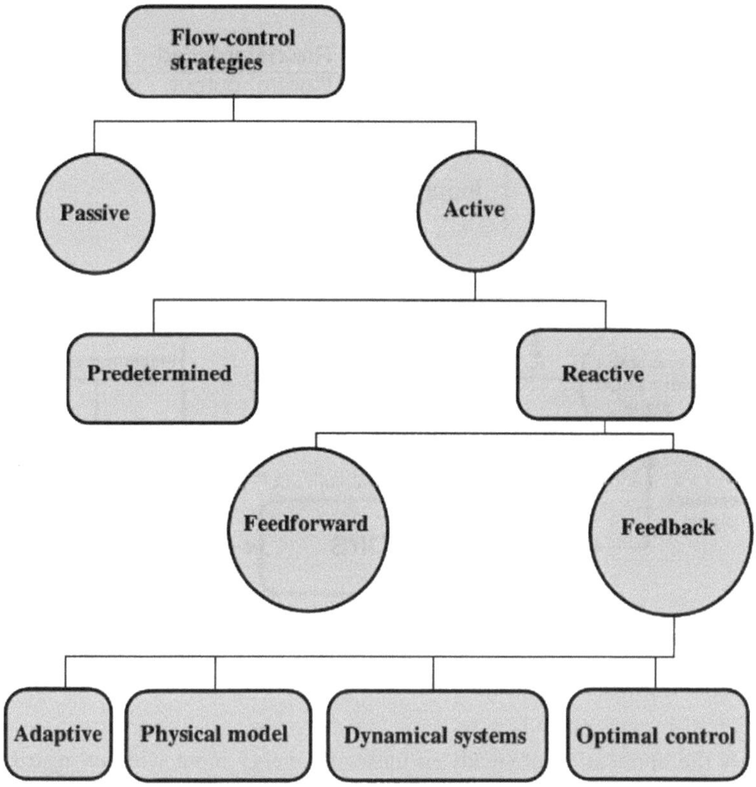

Fig. 4.15 Classification of flow control strategies [45]

with increased entrainment of high-speed fluid because of the turbulent mixing in reattachment of the separated region and formation of a laminar separation bubble. At higher angles of incidence, the bubble breaks down, either separating completely or forming a longer bubble. In either case, the form drag increases and the stall angles are reached. The ultimate goal of all this is to improve the performance of aerofoils by increasing the lift-to-drag ratio. However, induced drag is caused by the lift generated on a lifting surface with a finite span. Moreover, more lift is generated at higher incidence, but form drag also increases at these angles. An ideal method of control that is simple, inexpensive to build and operate, and does not have any trade-offs does not exist, and the skilled engineer has to make continual compromises to achieve a particular design goal [3]. It is easy to classify flow-control methods based on energy expenditure and the control loop involved. If control implementation requires energy expenditure, it is called active otherwise passive. Active control requires a control loop and is further divided into (a) predetermined or (b) reactive categories. Broad class of feedback control techniques include, open loop, closed loop controls which may be further classified into passive, active and reactive type as pictorially depicted in Fig. 4.15. Predetermined control

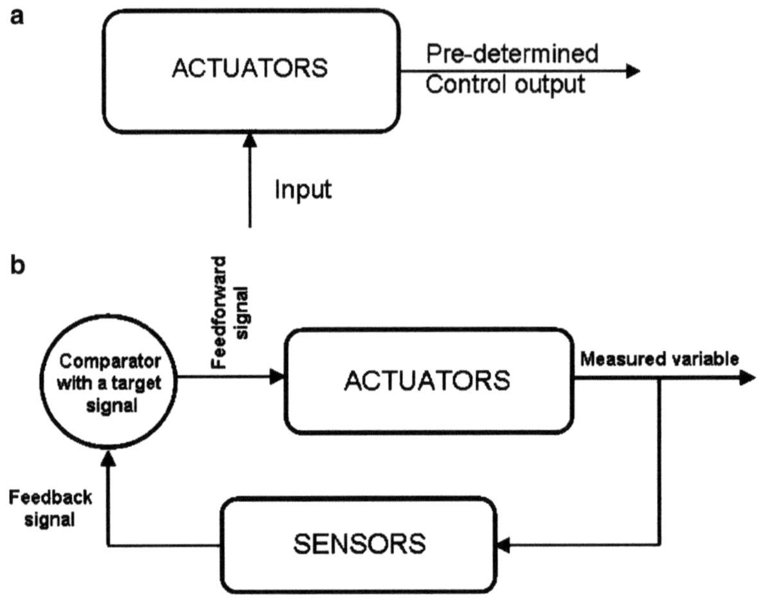

Fig. 4.16 Two different active flow control strategies: (**a**) Predetermined, open-loop control. (**b**) Reactive, feedback, closed-loop control

includes the application of steady or unsteady energy input without regard to the state of the flow. The control loop in this case is open, as shown in Fig. 4.16a, and no sensors are required. Because no sensed information is being fed forward, this open control loop is not a feedforward one. Control is a special class of active control in which the control input is continuously adjusted based on measurements. The control loop in this case can either be an open feedforward one (Fig. 4.16b) or a closed feedback loop (Fig. 4.16c). Classical control theory deals, mostly with reactive or interactive control techniques.

4.6 Flow Past External Bodies and Their Wake Control

Flow past bluff and streamlined bodies are widely encountered in a number of design. A variety of air-borne, sea bound, land vehicles, encounter fluid forcing on the body. The component of this force along the flow direction refers to the drag force, while its normal is the lift force. The distinction between streamlined and bluff bodies lies in their contribution to total drag. Typically, drag force comprises of both viscous and pressure drag. When substantial contribution of total drag encountered by the body is from pressure drag, it is termed as a bluff body. Such a contribution is highly likely if there is flow separation, and wake formation. Here, typically

Table 4.1 Regimes of fluid flow past a smooth circular cylinder [53]

Flow pattern (pictorial view)	Reynolds number $(Re = \frac{UD}{\nu})$	Description
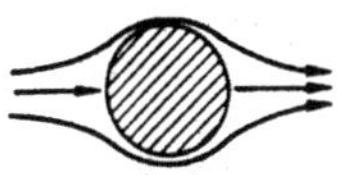	$Re < 5$	Regime of unseparated flow
	$5 to 15 \leq Re < 40$	A fixed pair of Föppl vortices in wake
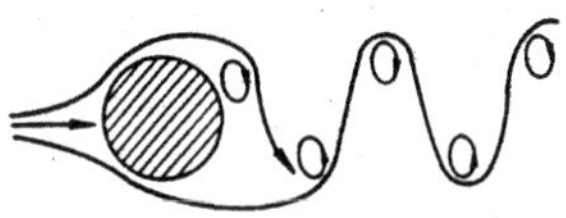	$40 \leq Re < 90$ and $90 \leq Re < 150$	Two regimes in which vortex street is laminar
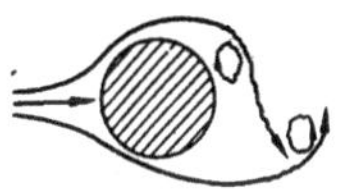	$150 \leq Re < 300$	Transition range to turbulence in vortex
	$300 \leq Re < 3 \times 10^5$	Vortex street is fully turbulent
	$3 \times 10^5 < Re < 3.5 \times 10^6$	Laminar boundary layer has undergone turbulent transition and wake is narrower and disorganized
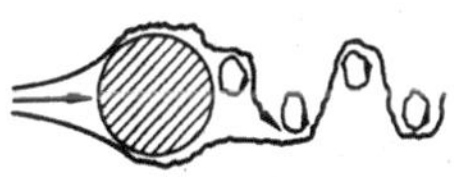	$3.5 \times 10^6 \leq Re$	Re - establishment of turbulent vortex street

alternative pair of vortices are shed from the body and this is popularly called Benard-Kármán vortices. Table 4.1 provides a detailed account of spatio temporal dynamics behind the body as a function of Reynolds number. Understanding and controlling these wakes is of paramount scientific and engineering importance. Bluff body wakes have been extensively studied by employing theoretical, numerical and experimental means [7, 8, 80, 81].

The wake region formed behind a bluff body is complex, as it involves the interaction of three types of shear-layers, viz., a boundary-layer, a separating shear-layer and a wake [8]. A wide variety of structures encountered in civil, mechanical and ocean engineering practice can be classified either as bluff or streamlined. Bluff body configurations are not highly desirable, as the form drag of these bodies makes a contribution of roughly two-thirds to the total drag. For a flexibly mounted bluff body, transverse vibrations are induced due to *resonance* or *phase-locking*, when the vortex shedding frequency (f_{vs}) is in the neighborhood of the natural frequency (f_n) of the structure [9, 53]. Vortex induced resonance is a well known phenomenon even at low to moderate Reynolds numbers. The challenge lies in controlling these vibrations caused by vortex shedding. At moderate to high Reynolds numbers, the wake region and the boundary-layer are turbulent, such flows encompass a wide spectrum of eddies from Kolmogorov's micro scale to eddies as big as the size of the body, known as the integral scale. Understanding and controlling these eddies of different length scales and time scales is central to achieving energy efficient systems and processes. Although a large body of literature is available, even after over a century of effort by researchers and engineers, the problem of bluff body flow still remains almost entirely in the empirical, descriptive realm of knowledge. These numerical simulations are not just theoretical solutions but experimental ones, parallel to those realized in the laboratory [82].

4.6.1 Governing Equations

Let Ω be an open bounded set in $\Re^2$ with boundary Γ. The state equations for the velocity (u) and pressure (p) in the non-dimensional form an incompressible, Newtonian fluid are,

$$\frac{\partial \mathbf{u}}{\partial t} + (\mathbf{u}.\nabla)\,\mathbf{u} = -\nabla p + \frac{1}{Re}\nabla^2 \mathbf{u} \tag{4.11}$$

$$\nabla.\mathbf{u} = 0 \quad \text{in} \quad \Omega \times [0,T] \tag{4.12}$$

$$\mathbf{u}(\mathbf{x},t)|_\Gamma = \mathbf{u}_b \tag{4.13}$$

$$\mathbf{u}(\mathbf{x},0) = \mathbf{u}_0 \;\; \text{in}\, \Gamma \tag{4.14}$$

The computational domain of interest can be tessellated with a non-homogeneous grid. In particular, a fine mesh can be employed in all the zones where the gradients are likely to be very high, such as no-slip surfaces and the wake region. Spatial discretization methods that can be employed are, methods include finite difference [83], finite element [10, 84], finite volume [11] etc. The spatial discretization is performed on a standard collocated grid using finite volume method. The detailed method of solution is available in Patankar [85].

4.6.2 Simulation of Wake Vortices

Navier-Stokes calculations involve multitude of fluid flow features in conjunction with flow past a cylinder. A visual feel of the vortex formation and wake is provided through streaklines as shown in Fig. 4.17a–c at two different Reynolds numbers (*Re*). It should be pointed out that, although *Re* is two orders higher, there are striking similarities between the fluid flow features of the turbulent wake for $Re = 10^4$, and predominantly laminar wake at $Re = 10^2$. Here, every shed eddy is ultimately connected through its own umbilical chord, which is revealed by the electrolytic precipitation in Fig. 4.17a. However, in the numerics, these are produced by tracing the non-inertial and mass-less numerical particles which are obtained by solving,

$$\frac{d\mathbf{x}_i}{dt} = \mathbf{u}_i \qquad \forall\, i : 1 \cdots P. \tag{4.15}$$

where, P refers to the total number of distinct particles released in the system (Fig. 4.17b). Understanding the vortex formation region is central to the control of laminar or turbulent wakes and associated chaotic scattering. Many rapid prototyping strategies hinge on understanding of wake properties.

Simulation of complex fluid flow patterns behind bluff bodies is a challenging task. Capturing the main features of the flow field and associated fluid dynamical parameters of design interest is the primary objective of any Computational Fluid Dynamics (CFD) study. Before proceeding to the problem under investigation, the first logical step would be to demonstrate the usefulness of the procedure through its application to a well established problem. Streamlines depict the instantaneous snapshot of the fluid flow behavior at a particular instant of time. Compared to velocity vectors, they provide a better visual appreciation of the fluid flow characteristics. The influence of rate of rotation ($\xi = \frac{U_c}{U_\infty}$) on the wake is assessed through the investigations at specific values of ξ. The case of two control cylinders kept stationary with $\xi = \frac{U_c}{U_\infty} = 0.0$ is investigated. The temporal evolution, for the case where both control cylinders are not rotating, is shown in Fig. 4.18. In the beginning ($\tau = 0.001$), the flow has a fore and aft symmetry (Fig. 4.18a). With the percolation of viscous influences, the asymmetry on the aft, sets in as shown in Fig. 4.18b. The formation of two symmetric attached eddies (popularly known as Föpal vortices), of opposite vorticity, can be noticed in Fig. 4.18c. These eddies are fed by circulation from the shear layers, they grow in size ($\tau = 20$), and gradually become asymmetric. In particular note the presence of a passage way with a fascinating transfer of momentum; in this case from the outer edge of the top vortex, via central region, engulfing the lower edge of the bottom vortex (Fig. 4.18e). The figure in particular, identifies two streamlines undergoing crossover. Similar to that found in Fig. 4.18e, using a thorough analysis of streamlines, Perry et al. [86] have shown that an *alley-way* passage is the precursor for the inception of vortex shedding. With further progress in time, vortex shedding begins and the wake becomes gradually tighter and more organized (Fig. 4.18f), which in turn induces periodic lift forces on the body. This periodic vortex shedding is also visible in

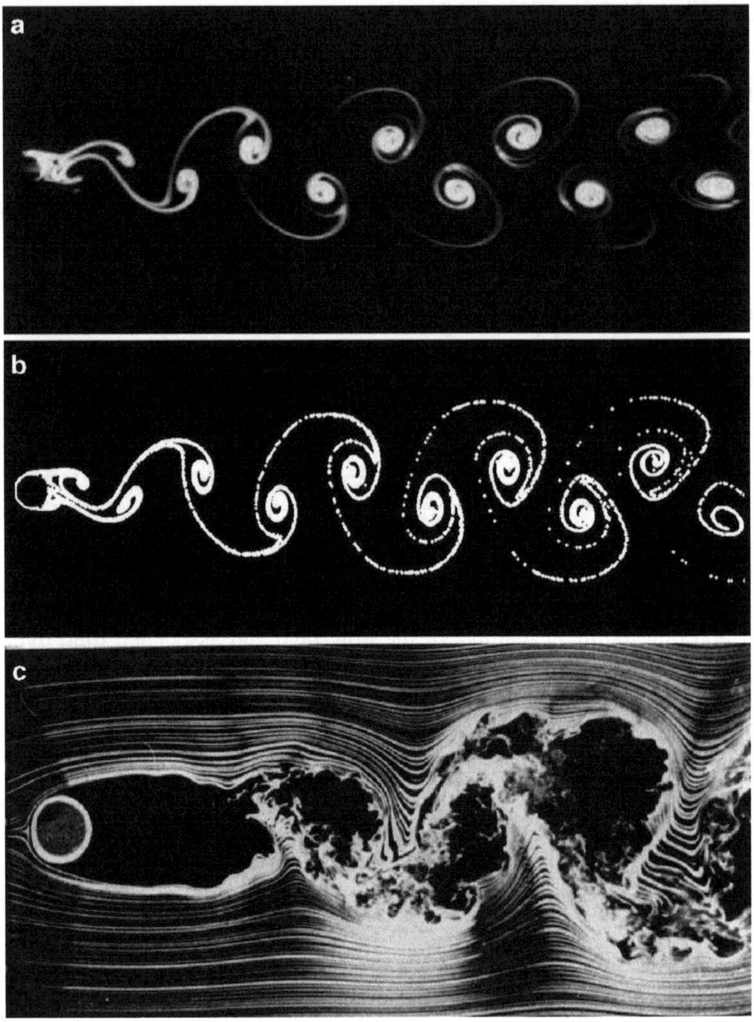

Fig. 4.17 Flow separation and vortex shedding behind a circular cylinder (**a**) $Re = 105$ [105], (**b**) Present simulation ($Re = 100$) and (**c**) $Re = 10^4$ [105]

the lift coefficient history. Figure 4.18g,h indicate the sinusoidal nature of the flow patterns in the wake region and the presence of an eddy (either at the top or bottom) in the formation region.

4.6.3 Different Wake Control Strategies

In the last decade, active flow control has assumed wider ramifications with a focus on optimal control strategies as it directly translates into energy savings. A variety

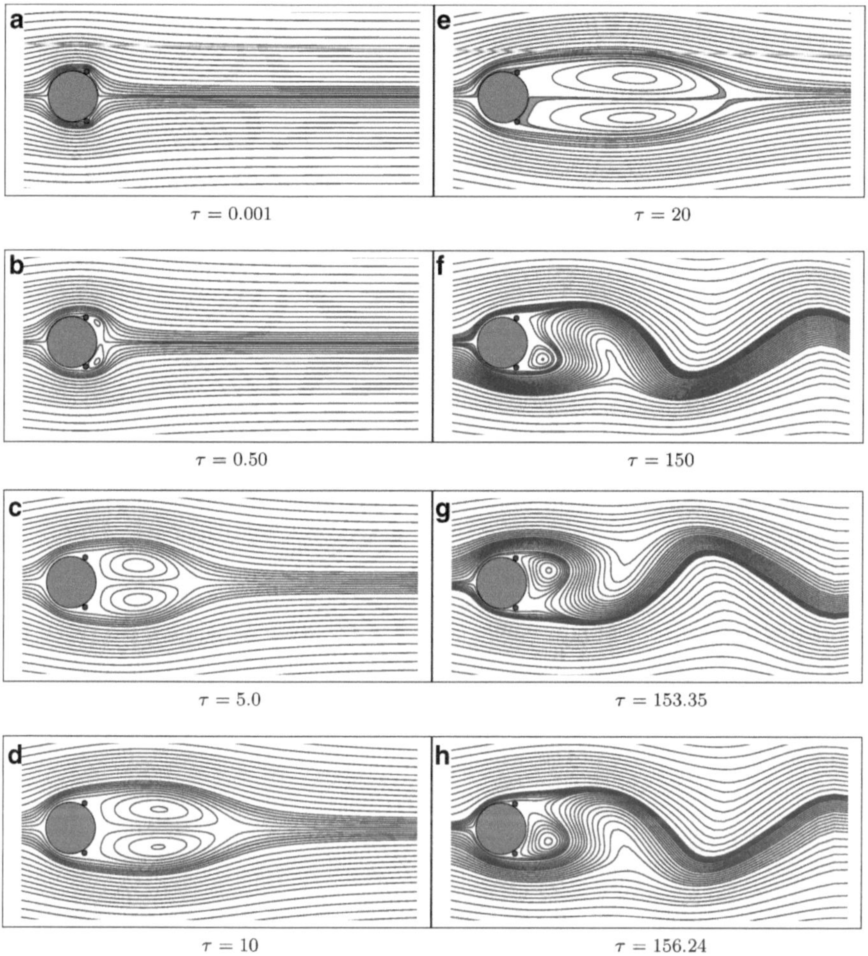

Fig. 4.18 Temporal evolution of streamline patterns for flow over a circular cylinder with control cylinders ($\xi = 0$) at $Re = 100$. The flow separation, twin eddy formation, inception of vortex shedding and periodic vortex formation can be noticed

of control algorithms (both reactive and non-reactive type) have emerged in the nonlinear dynamics and control community such as, chaos control [5], synchronization based control [52], pinning control [87], sporadic feedback control [88] etc. It is highly interesting to see how these control techniques (some of them being model control) can be integrated to exploit practically useful problems of engineering interest, such as those triggered by fluid mechanical instabilities. In the context of flow past a circular cylinder, vortex shedding due to von Kármán is the first global instability mode, which needs to be predicted and controlled before attempting the control of higher modes that involve turbulence. The latter also arises in the range of moderate to high Reynolds numbers. Several promising strategies such as pinning

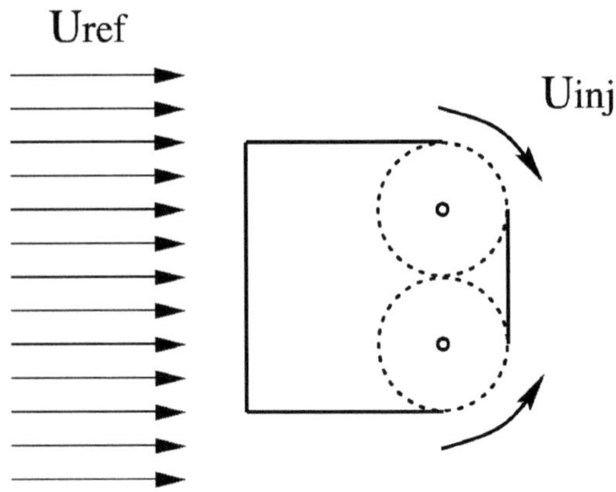

Fig. 4.19 Flow past a D cylinder with angular momentum injection by two rotating cylinders

control have been applied in the literature [87]. The idea of pinning can also be viewed in the context of synchronization, where the objective is to synchronize a chaotic system with the help of a target system [52, 89]. Chen and Aubry [90] have developed a closed loop algorithm to suppress vortex induced vibrations by means of DNS. Their strategy involved the use of Lorentz force in the system based on partial flow information available on the surface of the cylinder. However, for the purpose of control, in practical applications, blowing and suction [91], acoustic actuations [60] and cylinder rotations [57] are some of the popular active flow control strategies.

4.6.4 D-Cylinder Wake Control

Inspired by the pioneering work on chaos control by OGY [5], Patnaik and Wei [10] have proposed the stabilization of unstable periodic orbits through synchronization based control for the annihilation of wake vortices behind D-cylinder. Typical configuration of the D-cylinder with two embedded control cylinders is shown in Fig. 4.19.

Angular momentum injection into the wake is simulated. A streakline technique is utilized for the visualization of the spatiotemporal motion of the flow field. The plot depicts a gradual increment in the value of the injection parameter (ξ) for wake control. Some analytical and experimental studies are available for flow past a circular cylinder [10], establishing the link between vortex shedding and stability theory. It is established in the literature that the Kármán vortex shedding at low Reynolds numbers is a self-excited limitcycle oscillation of the near wake, resulting from a global instability. Chomaz et al. [93] have postulated that the entire wake region could be described by two zones of instability, viz. absolute and convective. There is a branch off from absolute to convective instability at some downstream location. We surmise that the efficacy of the angular momentum injection essentially

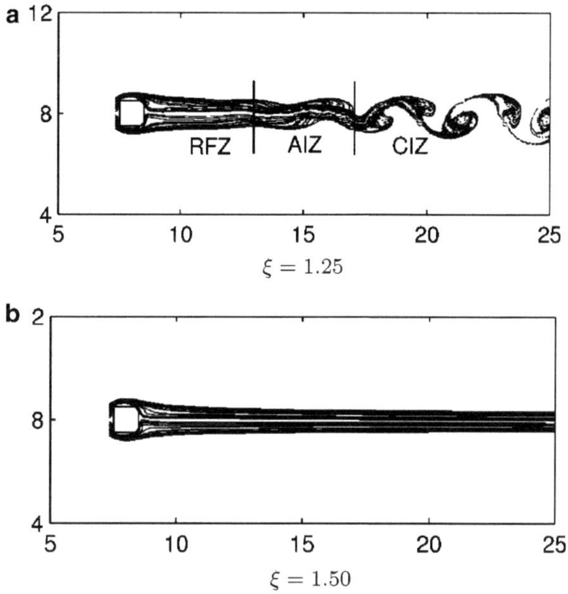

Fig. 4.20 Streakline plots depicting a gradual wake control by angular momentum injection. The flow domain can be thought of as having been divided into three zones, viz., RFZ: Recirculation Free Zone; AIZ: Absolute Instability Zone; CIZ: Convective Instability Zone

depends on the ability to effectively control the base suction behind the cylinder, which in turn leads to changes in the mean recirculation length, absolute instability zone (AIZ), and the convective instability zone (CIZ). To verify our idea, we set $\xi = 1.0$. Indeed, we see a waning in the Kármán vortex street as shown in Fig. 4.2b (figure have not taken from source). Such a reduction in the wake width is associated with an obvious increase in the shedding frequency. Further, it is noted that the AIZ is enlarged due to the increase in the base suction (a reduction in its absolute value) by the injection of angular momentum. Nevertheless, the eddies continue to be fed by circulation from the upstream shear layers. However, with a subsequent increase in the value of the angular momentum injection parameter $\xi = 1.25$, we observe a new zone behind the cylinder – the occurrence of a recirculation-free zone (RFZ) just behind the cylinder. As the AIZ is characterized by the eddy recirculation, the existence of the RFZ is verified by the ordered, parallel streaklines. Behind the RFZ, there is still an AIZ and a CIZ, although their sizes are significantly smaller (see Fig. 4.20c). The target state is achieved at $\xi = 1.5$, and is depicted in Fig. 4.20d. Here, only RFZ persists over the entire computational domain, while the two instability zones, AIZ and CIZ completely vanish.

Figure 4.21 illustrates the entire simulation process for $Re = 200$ and $Re = 400$. In the beginning, both the angular momentum injection parameter $\xi(t)$ and the total variance $\sigma^2(t)$ start from a quiescent state (zero values). As the fluid flow patterns evolve behind the D cylinder, the instability leads to the onset of shedding in the wake at around t $-$ 20. The synchronization coupling gives rise to a positive $\sigma^2(t)$ which in turn feeds back to the control equation and triggers the injection of angular momentum. After the initial buildup, the total variance subsides to a minimum

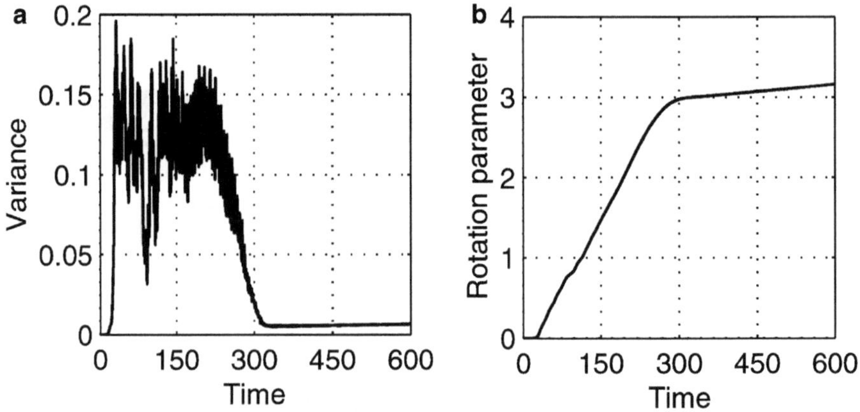

Fig. 4.21 Temporal evolution of the total variance and its associated rotation (injection) parameter at $Re = 400$

value at around t = 200 for $Re = 200$ (t = 300 for $Re = 400$), both $\xi(t)$ and $\sigma^2(t)$ gradually approach steady state values. Particularly, the latter reaches a value close to zero, signifying that it has reached the target synchronized state. Here, the value of constant C (chosen as 0.1) assists in the incorporation of gradual change for the rotation parameter. Interestingly, the injection parameter stabilizes at a value close to $\xi = 1.5$, which is in compliance with the streakline plot depicted in Fig. 4.20d for $Re = 200$.

Computational model control of Patnaik and Wei [10] was experimentally implemented by Beaudoin et al. [94, 95] and they have demonstrated effective control over the recirculation zones. They have experimentally realized these findings with smoke visualization as captured in Fig. 4.22, defecting uncontrolled and controlled wake features. The rotation of the cylinder clearly delays the separation leading to a smaller recirculation bubble on the base of the bluff body and a smaller recirculation length. In Fig. 4.23 they have shown the modification of the wake velocity profiles for higher free-stream velocity. The profiles exhibit the strong shear layer, which is clearly shifted downward when the cylinder is rotating (the black points, without control, are always above the gray points, with control). This is consistent with the fact that the separation point is also shifted downward. As the ratio U_{inj}/U_0 is much smaller for high free-stream velocities, the modification of the base flow is less spectacular than the one observed in the flow visualizations. Imparting rotations through a control algorithm requires energy expenditure. It is therefore imperative to ensure that the energy expenditure incurred is suitably subtracted from the energy savings, being accrued due to drag reduction. To this end, we perform a cost function analysis where the objective is to find out the value of optimum rotation parameter (ξ_{opt}). Beaudoin et al. [94,95] have constructed the cost functional to obtain the extremum seeking control as shown in Fig. 4.24.

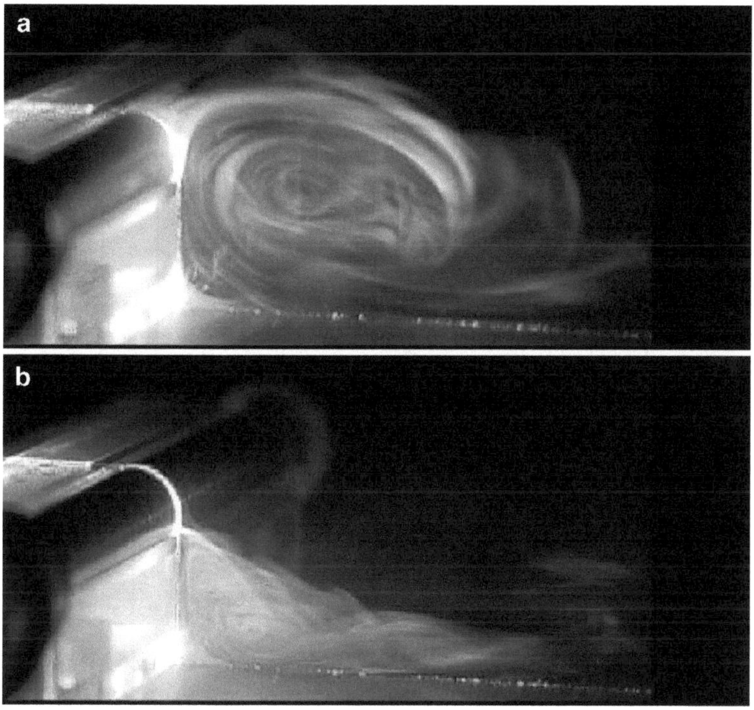

Fig. 4.22 Visualization of the separated flow downstream from the bluff body for $U_0 = 2ms^{-1}$. (**a**) Without control; (**b**) with control ($U_{inj}/U_0 = 2$)

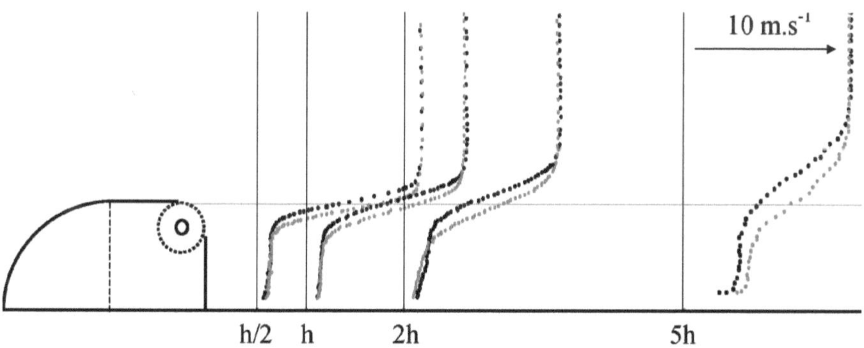

Fig. 4.23 Modulus of the mean velocity measured with hot wire anemometry for $U_0 = 12ms^{-1}$. *Black* dots: without control. *Gray* dots: with control ($U_{inj}/U_0 = 0.5$)

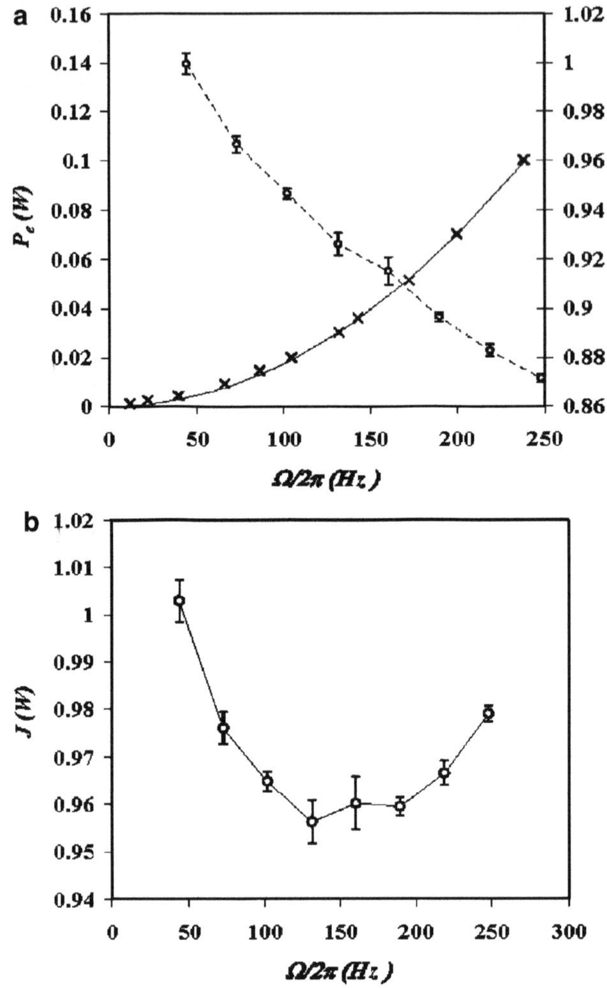

Fig. 4.24 (**a**) Measurements of the electrical power consumption P_e (*crosses*) fitted with a law $P_e = \alpha\Omega^2$ (*continuous line*); aerodynamic power P_a with $U_0 = 12m/s$ (*open circles*); (**b**) Global power function $J(\Omega, U_0)$

4.6.5 Momentum Injection Control

One of the attempts to control these vortex structures was initiated as early as 1910, by the legendary fluid dynamicist, Ludwig Prandtl, who demonstrated his ship of zero viscous resistance through flow around two counter rotating cylinders [96]. A large number of systematic experiments have been carried out by Modi [97] and his co-workers, who have employed rotating cylinders as an integral part of airfoils, a variety of bluff bodies such as buildings, moving vehicles etc. and the concept has been shown to be remarkably effective and successful in delaying separation [97,98]. Since momentum is injected into the shear layers by means of a moving wall (rotating control cylinders), in the literature it is also popularly known as Moving Surface Boundary-layer Control (MSBC).

4.6.5.1 Moving Surface Boundary-Layer Control (MSBC)

MSBC contributes to flow control in most of the following ways [97]:

- It retards the growth of the boundary-layer by minimizing relative motion between the surface and the free stream.
- It injects momentum into the boundary-layer.
- It creates a region of high suction and thereby accelerates the flow in its neighborhood and outside of the boundary-layer.
- It delays separation and interferes with the evolution of the wake.

Detailed historical progress and developments in the field of Moving Surface Boundary-layer Control (MSBC) have been summarized by Modi (see Ref. [97]).

4.6.5.2 The Control Algorithm

The algorithm employed by Muddada and Patnaik is depicted in the form a schematic in Fig. 4.25, where the recipe includes multiple-sensors, actuators and an algorithm that handles both sensors and actuators. In Fig. 4.25, the sensors are identified by c_1, c_2, c_3 etc., while the actuators are indicated by a_1, a_2, a_3 etc.The sensors monitor the state of both stream-wise and cross-stream velocity values as a function of time. The total variance $\sigma^2(t)$ as a function of time is constructed and the control equation specified in (4.17) is employed to decide the quantum of actuations. Here, the objective is to implement a synchronization based coupling strategy between the current state of the flow ($v(x,y,t)$) and the desired target state ($\tilde{v}(x,y,t)$) and is implemented through the Eqs. (4.16), (4.17). Here, the synchronization is taken to mean complete suppression of fluid flow oscillations. Therefore, the target state is a quiescent state given by $\tilde{v}(x,y,t)$ along the wake centerline. Total variance as a function of time is given by,

$$\sigma^2(t) = \frac{1}{\int_\Omega dxdy} \int_\Omega [v(x,y,t) - \tilde{v}(x,y,t)]^2 dxdy \tag{4.16}$$

where Ω denotes the region behind the circular cylinder. However, the above summation boils down to a summation in the discrete version, as the sensors are placed only along the line of symmetry. The total variance in turn decides the extent of actuations to be employed through the rotating elements. The rotation parameter

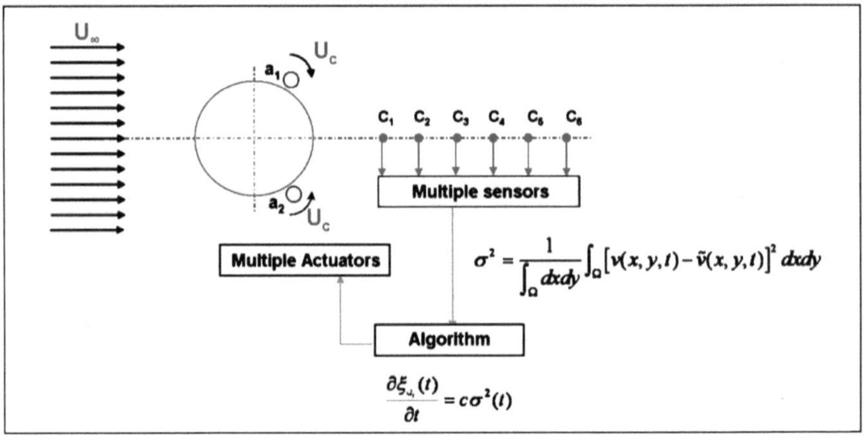

Fig. 4.25 Generic frame work for the closed-loop feed back control, involving actuators $(a_1, a_2, a_3, ...)$, multiple-sensors $(c_1, c_2, c_3 ...)$ and the control algorithm. The governing equations related to the control algorithm, and the multiple sensors are also indicated in the schematic. Here, the objective is to synchronize the current state of the system $(v(x,y,t))$ towards the desired target state $(\tilde{v}(x,y,t))$

(ξ) is the ratio of control cylinder velocity to the free stream velocity, $\xi = \frac{U_c(t)}{U_\infty}$ is given by the equation below:

$$\frac{d\xi(t)_{a_i}}{dt} = C\sigma^2(t) \tag{4.17}$$

The control equation (4.17) described above is a generic one that injects momentum and solved simultaneously with the Navier-Stokes equations. Here, the coefficient C is a proportionality constant and can be varied like a fuzz factor. This set of coupled equations render an opportunity to monitor the gradual control of wake instabilities. Momentum injection is initiated when the control cylinders start rotating which is actuated by applying power input. This generic flow control strategy of feedback sensors and actuators can indeed be applied irrespective of the type of flow (both laminar and turbulent). In the present study, we investigate only sub critical Reynolds number regime as it is a necessary precursor to any complicated investigations involving turbulence control.

Determining optimal control action is central to the construction of active flow control strategies. Here we analyze and compare two approaches employed in the present investigation. The system with twin controllers, present behind the main cylinder is considered. The controllers would inject momentum to achieve the annihilation of wake vortices. Thus, the wake formation region behind the circular cylinder is controlled through the rotatory action of the two cylinders. In the present context, designing an optimal control that incorporates proportional, integral and derivative control is required for two reasons: (a) To change the dynamics of the system – to make the system response both fast and stable. (b) To reduce the system sensitivity towards uncertainties due to internal and external disturbances. Designing suitable controllers and analyzing the resulting controlled system is therefore plausible through a PID controller. Often, the state of the flow with a known physical domain and initial configuration is chosen, and we can have admissible acting functions on the fluid through various devices (body forces, boundary values, temperature, etc.), then the issue of determining the optimal control action would depend on the system, in order to minimize turbulence within the flow.

The actuators are switched on (indicated by A), only after reaching steady periodic vortex shedding as shown in Fig. 4.26a. The rate of injection of momentum (through $\xi(t)$) gradually picks up and reaches a steady state value, while the temporal variance (σ^2(t)) constructed out of the sensor points droops down. The concomitant changes in the dynamics of the evolution of the wake and stabilization is characterized through the streaklines in Fig. 4.26b. Complete suppression of vortex shedding is achieved at the end of the control. However en route to the annihilation of vortex structures, the downstream is sinuous with persistent convective instabilities.

4.6.6 *Annihilation of Wake Vortices*

A wide variety of passive and active flow control techniques are typically employed for the annihilation of wake vortices. These can be classified into passive and active control techniques. Unlike the traditional control methods (such as twisted tapes, blowing and suction, acoustics, polymer additives etc.), smart control of turbulence involves sophisticated interactive control schemes, particularly those exploiting micro-machining technology and feedback.

Along with mass and momentum conservation equations, a control equation is designed and implemented to achieve desired flow control goals. A large number of sensors are chosen downstream of the body to report the state of the flow. The role of externally controllable actuators on the fluid flow patterns past a circular configuration is assessed. The generic framework for wake control is pictorially depicted in Fig. 4.25. Here, the multiple-feedback sensors report the dynamic state of the flow and an algorithm would synthesize the quantum of actuations (with zero-net mass injection) to be performed. Such flow manipulations are achieved by means of locating two rotating elements right behind the main cylinder. Practical realization of such control elements can be easily implemented by using hollow and bearing mounted configurations that can be either rotated clock-wise or counter-clock-wise. Similar feedback control was implemented in our earlier investigations on a D-cylinder [10]. However, any motor-driven rotating cylinder which is employed as an actuator will require energy expenditure. This issue is recently addressed by Beaudoin et al. [94, 95] by constructing an objective function to quantify the associated costs. In their investigations, the net reduction in drag is assessed vis-á-vis the associated energy penalty for the actuators.

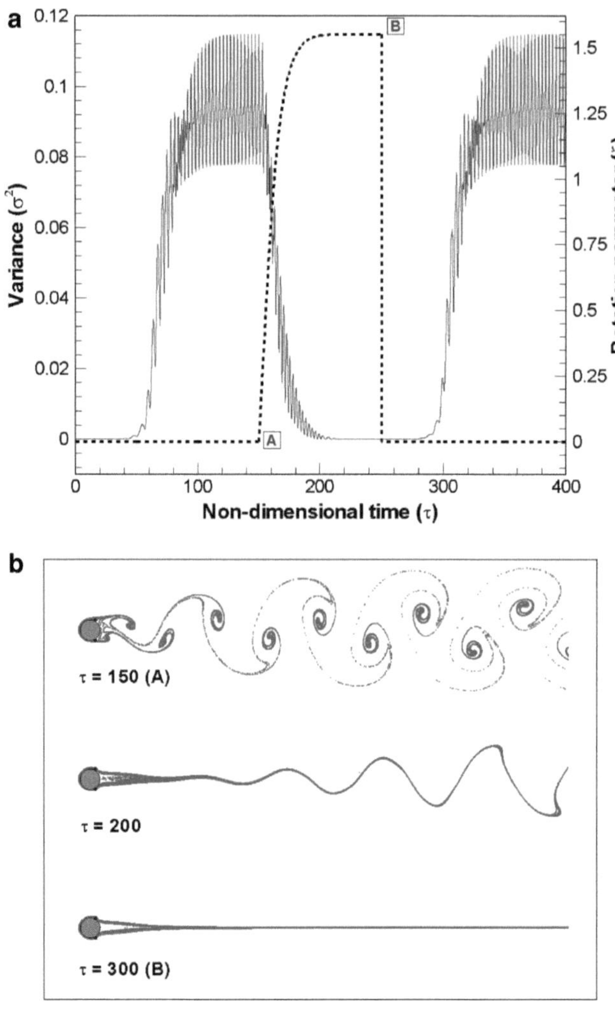

Fig. 4.26 (**a**) Temporal evolution of total variance (*solid line*) and its associated rotation parameter (**dashed line**), (**b**) Streakline patterns at different time levels during control. Point A refers to control being switched *on* and point B refers to control *off* in (**a**)

Flow visualization was achieved by Aluminum powder suspensions in conjunction with slit lighting to visualize the streamlines. The air-water interface thus captured with the aid of *Aluminum* dust as the passive tracer depicts the effectiveness of the momentum injection and the role of actuators in the reduction of wake width as in Fig. 4.27. The correlation between numerical predictions (Fig. 4.27) and flow visualization results is indeed satisfactory considering the complex character of the flow. In the absence of any rotations $\left(\frac{U_c}{U_\infty} = 0.0\right)$, the two actuators remain immersed in the separated flow field. This results in a wider

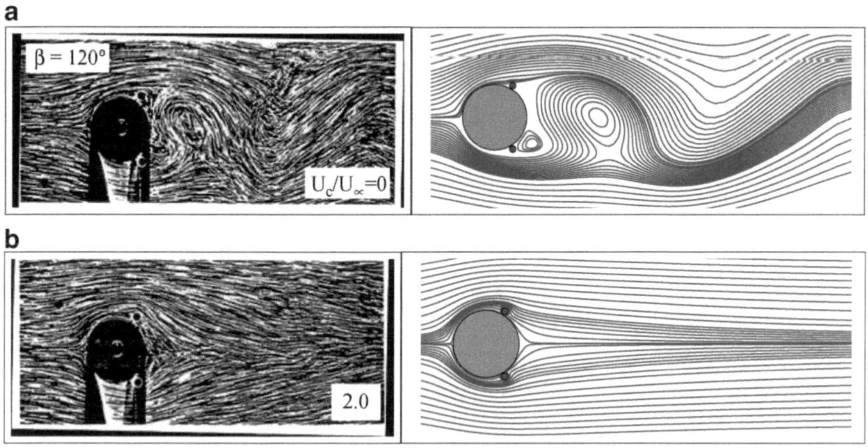

Fig. 4.27 Validation of numerical visualization for $Re = 100$, against experimental result [106]. Here, the two control cylinders are located at $\beta = 120°$ to the flow direction. The rotation parameter $\xi = \frac{U_c}{U_\infty}$ represents rate of momentum injection into the wake region

wake, and the downstream is dominated by eddy formation and shedding. The figure depicts a large vortex, which is about to get washed away into the wake. As noted by Perry et al. [86], a saddle point type of instability associated with the newly formed vortex is discernible. The overall configuration of the wake reverses itself in an alternating fashion in compliance with the vortex shedding process. The shed vortices grow rapidly as they are convected downstream, diffusing vorticity in the surrounding region at a faster rate. When the control cylinders are cranked with even higher rotation rates, say $\xi = 2.0$, the wake gradually becomes narrower and the vortex formation is completely suppressed. For this rotation parameter, the boundary-layer remains totally attached to the main cylinder and vortex shedding ceases completely. The flow domain becomes completely *inviscid like* in its appearance. In the near wake of the cylinder, a *V-cone* type formation is seen both in the experimental and numerical visuals.

4.6.7 Modeling the Components of Reynolds Stress Tensor

The process of Reynolds averaging applied on the Navier-Stokes equations results in apparent stresses, from the convective nonlinearity. Since the additional variables require closure, modeling is mandatory. In most of the engineering designs that need turbulent fluid flow analysis, it is adequate to investigate the influence of small scale eddies on the mean flow quantities.

4.6.7.1 $k-\epsilon$ Model

The spatial and temporal distribution of k and ϵ is determined from differential form of the transport equations for these quantities, considering the history and transport effects of turbulence are [99],

$$\frac{\partial k}{\partial t}+\frac{\partial\left[\overline{u_i}k\right]}{\partial x_i}=\frac{\partial}{\partial x_i}\left(\left(\nu+\frac{\nu_t}{\sigma_k}\right)\frac{\partial k}{\partial x_i}\right)+P_k-\epsilon, \tag{4.18}$$

$$\frac{\partial \epsilon}{\partial t}+\frac{\partial\left[\overline{u_i}\epsilon\right]}{\partial x_i}=\frac{\partial}{\partial x_i}\left(\left(\nu+\frac{\nu_t}{\sigma_\epsilon}\right)\frac{\partial \epsilon}{\partial x_i}\right)+C_{\epsilon 1}P_k\frac{\epsilon}{k}-C_{\epsilon 2}\frac{\epsilon^2}{k}, \tag{4.19}$$

with P_k referring to production of turbulent kinetic energy, which is given by,

$$P_k=\nu_t S^2 \quad \text{where} \quad S=\sqrt{\frac{1}{2}\left[\frac{\partial \bar{u}_i}{\partial x_j}+\frac{\partial \bar{u}_j}{\partial x_i}\right]^2} \tag{4.20}$$

S refers to the symmetric mean strain rate tensor. The model constants used in the above equations are: $c_\mu=0.09$, $C_{\epsilon 1}=1.44$, $C_{\epsilon 2}=1.92$, $\sigma_k=1.0$ and $\sigma_\epsilon=1.3$.

4.6.7.2 $k-\epsilon$ Model of Kato and Launder

The standard $k-\epsilon$ model is modified by Kato and Launder to control and regulate the over production of the turbulent kinetic energy in the stagnation regions. Kato and Launder [100] introduced the following ad hoc model relation for the production term in the turbulent kinetic energy equation:

$$P_k=\nu_t S\Omega \quad \text{instead of} \quad P_k=\nu_t S^2 \tag{4.21}$$

The quantity Ω is an antisymmetric rotation tensor and is proportional to the magnitude of the local vorticity and is given as,

$$\Omega=\sqrt{\frac{1}{2}\left[\frac{\partial \bar{u}_i}{\partial x_j}-\frac{\partial \bar{u}_j}{\partial x_i}\right]^2}; \quad S=\sqrt{\frac{1}{2}\left[\frac{\partial \bar{u}_i}{\partial x_j}+\frac{\partial \bar{u}_j}{\partial x_i}\right]^2} \tag{4.22}$$

In simple shear flows, the behavior of the production term remains unchanged as $\Omega \approx S$, while there is an advantage in the stagnation regions where spurious turbulence production is eliminated as $\Omega \approx 0$ in that region. Here, the model constants used are same as that in the standard $k-\epsilon$ model. However, these model constants need not be recalibrated owing to the fact that production term is identical both in standard $k-\epsilon$ and Kato and Launder's modification for simple shear flows.

The location of the centre of mean vortex and length of formation region (L_f) is quantified by plotting normalized mean streamwise velocity ($\bar{u}_{cl}$) against the wake

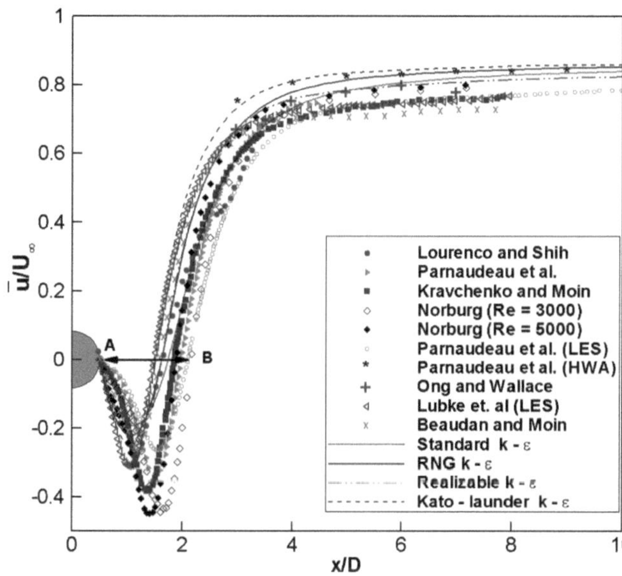

Fig. 4.28 Mean streamwise velocity along the wake center line for $Re = 3{,}900$. Lourenco and Shih [107], Parnaudeau et al. [107], Kravchenko and Moin [108], Norburg [107], Ong and Wallace [107], Lübcke et al. [109], Beaudan and Moin [109]

centreline (see Fig. 4.28). The plot signifies the predictive capability of different models vis-á-vis other numerical and experimental investigations. All the models tested have a better comparison with both PIV and LES data in the far wake region. However, there is a clear under prediction on the location of centre of mean vortex, length of the wake formation region (L_f) and the corresponding magnitude of velocity.

This is further corroborated through the time averaged Reynolds stress contours shown in Fig. 4.29. Here, the Reynolds stress contours are akin to butterfly wings both in PIV studies and LES simulations, which could be predicted reasonably well by all the turbulence models. The model predictions show that the wings are closer to the cylinder surface than the PIV data or LES simulations. The turbulence production term in the k transport equation is given by $P_K = \tau_{ij}^R \bar{S}$. There is an excessive production of turbulent kinetic energy in the standard $k - \epsilon$, RNG $k - \epsilon$ and Realizable $k - \epsilon$ models. This over prediction of turbulence generation term is nicely controlled in the modification introduced by Kato and Launder. Hence, on the fore of the main cylinder and in the stagnation region, the model by Kato and Launder has a better prediction of turbulence production due to improved accuracy in the Reynolds stress prediction. In the stagnation zone and in its proximity, the shear stress production of TKE is small and normal stress production is high. In an isotropic eddy viscosity model, such as standard $k - \epsilon$, normal stresses are not predicted that well, leading to excessive generation of turbulent kinetic energy.

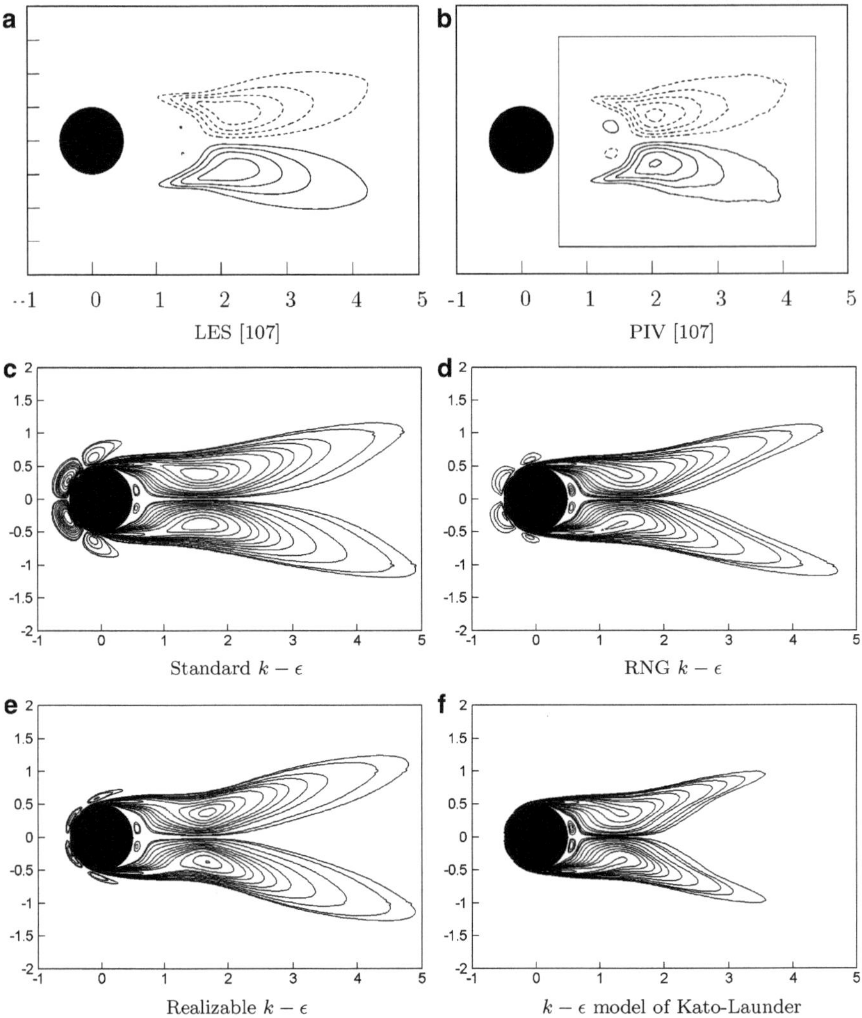

Fig. 4.29 Time averaged Reynolds stress contours at $Re = 3{,}900$. The contours resemble butterfly wings and the model by Kato and Launder has the best predictive capability among these two-equation models investigated. The prediction by other turbulence models is reasonable, although LES simulations have a better comparison against PIV data

The Reynolds stress contours are predicted by different models for a momentum injection parameter $(\xi) = 4.0$ (see Fig. 4.30). The laminar boundary layer region on the upstream is poorly predicted by the standard $k-\epsilon$ model, while Kato and Launder's version has the best prediction among the models tested.

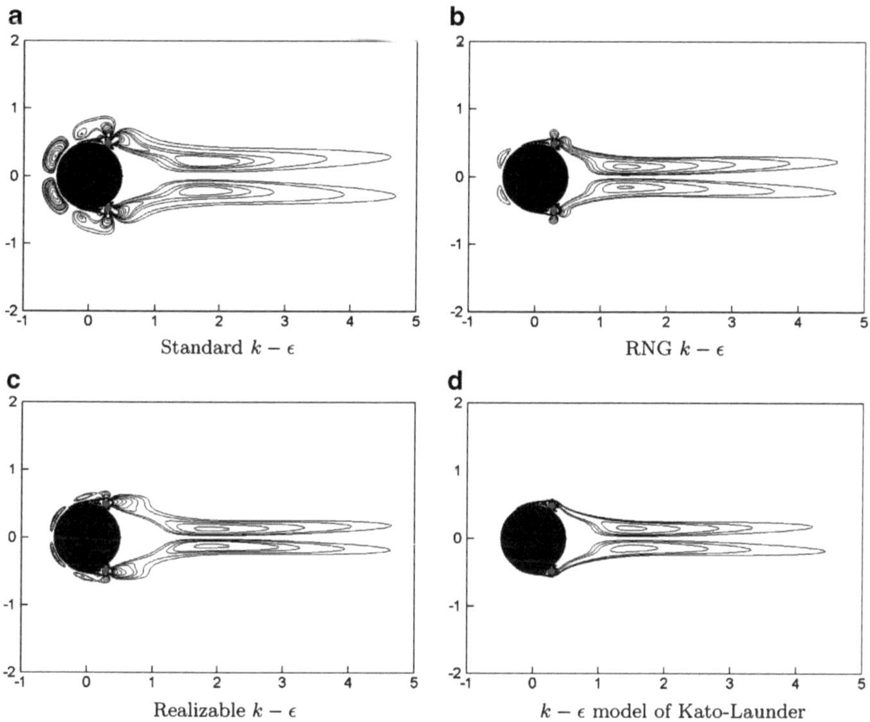

Fig. 4.30 Time averaged Reynolds stress contours for $\xi = 4.0$ and $Re = 3,900$ for four different eddy viscosity based models. The models have almost similar predictive capability in the wake region, although there are differences in the upstream

4.7 Conclusions and Future Directions

The role of chaos in fluid turbulence is pervasive and inter twined. Unfortunately, the link the two is limited by our current state of understanding. The nonlinearities in fluids, trigger instabilities and lead to chaos. In the opinion of the authors control of chaos would result in better understanding of turbulence and its control. In this review an attempt is made to bridge the gap between the developments in chaos control techniques and their applicability to turbulent fluid motion. Developments in smart sensors and actuators would lead to their integration into active flow control techniques to minimize the objective functional. The traditional engineering goals such as, drag reduction, enhanced mixing, delayed stall, reduction in noise levels etc. can perhaps be achieved by modern tools such as, microelectro mechanical systems.

Turbulence is often seen as, the last great unsolved problem of classical physics. The struggle by the scientific community to conquer turbulence has been long and arduous with lots of sweat, few victories, and much frustration. The way turbulence is being viewed as a complex physical phenomenon has changed over the years.

Indeed, key ideas in the field continue to rise and fall [113]. Turbulence is essentially random fluctuations superimposed on mean flow and is there fore, deterministic in the statistical sense. It has a combination of both coherent and incoherent motions. Turbulent fluid motion consists of a wide spectrum of length scales and time scales that are random, and there is zero energy in any one particular frequency or wave number. It implies, the probability density function and spectrum of any flow variable are continuous and finite [112]. Historically, there are perhaps five methods in approaching the central ologma of turbulence, visualization, first principles, the statistical approach, coherent structures, and other modern tools.

Chaos control strategies such as, synchronization based coupling, stabilization of unstable periodic orbits by choosing one of the orbits etc. are extensively applied to general non linear dynamics problems. However, the essence behind these ideas have not percolated to fluid flow systems in a useful fashion. The major stumbling block in fluids are geometric nonlinearity associated with advection term and multi-degree of freedom.

Implementation of chaos control methodologies would depend on progress in smart sensing, actuation and control algorithms, which form the essential back bone. How and when to interact with the flow is always a difficult question to answer due to sensitivity associated with instabilities and their generation and amplification mechanisms. The authors believe that, an understanding of how instabilities trigger chaos can unravel the connection between chaos and turbulence. Application of suitable chaos control methodologies would hinge on the above inter twining between the two.

Acknowledgements The authors are deeply indebted to some anonymous professors from NUS, Singapore and ASU, USA for their insightful comments on an earlier version of this chapter.

References

1. Gleick, J.: Chaos: Making a New Science. Cardinal, Sphere Books Ltd., London (1988)
2. Tél, T., Gruiz, M.: Chaotic Dynamics: An Introduction Based on Classical Mechanics. Cambridge University Press, New York (2006)
3. Gad-El-Hak, M.: Flow Control: Fundamentals and Practices. Spinger, New York (1998)
4. Lumley, J., Blossey, P.: Control of Turbulence. Ann. Rev. Fluid Mech. **30**, 311–327 (1998)
5. Ott, E., Grebogi, C., Yorke, J.A.: Controlling chaos. Phys. Rev. Lett. **64**(11), 1196 (1990)
6. Aranson, I.S., Kramer, L.: The world of the complex Ginzburg-Landau equation. Rev. Mod. Phys. **74**, 99–143 (2002)
7. Sreenivasan, K.R.: Fluid turbulence. Rev. Mod. Phys. **71**(2), S383–S395 (1999)
8. Williamson, C.H.K.: Vortex dynamics in the cylinder wake. Ann. Rev. Fluid Mech. **28**, 477–539 (1996)
9. Patnaik, B.S.V., Narayana, P.A.A., Seetharamu, K.N.: Numerical simulation of laminar flow past a transversely vibrating circular cylinder. J. Sound Vib. **228**(3), 459–475 (1999)
10. Patnaik, B.S.V., Wei, G.W.: Controlling wake turbulence. Phys. Rev. Lett. **88**, 054502 (2002)
11. Muddada, S., Patnaik, B.S.V.: An active flow control strategy for the suppression of vortex structures behind a circular cylinder. Eur. J. Mech. B Fluids **29**, 93–104 (2010)

12. Muddada, S., Patnaik, B.S.V.: An assessment of turbulence models for the prediction of flow past a circular cylinder with momentum injection. J. Wind. Eng. Ind. Aerod. **98**, 575–591 (2010)
13. Pope, S.B.: The Turbulent Flows. Cambridge University Press, New York (2000)
14. Egolf, D.A., Melnikov, I.V., Pesch, W., et al.: Mechanisms of extensive spatiotemporal chaos in Rayleigh-Bernard convection. Nature **404**(6779), 733–736 (2000)
15. Garfinkel, A., Spano, M.L., Ditto, W.L., et al.: Controlling cardiac chaos. Science, **257**(5074), 1230–1235 (1992)
16. Gollub, J.P., Swinney, H.S.: Onset of turbulence in a rotating fluid. Phys. Rev. Lett. **35**,927–930(1975)
17. Moon, F.C.: Chaotic Vibrations. Wiley, New York (1987)
18. Rössler, O.E., Wegmann, K.: Chaos in the Zhabotinskii reaction. Nature, **271**, 89–90 (1978)
19. Olsen, L.F., Schaffer, W.M.: Chaos vs. noisy periodicity: Alternative hypotheses for childhood epidemics. Science, **249**, 499–504 (1990)
20. Arnsdorf, M.F.: Arrhythmogenesis, the electrophysiologic matrix, and electrophysiologic chaos. Curr. Opin. Cardiol. **6**(1), 3–10 (1991)
21. Peitgen, H.-O., Jürgens, H., Saupe, D.: Chaos and Fractals: New Frontiers of Science. Springer, New York (2004)
22. Fefferman, C.L.: Existence and smoothness of the Navier-Stokes equation, Official problem description, Clay Mathematics Institute (http://www.claymath.org/millennium/Navier-Stokes_Equations/navierstokes.pdf)
23. Reynolds, O.: On the dynamical theory of incompressible viscous flows and the determination of the criterion. Philos. Trans. R. Soc. London Ser. A **186**, 123–161 (1894)
24. Tritton, D.J.: Physical Fluid Dynamics. Oxford University Press, New York (2007)
25. Davidson, P.A.: Turbulence: An introduction for Scientists and Engineers. Oxford University Press, New York (2007)
26. Gad-el-Hak, M., Her Mann Tsai.: Transition and Turbulence control. Lecture Notes Series, Institute of Mathematical Sciences, vol. 8. National Univercity of Singapore (2006)
27. Rodi, W.: On the simulation of turbulent flow past bluff bodies. J. Wind Engg. Ind. Aerodyn. **46–47**, 3–19 (1993)
28. Launder, B.E., Spalding, D.B.: The numerical computation of turbulent flow. Comput. Meth. Appl. Mech. Eng. **3**(2), 269–289 (1974)
29. Kline, S.J., Robinson, S.K.: Quasi-coherent structures in the turbulent boundary layer: part I. Status report on a community - wide summary of the data. In: Kline, S.J., Afgan, N.H. (eds.), Near-wall Turbulence, pp. 200–217. Hemisphere, New York (1990)
30. Robinson, S.K.: Coherent motions in the turbulent boundary layer. Annu. Rev. Fluid. Mech. **23**, 601–639 (1991)
31. Sreenivasan, K.R.: The turbulent boundary layer. In: Gad-el-Hak, M. (ed.), Frontiers in Experimental Fluid Mechanics, chap. 4, pp. 159–209. Springer, Berlin (1989)
32. Cantwell, B.J.: Organized motion in turbulent flow. Annu. Rev. Fluid. Mech. **75**, 457–515 (1981)
33. Gad-el-Hak, M., Bandyopadhyay, P.R.: Reynolds number effects in wall-bounded turbulent flows. Appl. Mech. Rev. **47**, 307–365 (1994)
34. Gad-el-Hak, M., Blackwelder, R.F., Riley, J.J.: On the growth of turbulent regions in laminar boundary layers. J. Fluid Mech. **110**, 73–95 (1981)
35. Thorpe, S.A.: The turbulent Ocean. Cambridge University Press, Cambridge (2005)
36. Theodorsen, T.: Mechanism of turbulence. In: Proceedings of Second Midwestern Conference on Fluid Mechanics, Ohio State University, USA (1952)
37. Kline, S.J., Reynolds, W.C., Schraub, F.A., et al.: The structure of turbulent boundary layers. J. Fluid Mech. **30**, 741–773 (1967)
38. Moin, P., Mahesh, K.: Direct numerical simulation: A tool in turbulence research. Annu. Rev. Fluid. Mech. **30**, 539–578 (1998)
39. Chernyshenko, S.F., Baig, M.F.: Streaks and vortices in near-wall turbulence. Phil. Trans. R. Soc. A. **363**, 1097–1107 (2005)

40. Arecchi, F.T., Boccaletti, S., Ciofini, M., et al.: The control of chaos: Theoretical schemes and experimental realization. (Int. J. Bifurcat. Chaos **8**(8), 1643–1655 (1998)
41. Auerbach, D., Cvitanovic, P., Eckmann, J.-P., et al.: Exploring chaotic motion through periodic orbits. Phys. Rev. Lett. **58**, 2387–2390 (1987)
42. Ditto, W.L., Rauseo, S.N., Spano, M.L.: Experimental control of chaos. Phys. Rev. Lett. **65**, 3211–3214 (1990)
43. Hunt, E.R.: Stabilizing high-period orbits in a chaotic system: The diode resonator. Phys. Rev. Lett. **67**, 1953–1955 (1991)
44. Roy, R., Murphy, T.W., Maier, T.D. Jr., et al.: Dynamical control of a chaotic laser: Experimental stabilization of a globally coupled system. Phys. Rev. Lett. **68**, 1259–1262 (1992)
45. Gad-el-Hak, M.: Flow Control: Passive, Active, and Reactive Flow Management. Cambridge University Press, Cambridge (2000)
46. Pyragas, K.: Continuous control of chaos by self-controlling feedback. Phys. Lett. A **170**(6), 421–428 (1992)
47. Xiao, J., Hu, G., Yang, J., et al.: Controlling turbulence in the complex Ginzburg-Landau equation. Phys. Rev. Lett. **81**, 5552–5555 (1998)
48. Boccaletti, S., Bragard, J., Arecchi, F.T.: Controlling and synchronizing space time chaos. Phys. Rev. E **59**, 6574–6478 (1999)
49. Junge, L., Parlitz, U.: Synchronization and control of coupled Ginzburg-Landau equations using local coupling. Phys. Rev. E **61**, 3736–3742 (2000)
50. Roussopoulos, K.: Feedback control of vortex shedding at low Reynolds number. J. Fluid Mech. **248**, 267–296 (1993)
51. Gillies, E.A.: Multiple sensor control of vortex shedding. AIAA Journal **39**(4), 748–750 (2000)
52. Wei, G.W.: Synchronization of single-side locally averaged adaptive coupling and its application to shock capturing. Phys. Rev. Lett., **86**(16), 3542–3545 (2001)
53. Blevins, R.D.: Flow Induced Vibrations. Van Nostrand Reinhold, New York (1990)
54. Park, D.S., Ladd, D.M., Hendricks, E.W.: Feedback control of von Karman vortex shedding behind a circular cylinder at low Reynolds numbers. Phys. Fluid. **6**, 2390–2405 (1994)
55. Gunzburger, M.D., Lee, H.C.: Feedack control of vortex shedding. Tran. ASME J. Appl. Mech. **63**, 828–835 (1996)
56. Min, C., Choi, H.: Suboptimal feedback control of vortex shedding at low reynolds numbers. J. Fluid Mech., **401**, 123–156 (1999)
57. Tokumaru, P.T., Dimotakis, P.E.: Rotary oscillation control of a cylinder wake. J. Fluid Mech. **224**, 77–90 (1991)
58. Warui, H.M., Fujisawa, N.: Feedback control of vortex shedding from a circular cylinder by cross-flow cylinder oscillations. Exp. Fluids **21**, 49–56 (1996)
59. Ffowcs Williams, J.E., Zhao, B.C.: The active control of vortex shedding. J. Fluid. Struc. **3**, 115–122 (1989)
60. Fujisawa, N., Takeda, G.: Flow control around a circular cylinder by internal acoustic excitation. J. Fluid. Struc. **17**, 903–913 (2003)
61. Chen, Z., Fan, B., Zhou, B., Aubry, N.: Control of vortex shedding behind a circular cylinder using electromagnetic forces. Mod. Phys. Lett. B, **19**(28/29), 1627–1630 (2005)
62. Posdziech, O., Grundmann, R.: Electromagnetic control of seawater flow around circular cylinders. Eur. J. Mech. B Fluids **20**, 255–274 (2001)
63. Cattafesta, L.N., Garg, S., Shukla, D.: Development of piezoelectric actuators for active flow control. AIAA Journal **39**(8), 1562–1568 (2001)
64. Kurimoto, N., Suzuki, Y., Kasagi, N.: Active control of lifted diffusion flumes with arrayed micro actuators. Exp. Fluids **39**, 995–1008 (2005)
65. Gerhard, J., Pastoor, M., King, R., et al.: Model-based control of vortex shedding using low-dimensional Galerkin models. AIAA Paper, 2003–4261 (2003)
66. Carpenter, P.W., Kudar, K.L., Ali, R., et al.: A deterministic model for the sublayer streaks in turbulent boundary layers for application to flow control. Phil. Trans. R. Soc. A **365**, 2419–2441 (2007)

67. Kroo, I.: Drag due to lift: concepts for prediction and reduction. Annu. Rev. Fluid Mech. **33**, 587–617 (2001)
68. Sobieczky, H., Seebass, A.R.: Supercritical airfoil and wing design. Annu. Rev. Fluid Mech. **16**, 337–363 (1984)
69. Braslow, A.L.: A history of suctian-type laminar-flow control with emphasis on flight research. In: Monographs in Aerospace History, no. 13. NASA History Division, Office of Policy and Plans, NASA Heaquarters, Washington, DC (1999)
70. Carpenter, P.W.: Status of transition delay using compliant walls. In: Bushnell, D.M., Hefner, J.N. (eds.), Viscous drag reduction in boundary layers. Progr. Astronaut. Aeronaut. vol. 123, pp. 79–113. Washington, DC: AIAA (1990)
71. Carpenter, P.W., Lucey, A.D., Davies, C.: Progress on the use of compliant walls for laminar-flow control. J. Aircraft. **38**, 504–512 (2001)
72. Walsh, M.J., Weinstein, L.M.: Drag and heat transfer on surfaces with small longitudinal fins. In: 11th AIAA, Fluid and Plasma Dynamics Conference, Seattle, Washington, **1161** (1978)
73. Bechert, D.W., Bruse, M., Hage, W., et al.: Experiments on drag-reducing surfaces and their optimization with an adjustable geometry. J. Fluid Mech. **338**, 59–87 (1997)
74. Sirovich, L., Karlsson, S.: Turbulent drag reduction by passive mechanisms. Nature, **388**, 753–755 (1997)
75. Carpenter, P.W.: The right sort of roughness. Nature **388**, 713–714 (1997)
76. Choi, K.S.: Fluid dynamics: the rough with the smooth. Nature **440**, 754 (2006)
77. Abergel, F., Temam, R.: On some control problems in fluid mechanics. Theor. Comp. Fluid Dyn. **1**, 303–325 (1990)
78. Medjo, T.T., Tenam, R., Ziane, M.: Optimal and robust control of fluid flows: Some theoretical and computational aspects. Trans. ASME App. Mech. Rev. **61**(1), 23 (2008)
79. Guan, S., Zhou, Y.C., Wei, G.W., et al.: Controlling flow turbulence. Chaos **13**(1), 64–70 (2003)
80. Zdravkovich, M.M.: Flow around circular cylinders: Fundamentals, vol. 1, pp. 672. Oxford University Press, New York (1997)
81. Williamson, C.H.K., Govardhan, R.: Vortex-induced vibrations. Annu. Rev. Fluid Mech. **36**, 413–455 (2004)
82. Roshko, A.: Perspectives on bluff body aerodynamics, J. Wind. Eng. Ind. Aerod. **49**, 79–100 (1993)
83. Noto, K., Fujimoto, K.: Formulation and numerical methodology for three-dimensional wake of heated circular cylinder. Numer. Heat Tran. A Appl. **49**(2), 129–158 (2006)
84. Patnaik, B.S.V., Narayana, P.A.A., Seetharamu, K.N.: Numerical simulation of vortex shedding past a circular cylinder under the influence of buoyancy. Int. J. Heat Mass Tran. **42**(18), 3495–3507 (1999)
85. Patankar, S.V.: Numerical Heat Transfer and Fluid Flow. Hemisphere, Washington (1980)
86. Perry, A.E., Chong, M.S., Lim, T.T.: The vortex shedding process behind two-dimensional bluff bodies. J. Fluid Mech. **116**, 77–90 (1982)
87. Tang, G.N., Guan, S., Hu, G.: Controlling flow turbulence with moving controllers. Eur. Phys. J. B. **48**, 259–264 (2005)
88. Tang, G.N., Hu, G.: Controlling flow turbulence using local pinning feedback. Chin. Phys. Lett. **23**(6), 1523–1526 (2006)
89. Guan, S., Wei, G.W., Lai, C.H.: Controllability of flow turbulence. Phys. Rev. E. **69**, 066214 (2004)
90. Chen, Z., Aubry, N.: Closed-loop control of vortex-induced vibration. Comm. Nonlinear Science and Num. Simulation **10**, 287–297 (2005)
91. Li, Z., Navon, I.M., Hussaini, M.Y., et al.: Optimal control of cylinder wakes via suction and blowing. Comp. Fluids **32**, 149–171 (2003)
92. Arístegui, J., Tett, P., Hernández-Guerra, A., et al.: The influence of island-generated eddies on chlorophyll distribution: a study of mesoscale variation around Gran Canaria. Deep Sea Research Part I: Oceanographic Research Papers **44**, 71–96 (1997)

93. Chomaz, J.M., Huerre, P., Redekopp, L.G.: Bifurcations to local and global modes in spatially developing flows. Phys. Rev. Lett. **60**, 25–28 (1988)
94. Beaudoin, J.F., Cadot, O., Aider, J.L., et al.: Bluff body drag reduction by extremum seeking control. J. Fluids Struc. **22**, 973–978 (2006)
95. Beaudoin, J.F., Cadot, O., Aider, J.L., et al.: Drag reduction of a bluff body using adaptive control methods. Phys. Fluids **18**(8), 085107 (2006)
96. Betz, A.: History of boundary layer control in Germany. In: Lachmann, G.V. (ed.), Boundary layer and flow control, pp. 1–20. Pergamon Press, New York (1961)
97. Modi, V.J.: Moving surface boundary-layer control: A review. J. Fluids Struc. **11**, 627–663 (1997)
98. Mokhtarian, F., Modi, V.J.: Fluid dynamics of airfoils with moving surface boundary-layer control. J Aircraft **25**, 163–169 (1988)
99. Launder, B.E., Sharma, B.I.: Application of the energy-dissipation model of turbulence to the calculation of flow near a spinning disc. Lett. Heat and Mass Trans. **1**, 131–138 (1974)
100. Rodi, W.: Comparison of LES and RANS calculations of the flow around bluff bodies. J. Wind Engg. Ind. Aerodyn. **69**, 55–75 (1997)
101. Manneville, P.: Instabilities, Chaos and Turbulence. Imperial College Press, London (2004)
102. Saddoughi, S.G., Veeravalli, S.V.: Local isotropy in turbulent boundary layers at high Reynolds number. J. Fluid Mech. **268**, 333–372 (1994)
103. Tél, T., Alessandro de Moura, Grebogi, C., et al.: Chemical and biological activity in open flows: A dynamical system approach. Phys. Rep. **413**, 91–196 (2005)
104. Brown, G.L., Roshko, A.: On the density effects and large structure in turbulent mixing layers. J. Fluid Mech. **64**, 775–816 (1974)
105. Van Dyke, M.: An Album of Fluid Motion. Parabolic Press, Stanford (1988)
106. Private communication from Prof. T. Yokomizo
107. Parnaudeau, P., Carlier, J., Heitz, D., et al.: Experimental and numerical studies of the flow over a circular cylinder at Reynolds number 3900, Phys. Fluids **20**, 085101 (2008)
108. Kravchenko, A.G., Moin, P.: Numerical studies of flow over a circular cylinder at $Re_D = 3900$, Phys. Fluids **12**(2), 403–417 (2000)
109. Lübcke, H., Schmidt, St., Rung, T., et al.: Comparison of LES and RANS in bluff-body flows. J. Wind. Eng. Ind. Aerod. **89**, 1471–1485 (2001)
110. Kolmogorov, A.N.: Dissipation of energy in the locally isotropic turbulence. In: Friedlander, S.K., Topping, L. (eds.) Turbulence: Classic Papers on Statistical Theory, pp. 159–161. Interscience, New York (1961)
111. Head, M.R., Bandyopadhyay, P.: New aspects of turbulent boundary-layer structure. J. Fluid Mech. **107**, 297–338 (1981)
112. Hunt, J.C.R., Carruthers, D.J., Fung, J.C.H.: Rapid distortion theory as a means of exploring the structure of turbulence. In: Sirovich, L. (ed.) New Perspectives in Turbulence, pp. 55–103. Springer, Berlin (1991)
113. Liepmann, H.W.: The rise and fall of ideas in turbulence. Am. Sci. **67**(2), 221–228 (1979)

Chapter 5
Detection and Characterization of Cracks in Beams via Chaotic Excitation and Statistical Analysis

Chandresh Dubey and Vikram Kapila

5.1 Introduction

In one common approach to detect the presence of cracks in beam type structures, the structure is made to vibrate to extract some quantitative features that increase or decrease based on the crack depth. For example, numerous experimental and simulation studies analyze variations in natural frequency of the structure to detect crack depth. Unfortunately, the ability to effectively characterize crack depth in this manner is limited by the very small variation in the natural frequency resulting from the change in crack depth [15]. Harmonic excitation is the most common method used in vibration-based crack detection studies [1, 14, 22].

In recent years, many researchers have investigated the use of chaotic excitation signals to develop vibration-based crack detection techniques due to their applicability over a wide frequency spectrum [12, 13, 16]. Although the influence of chaotic excitation on statistics-based characterizing parameters has been investigated [12], the results are not very satisfactory. In fact, statistics-based characterizing parameters are significantly influenced by the choice of chaotic signal used for excitation. Statistics-based characterizing parameters have been shown to yield satisfactory results in predicting the severity and location of crack in [22], which used a harmonic signal to excite a plate structure in the vicinity of its natural frequency.

In this chapter, we consider the chaotic excitation of [16] and statistics-based characterizing parameters examined in [22] for crack detection in beams. Specifically, similar to [16], we use a SDOF approximation of a cracked beam excited by a chaotic input. However, unlike the Hausdorff distance used in [16] as a characterizing parameter for crack detection, we consider several statistical characteristics to quantify the crack. Moreover, we extend this framework to the continuous

C. Dubey · V. Kapila (✉)
Polytechnic Institute of NYU, 6 Metrotech Center Brooklyn New York
e-mail: cdubey01@students.poly.edu; vkapila@poly.edu

S. Banerjee et al. (eds.), *Applications of Chaos and Nonlinear Dynamics in Engineering – Vol. 1*, Understanding Complex Systems, DOI 10.1007/978-3-642-21922-1_5,

model of structure to predict not only the extent of crack but also its location along the beam. In contrast to [16], our proposed technique does not necessitate the reconstruction of the chaotic attractor, instead all relevant information is extracted directly from the time series of vibration response. We present several alternatives to allow the detection of (i) only the presence of crack using one time series data; (ii) the location and severity of crack using time series data measured at uniformly distributed locations along the beam span; and (iii) approximate location and severity of crack using only two time series data.

This chapter is organized as follows. First, in Sect. 5.2, we explain various inputs used in this effort for exciting the SDOF model of a cracked beam. Second, in Sect. 5.3, we build on the SDOF mathematical model of a cracked beam [2, 5, 16] to include a chaotic excitation. Third, in Sect. 5.4, we define and explain various statistical characterizing parameters that are to be used to detect crack. Fourth, in Sect. 5.5, we provide results for the SDOF model of a cracked beam. Fifth, in Sect. 5.6, we review a mathematical model of a cantilever beam with crack [17, 21]. Sixth, in Sect. 5.7, we provide a finite element (FE) model that is used to simulate the dynamics of a cracked beam with chaotic displacement at its support. Seventh, in Sect. 5.8, we illustrate the results obtained by analyzing time series produced by simulating the continuous structure and we present various alternative methods to characterize cracks. Eighth, in Sect. 5.9, we provide the experimental verification of the numerical simulation. Finally, in Sect. 5.10, some concluding remarks are given.

5.2 Input Signals

A variety of input signals has been studied for vibration-based crack detection in the literature [1, 8, 12–14, 16, 22]. In this effort, we first evaluate the effectiveness of three types of input signals in crack detection with a SDOF model.

5.2.1 Random Input

Random excitation is applied to a SDOF model of a cracked beam to examine the system behavior. Note that chaotic inputs, which are of primary interest in this effort, appear random on a cursory glance (although a careful analysis of their time series reveals hidden structure and pattern). This motivated our interest in studying the use of random signals for vibration-based crack detection. Using one random input, we excite a SDOF beam model with different crack depths and analyze the resulting time series data to determine whether a crack characterization methodology could be developed. Figure 5.1 shows a zero mean random signal used in our study.

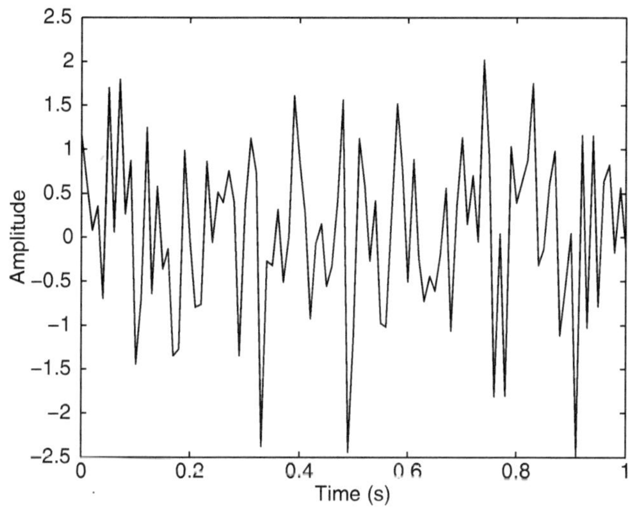

Fig. 5.1 Random input signal

5.2.2 *Sinusoidal Input*

Harmonic or sinusoidal inputs have been widely used for vibration-based crack detection in the literature [1, 14, 19, 22]. A key advantage of using sinusoidal excitation for crack detection is that the signal is easy to produce and apply to the structure. In the current literature, it has been observed that for harmonic excitation based crack detection, the excitation frequency ought to be in the vicinity of system natural frequency [22]. Such a harmonic input drives the system close to its resonance condition which additionally aggravates the effect of defects on system response. Figure 5.2 shows a typical sinusoidal signal of unit amplitude used in our study.

5.2.3 *Chaotic Input*

The chaotic solution of a forced Duffing's equation [20] is applied as a displacement excitation to a cracked beam in this study. The forced Duffing's oscillator can be expressed as

$$\ddot{y} + c\dot{y} - k_1 y + k_2 y^3 = F\cos(\omega t), \tag{5.1}$$

where c, k_1, and k_2 correspond to some physical parameters of the Duffing oscillator, F is the amplitude of the sinusoid driving the oscillator, and ω is the angular frequency of the forcing sinusoid. As evidenced from (5.1), Duffing's equation has a nonlinear stiffness term proportional to y^3. By defining state variables, $y_1 \triangleq y$,

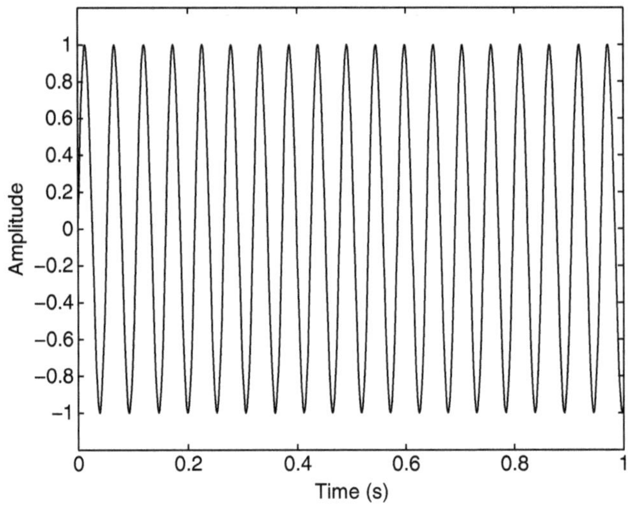

Fig. 5.2 Sinusoidal input signal

$y_2 \triangleq \dot{y_1} = \dot{y}$, and $y_3 \triangleq \omega t$, (5.1) can be rewritten as the following set of three first-order ordinary differential equations

$$\begin{aligned}\dot{y_1} &= y_2,\\ \dot{y_2} &= F\cos(y_3) + k_1 y_1 - k_2 y_1^3 - c y_2,\\ \dot{y_3} &= \omega.\end{aligned} \tag{5.2}$$

In a nonlinear system, there may be several attracting solutions corresponding to different initial conditions and system parameters. For instance, in (5.2), the time series of y_1 and the phase space trajectory produced by y_1 (displacement) versus y_2 (velocity) exhibit a periodic solution if the parameters are set to $c = 0.05$, $k_1 = -0.53$, $k_2 = 1$, $F = 7.5$, and $\omega = 1$ with initial values of $y_1(0) = 0$, $y_2(0) = 0.4$, and $y_3(0) = 0$ [16]. Alternatively, if the system parameter k_1 is changed to $k_1 = 0$, the Duffing's equation exhibits a chaotic behavior as evidenced from Fig. 5.3. It can be seen from Fig. 5.3 that chaotic signals exhibit complex and sophisticated attractor geometries in phase space and that their time domain behavior is non-periodic.

5.3 Modeling of a Cracked Beam as a SDOF System with Displacement Input

We begin by modeling a cracked beam using a SDOF model that allows us to capture the severity of crack at any location along the beam as shown in Fig. 5.4. Consideration of this model requires several assumptions. First, the crack is open to the surface and it is a fine hair crack. Second, only the stiffness of the beam is

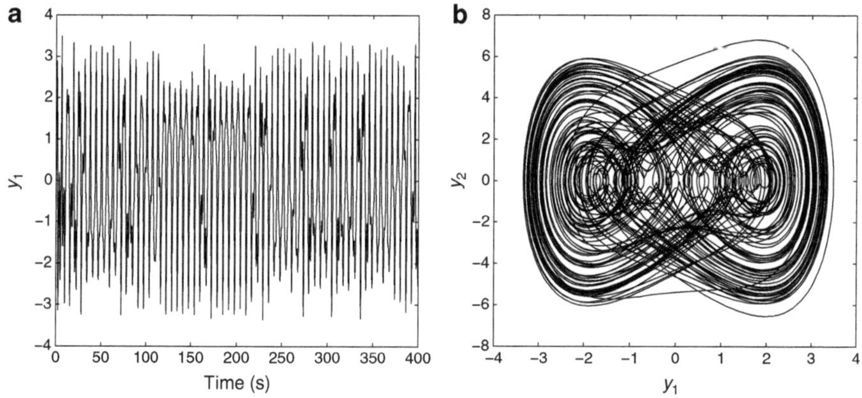

Fig. 5.3 Chaotic solution of the Duffing equation (5.2) with $c = 0.05$, $k_1 = 0$, $k_2 = 1$, $F = 7.5$, $\omega = 1$, $y_1(0) = 0$, $y_2(0) = 0.4$, and $y_3(0) = 0$. (**a**) Time series of y_1 and (**b**) Phase portrait projected onto the (y_1, y_2) plane

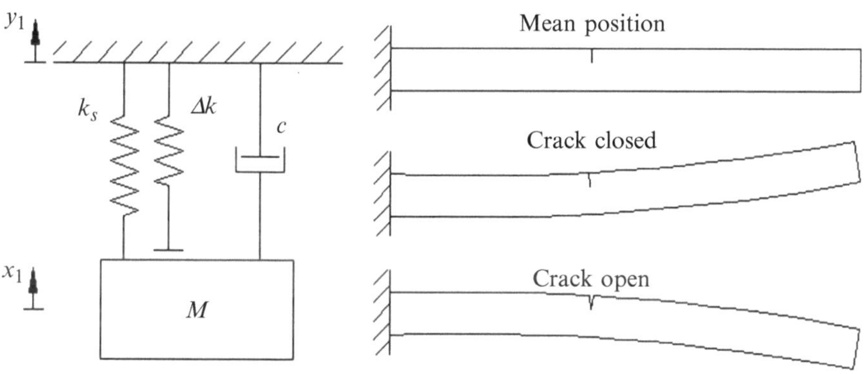

Fig. 5.4 SDOF model of continuous beam with crack

affected by the presence of crack, i.e., the beam stiffness is lowered as the crack opens. The amount of change in beam stiffness Δk is related to the depth of the crack. Assuming the presence of a crack on the top surface of the beam, as the beam vibrates in its first mode, the crack causes the beam to experience stretching and compression in addition to the nominal stretching and compression experienced in the absence of crack. During compression of the top surface, the structure is deemed as being continuous with effective stiffness of k. However, during stretching of the top surface, the effective stiffness of the structure reduces as follows

$$k_s = k - \Delta k, \tag{5.3}$$

where k is the stiffness of the beam without crack, k_s is the stiffness during stretching, and Δk is the stiffness difference [16]. Therefore, a beam with crack is considered as a piecewise linear system which behaves nonlinearly due to the presence of crack. For a SDOF model with a relatively small crack, the ratio of Δk to k is equal to the ratio of the crack depth a to the thickness h of the beam [2, 16], i.e.,

$$\frac{\Delta k}{k} = \frac{a}{h}, \qquad \text{for } a \ll h. \tag{5.4}$$

Next, as shown in Fig. 5.4, an excitation input (random, harmonic, or chaotic) is applied at the base of the SDOF system as a displacement. For the case of chaotic input, we use y_1 solution of the Duffing's equation to excite the base. The equations of motion for this piecewise continuous SDOF system are [16]

$$\begin{aligned} M\ddot{x}_1 + c\dot{x}_1 + kx_1 &= c\dot{y}_1 + ky_1, &\quad \text{for } x_1 \geq 0, \\ M\ddot{x}_1 + c\dot{x}_1 + k_s x_1 &= c\dot{y}_1 + k_s y_1, &\quad \text{for } x_1 < 0, \end{aligned} \tag{5.5}$$

where M is the mass of cantilever beam, c is the damping coefficient, and x_1 is the displacement of beam from the undeformed position.

5.4 Statistical Parameters

5.4.1 *Standard Deviation* (σ)

The standard deviation is a measure of the dispersion of a data set or a distribution [18]. Essentially, standard deviation characterizes the amount of variation a data set experiences relative to its mean or average. It may be thought of as the average deviation of data points from their mean. A data set with a low standard deviation indicates that the data points are in close proximity to one another, whereas a data set with a high standard deviation indicates that the data points are spread over a wide range of values. Standard deviation for a data set is calculated using

$$\sigma = \left(\frac{1}{n-1} \sum_{i=1}^{n} (x_i - \bar{x})^2 \right)^{1/2}, \tag{5.6}$$

where x_i, $i = 1, \ldots, n$, denote individual data points, $\bar{x}$ denote the mean of data set, and n denotes the number of elements in the data set. Figure 5.5a shows the effect of low and high standard deviation on normal distribution curves with the same average in both cases.

5.4.2 *Skewness*

Skewness denotes the degree of asymmetry or departure from symmetry of a data set or distribution from its mean [18]. Skewness of a given data set can be negative

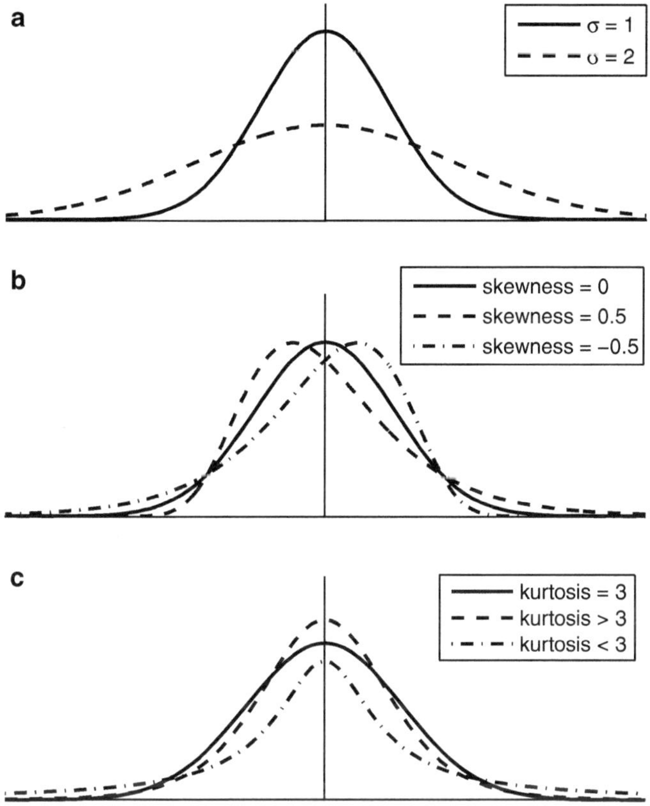

Fig. 5.5 (**a**) Standard deviation, (**b**) skewness, and (**c**) kurtosis

or positive as illustrated in Fig. 5.5b. If skewness is zero (i.e., the distribution is symmetric) then the mean and median of the data set are equal. Skewness for a data set is calculated using

$$\text{skewness} = \frac{\sum_{i=1}^{n}(x_i - \bar{x})^3}{\sigma^3(n-1)}. \tag{5.7}$$

5.4.3 *Kurtosis*

Kurtosis characterizes the "peakedness" or "flatness" of a distribution of a data set with reference to a normal distribution [18]. A data set with normal distribution has kurtosis of about 3 as shown in Fig. 5.5c. Distribution of a given data set may have a kurtosis below 3 (*platykurtic*) or above 3 (*leptokurtic*). Frequently kurtosis is viewed as tail heaviness of a distribution. Leptokurtic distribution has heavier tail than normal distribution, i.e., it has many data points away from the mean, indicating presence of many outliers, than a normally distributed data. Conversely, a platykurtic distribution has thinner tail than normal distribution, i.e., it has few data points away from the mean than a normally distributed data. Kurtosis of a distribution of a data set is calculated using

$$\text{kurtosis} = \frac{\sum_{i=1}^{n}(x_i - \bar{x})^4}{\sigma^4(n-1)}. \tag{5.8}$$

5.5 SDOF Results

In this section, we subject the SDOF model of Sect. 5.3 to the random, harmonic, and chaotic inputs described in Sect. 5.2. The resulting time series output is recorded and used to evaluate the values of statistical parameters such as standard deviation, skewness, and kurtosis. This process is repeated for varying crack depths. The results are graphed and examined to determine the existence of any trends that may indicate the presence and severity of crack. Table 5.1 shows the physical parameters of the problem data used in our simulations. Moreover, in all cases, we also consider one stiffness value above and one below the nominal stiffness. This allows us to investigate the effect of various inputs on randomly selected specimen since in real world situations the natural frequency of specimen may not be know *a priori*.

5.5.1 SDOF Results of Statistical Characterizing Parameters for Random Input

The results of our analysis for a SDOF model vibrated using a normally distributed random signal are summarized in Fig. 5.6. Specifically, Fig. 5.6 shows the changes in statistical parameters (such as, standard deviation, skewness, kurtosis) versus the normalized crack depth for the three different cases of stiffness. We observe that there is no discernible trend relating standard deviation (or kurtosis) versus crack depth ratio. Although skewness exhibits a constant increasing trend with increasing crack depth, a sudden change at the crack depth ratio of 0.10 renders it unusable.

Table 5.1 SDOF simulation parameters

Property	Value
Mass	0.18 kg
Nominal stiffness	295 $\frac{\text{N}}{\text{m}}$
Damping	0.03 $\frac{\text{Ns}}{\text{m}}$

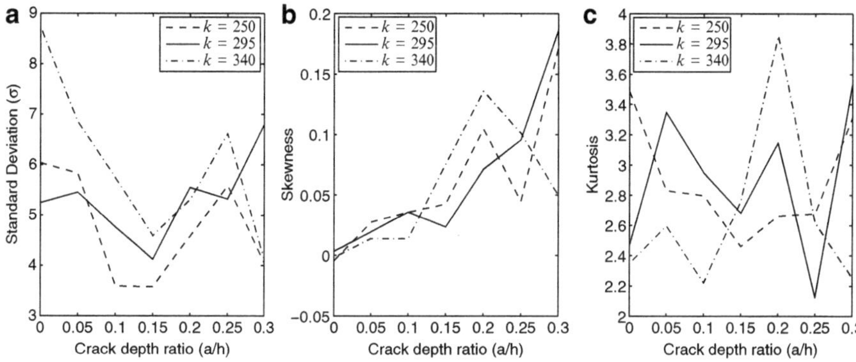

Fig. 5.6 SDOF results of statistical characterizing parameters for random input. Crack depth ratio versus (**a**) standard deviation, (**b**) skewness, and (**c**) kurtosis

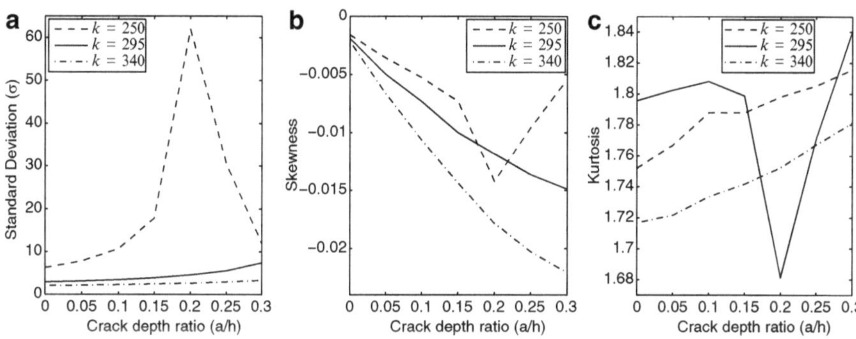

Fig. 5.7 SDOF results of statistical characterizing parameters for sinusoidal input with $\omega = 35$ rad s^{-1} $< \omega_n$. Crack depth ratio versus (**a**) standard deviation, (**b**) skewness, and (**c**) kurtosis

5.5.2 *SDOF Results of Statistical Characterizing Parameters for Sinusoidal Input*

The results of our analysis for a SDOF model vibrated using sinusoidal inputs are summarized in Figs. 5.7 and 5.8. Specifically, Fig. 5.7 shows the behavior of statistical characterizing parameters when the frequency ω of the sinusoidal input is smaller than the natural frequency ω_n of the SDOF model, i.e., $\omega < \omega_n$. For the two higher values of ω_n (corresponding to $k = 295$ and $k = 340$), with the increasing crack depth, the standard deviation σ increases while the skewness decreases. However, this trend changes abruptly for ω_n corresponding to $k = 250$. Moreover, kurtosis is found to be increasing for $k = 250$ and $k = 340$ but behaves abruptly for $k = 295$.

Next, Fig. 5.8 shows the behavior of statistical characterizing parameters when $\omega > \omega_n$. In this case, for the two higher values of ω_n (corresponding to

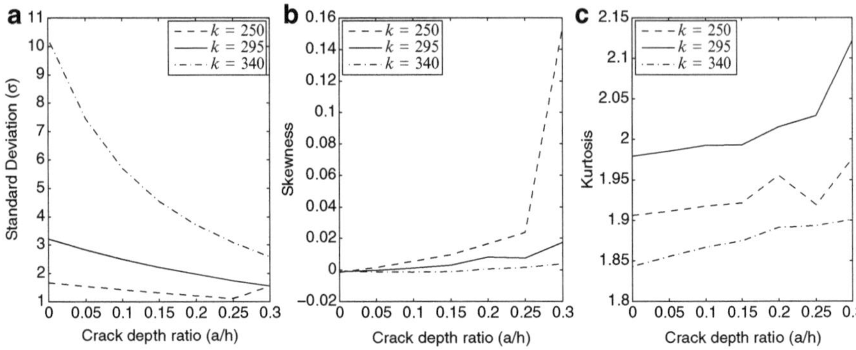

Fig. 5.8 SDOF results of statistical characterizing parameters for sinusoidal input $\omega = 45$ rad s^{-1} $> \omega_n$. Crack depth ratio versus (**a**) standard deviation, (**b**) skewness, and (**c**) kurtosis

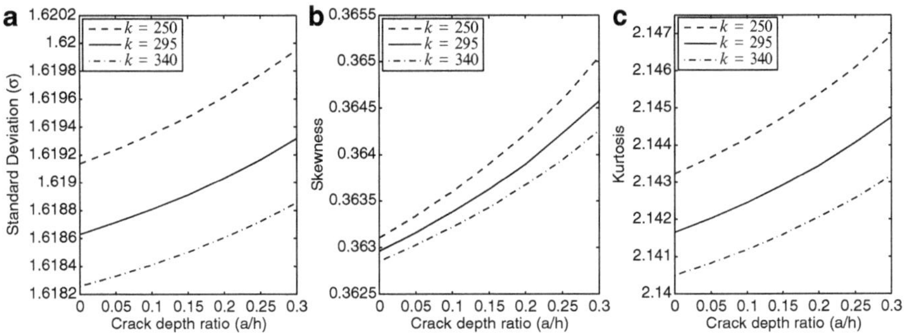

Fig. 5.9 SDOF results of statistical characterizing parameters for chaotic input. Crack depth ratio versus (**a**) standard deviation, (**b**) skewness, and (**c**) kurtosis

$k = 295$ and $k = 340$), with the increasing crack depth, the standard deviation σ decreases while the skewness increases. However, this trend changes abruptly for ω_n corresponding to $k = 250$. Moreover, kurtosis is found to be increasing for $k = 295$ and $k = 340$ but behaves abruptly for $k = 250$. The above analysis shows that the use of harmonic input for crack detection yields acceptable results if the natural frequency of the structure is known, thereby allowing for an adjustment in the excitation signal frequency. Alternatively, if the natural frequency of the structure is not known, a sine sweep would have to be performed to discern a trend and make conclusions about crack depth.

5.5.3 SDOF Results of Statistical Characterizing Parameters for Chaotic Input

The results of our analysis for a SDOF model vibrated using the chaotic input of Sect. 5.2.3 are summarized in Fig. 5.9. Note that all three statistical characterizing parameters, namely σ, skewness, and kurtosis exhibit a continuous growth with increasing crack depth. Moreover, note that this trend is observed for three test structures of different natural frequencies, each excited with exactly the same chaotic input. This result indicates that we do not need to have the exact knowledge of the natural frequency of the structure that is being investigated for crack detection. This of course stems from the fact that the chaotic excitation signal possesses a broad frequency content.

5.6 Continuous Model

In this section, we consider extending the results obtained in Sect. 5.5 for SDOF case to the continuous model case. We begin by developing a continuous model of the dynamical behavior of the beam with surface crack in two parts.

Referring to Fig. 5.4, during the upward motion of the beam, i.e, when the beam moves above the neutral position, the crack is closed and beam behaves as a typical continuous beam. However, during the downward motion of the beam, i.e, when the beam moves below the neutral position, the crack is open and the resulting dynamics is significantly different from a typical continuous beam. The typical continuous beam model resulting for the case of a closed crack is reviewed in Appendix 1. Furthermore, the modification to the beam model resulting for the case of an open crack involves modeling the crack with a rotational spring whose stiffness is related to the crack depth. This modified model is reviewed in Appendix 2.

5.7 Simulation of the Continuous Model of a Cracked Beam in ANSYS

In this effort, we used the ANSYS® software [10] to simulate the dynamics of a cracked beam under external excitation. We modeled the beam as a 2-D elastic object using a *beam3* element [10] which has tension, compression, and bending capabilities. The crack is simulated by inserting a torsional spring at the location of the crack and using the mathematical model described in Appendix 2. The torsional spring is modeled using a *combin14* element [10] which is a spring-damper element used in 1-D, 2-D, and 3-D applications. In our FE model, we used the *combin14* element as a pure spring with 1-D (i.e., torsional) stiffness since the model of Appendix 2 does not consider damping. Figure 5.10 shows the FE model with a crack located at $0.2L$. Moreover, the physical characteristics of the beam used in our FE model are provided in Table 5.2. Before using our FE model to produce the cracked beam response under external excitation, we first verify its accuracy. To do so, the mathematical model of Appendix 2 is implemented as a MATLAB® [7]

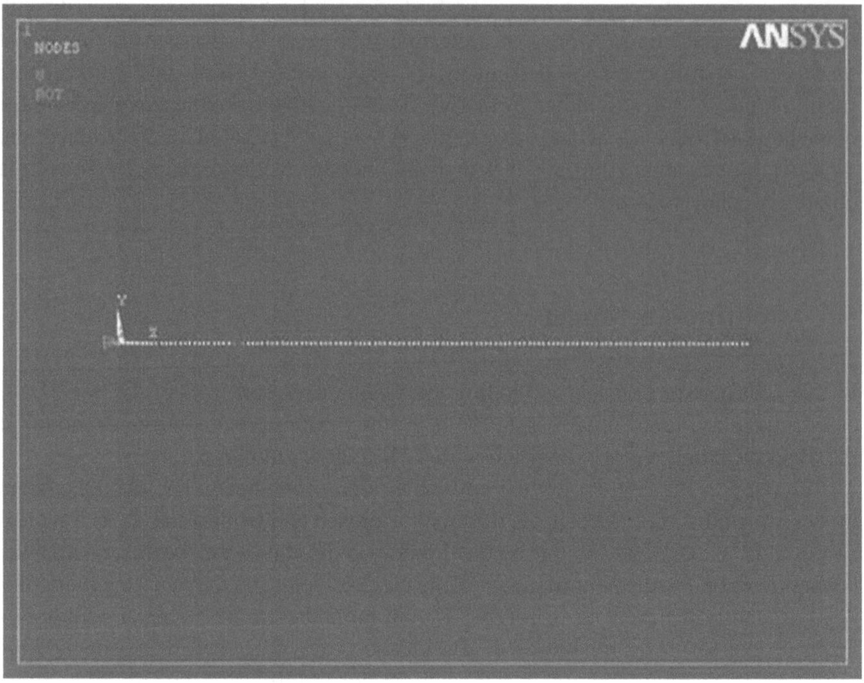

Fig. 5.10 Finite element model of a beam with one crack

Table 5.2 Physical characteristics of beam used in the FE model

Property	Value
Material	Plexiglass
Length	500 mm
Width	50 mm
Thickness	6 mm
Modulus of elasticity	3300 MPa
Density	1190 kg m^{-3}
Poisson's ratio	0.35

algorithm, which allowed us to compute the natural frequency of the cracked beam for different crack depths. The natural frequencies obtained from the MATLAB® algorithm are compared vis-à-vis the natural frequencies determined from the FE simulations performed using ANSYS®. The results of this model validation study are provided in Fig. 5.11, which shows the natural frequencies for the first two vibration modes computed using both the MATLAB® and ANSYS®. Note that with increasing crack depth natural frequency of the beam decreases, which is validated by the given results.

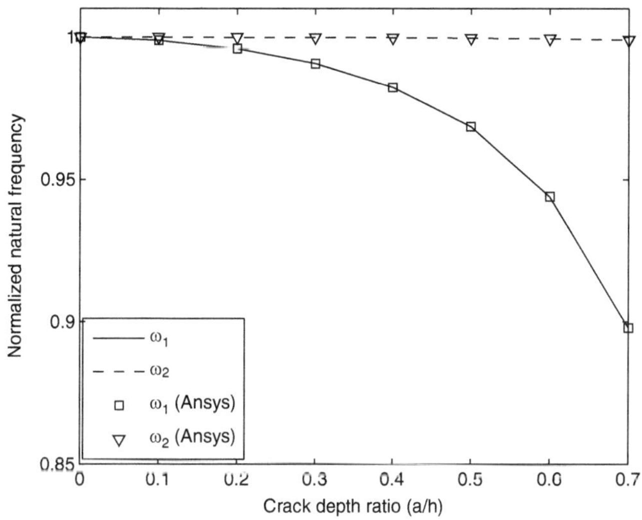

Fig. 5.11 Natural frequency vs. crack depth

5.8 Continuous Model Results

After validating the FE model of Sect. 5.7, we subject it to a base displacement using the time series y_1 corresponding to the chaotic input described in Sect. 5.2.3. In particular, using MATLAB®, we simulate the chaotic oscillator of (5.2) and save 15 000 time steps of time series corresponding to y_1, which is used as an input to excite the FE model. The FE simulation is used to produce and record spatio-temporal responses for each node (corresponding to discretized locations along the beam span). The resulting data is imported in MATLAB® for a detailed time series analysis, whose results are grouped in two parts as explained below.

5.8.1 Detection of the Presence of Crack

We begin by analyzing the time series data corresponding to the response of only the beam tip to detect the presence of any cracks along the beam span. Figure 5.12 provides plots of the normalized crack depth a/h versus two statistical characterizing parameters, namely σ and kurtosis. Note that both σ and kurtosis exhibit a growth trend with increasing crack depth. However, increasing crack depth produces a significantly greater growth in kurtosis compared to σ. Moreover, note that the rate of growth in σ and kurtosis increases when the crack is near the fixed end of the beam and decreases as the crack location shifts to the free end of the beam. Finally, a sound beam starts out with a nominal kurtosis value

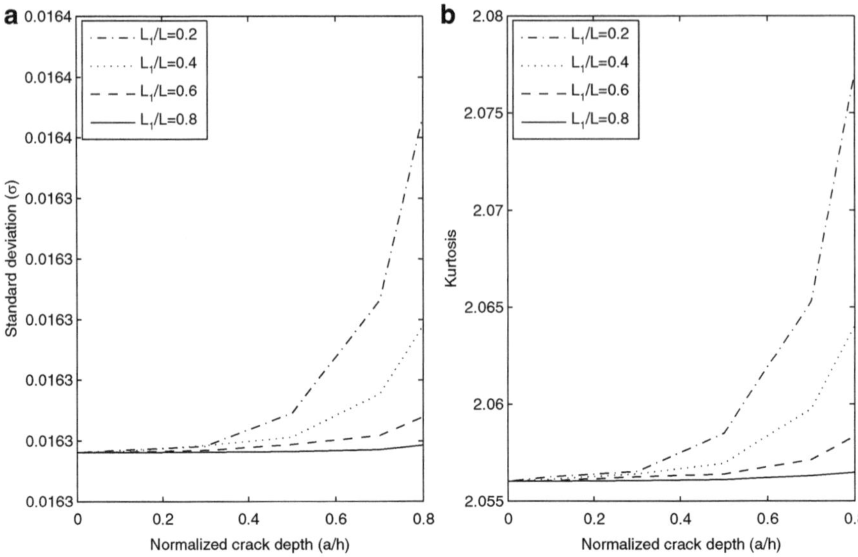

Fig. 5.12 Variation in standard deviation σ and kurtosis with crack severity for different crack locations

(in our example, 2.056) and the presence of crack anywhere along the beam causes the kurtosis value to increase indicating the presence of crack. For example, if the kurtosis calculated from the tip time series data is found to be 2.06, we can conclude presence of crack since the kurtosis for the healthy beam is 2.056. Next, for the kurtosis value of 2.06, according to Fig. 5.12, we may identify that the normalized crack depth may be $a/h = 0.54$ at $L_1/L = 0.2$, or it may be $a/h = 0.705$ at $L_1/L = 0.4$, or something else in between the two indicated results. Thus, we conclude that it is not possible to determine either the crack depth or crack location using only the beam tip response analysis. However, as discussed above, the beam tip response analysis can allow crack detection and can be used in quality control applications to reject damaged components.

5.8.2 Detection of the Location and Severity of Crack

In this subsection, we use the time series data of the beam response along its span to develop a method to predict the severity and approximate location of the crack on the beam surface. We begin by conducting a FE simulation of a beam with a known crack depth and a known crack location. As previously discussed in Sect. 5.8, we use the time series y_1 of chaotic oscillator (5.2) as the input excitation and record the time series of beam response at every 5 mm distance along the beam length. This simulation is repeated for three different crack locations to test the consistency of

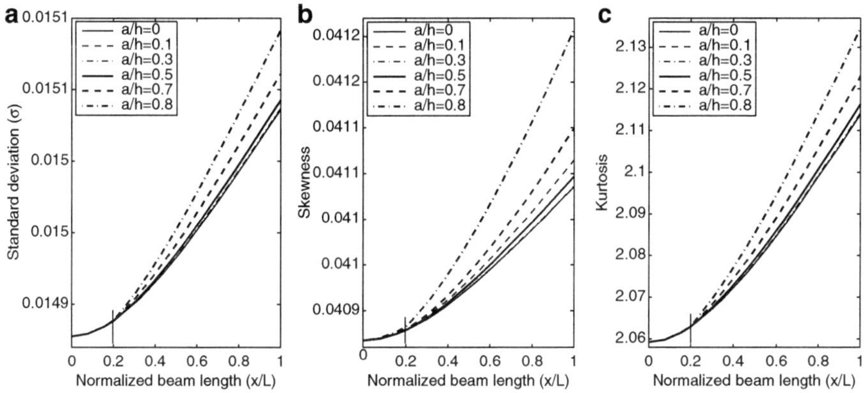

Fig. 5.13 Variation in standard deviation σ, skewness, and kurtosis with varying crack severity for a crack located at $0.2L$

our results. Next, all three statistical characterizing parameters, namely σ, skewness, and kurtosis are computed for the time series at each location for each crack depth.

Figure 5.13 provides plots of the three statistical characterizing parameters for cracks of varying depth located at $L_1 = 0.2L$. The standard deviation σ, skewness, and kurtosis values are plotted against various locations along the beam. The standard deviation σ versus normalized beam length plot shows that the slope of the resulting curve increases with increasing crack depth, i.e., the beam without crack has the smallest slope and the beam with the most severe crack has the largest slope. Moreover, as seen from Figs. 5.13–5.15, the slope of each curve changes abruptly at the location of crack; thus this behavior can be used to determine the approximate location of crack. A careful examination of the skewness plot reveals that the various curves corresponding to different crack depths do not follow the kind of ordering observed in the standard deviation σ plots. Thus, in the following analysis, we will no longer consider the skewness plots. Next, the kurtosis plots are found to behave in a similar manner as the standard deviation σ plots and can be used to predict crack location and severity. In fact, the kurtosis plots exhibit an order of magnitude larger changes compared to standard deviation σ, rendering them more suitable for detecting crack location and severity. Finally, Figs. 5.14 and 5.15 show the plots of statistical characterizing parameters with crack location $L_1 = 0.4L$ and $L_1 = 0.6L$, respectively.

5.8.3 *Crack Depth and Location Prediction Using Reduced Measurement*

As seen in the previous subsection, we can detect the severity and location of a crack provided that the spatio-temporal data is recorded for a series of locations along the

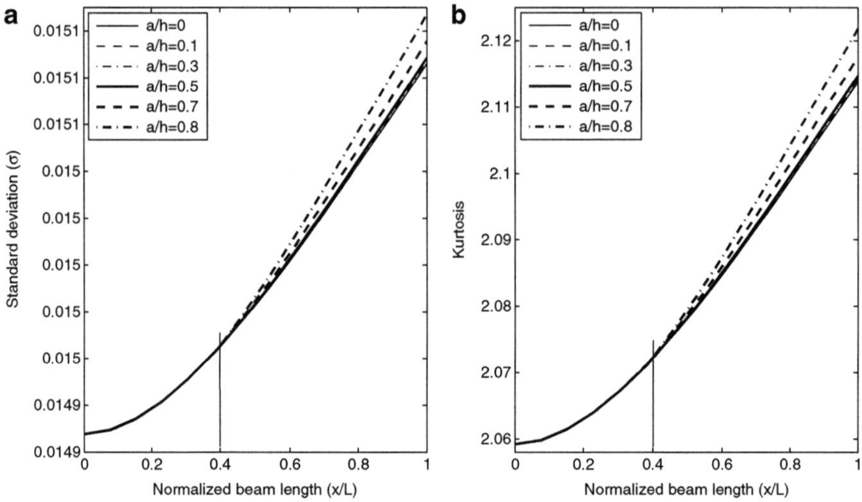

Fig. 5.14 Variation in standard deviation σ and kurtosis with varying crack severity for a crack located at $0.4L$

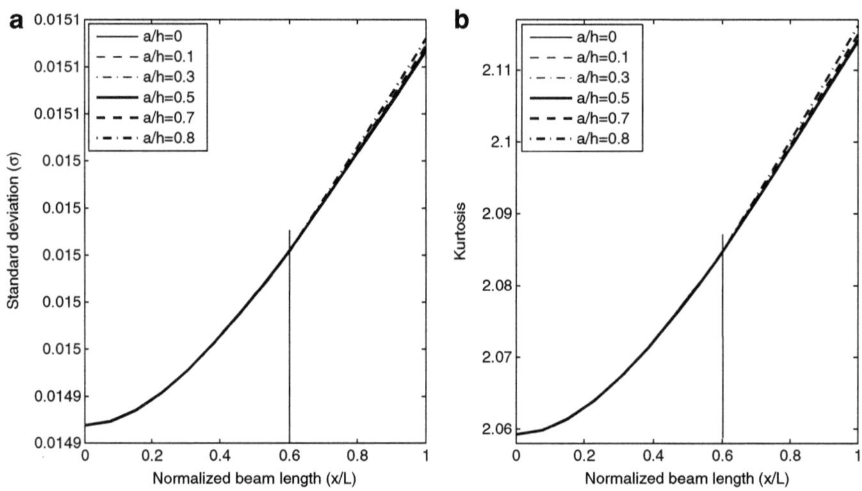

Fig. 5.15 Variation in standard deviation σ and kurtosis with varying crack severity for a crack located at $0.6L$

beam length. Although it may be feasible to do so in a laboratory environment, in a real-world setting this would be prohibitive due to the need to instrument the structure with numerous sensors along its span. A careful examination of the kurtosis plots of Figs. 5.13–5.15 suggests one approach to overcome this problem. Specifically, note that for each crack depth and each crack location, the kurtosis yields a distinct slope past the crack location. For example, Fig. 5.16a shows the

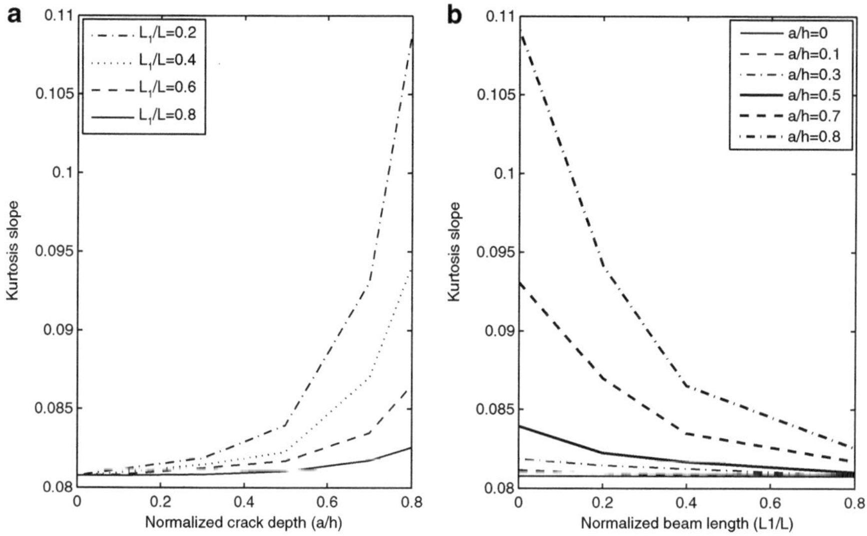

Fig. 5.16 (**a**) Kurtosis slope versus normalized crack depth for fixed crack locations and (**b**) kurtosis slope versus normalized length for fixed crack depths

variation of kurtosis slope versus the crack depth, thus yielding *constant crack location curves*. Similarly, Fig. 5.16b shows the variation of kurtosis slope versus the crack location, thus yielding *constant crack depth curves*. To use the diagrams in Fig. 5.16, we record only two time series along the beam span, preferably near the beam tip. The recorded time series can be analyzed to obtain the corresponding kurtosis values. Furthermore, the knowledge of location where the time series data was recorded allows us to compute the kurtosis slope. Next, superimposing the computed kurtosis slope value on Fig. 5.16a yields the approximate crack location whereas doing the same on Fig. 5.16b yields the approximate crack depth. Having obtained the approximate knowledge of the crack location, we can focus on either collecting additional time series data in the identified region of crack or we can make efficient use of NDT techniques by focusing only on the identified region of crack.

5.9 Experimental Verification

In this section, we briefly describe an experimental study that was used to validate the results obtained from numerical and simulations studies for the SDOF models of a cracked cantilever beam. In Sect. 5.9.1, we provide a schematic of our experimental setup and briefly describe its main components. In addition, we explain our approach to specimen preparation and introduction of cracks by means of a saw

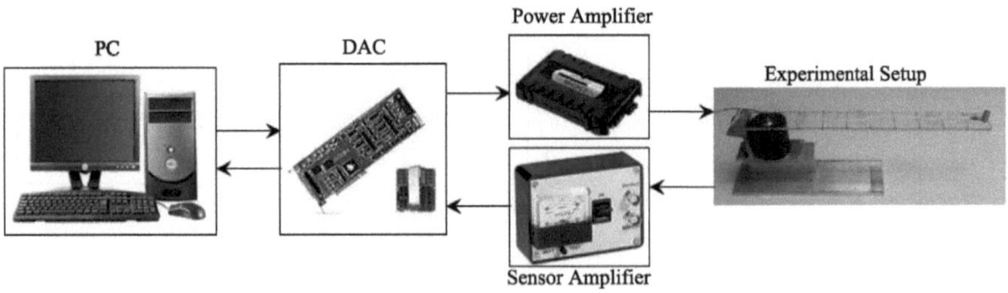

Fig. 5.17 Experimental setup

Table 5.3 Specimen properties

Property	Value
Material	Plexiglass
Length	500 mm
Width	50 mm
Thickness	6 mm
Modulus of elasticity	3300 MPa
Density	1190 kg m^{-3}
Poisson's ratio	0.35

cut. In Sect. 5.9.2, we provide an analysis of the experimental data. Note that the purpose of this experimental study has been to validate the general trend of various crack characterizing parameters studied in earlier in this chapter.

5.9.1 Experimental Setup and Specimen Preparation

A schematic of the experimental setup used in our study is given in Fig. 5.17. An aluminum base is used to hold both the shaker (Brüel & Kjær Type 4810) and test specimen in place. The setup allows us to produce a base excitation by directly clamping beam on shaker. An accelerometer (Omega ACC 103) is mounted at the tip of the specimen using mounting bee wax. Our software environment consists of Matlab, Simulink, and Real Time Workshop in which the Duffing's chaotic oscillator is propagated to obtain the time series corresponding to the y_1 signals of (5.2). Next, an analog output block in the Simulink program outputs the y_1 signal to a digital to analog converter of Quanser's Q4 data acquisition and control board which in turn is fed to a 12 volt amplifier (Kenwood KAC-8202) to drive the shaker. The accelerometer output is processed by an amplifier (Omega ACC PSI) and interfaced to an analog to digital converter of the Q4 board for feedback to the Simulink program. For specimen of different crack depth, the accelerometer measurement is recorded and used to produce the output response time series, which is used to perform our analysis.

Properties of the specimen used in our experiments are given in Table 5.3 and are same as the properties given in Table 5.3.

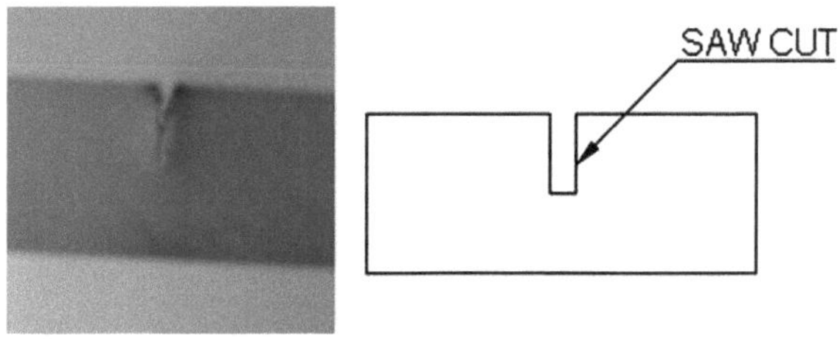

Fig. 5.18 Emulating crack in beam by saw cut

To emulate a fine hair crack, we used a 0.1 mm saw to introduce cracks of several different desired depths. As noted in [6], sawed and cracked beams yield different natural frequencies wherein the frequency difference is dependent on the width of the cut. Thus, it follows that the frequency characteristics of sawed and cracked beams may differ significantly for larger crack width and render the natural frequency based crack detection methods ineffective. The results of this effort are not significantly affected since, instead of relying on changes in natural frequency, our crack detection approach relies on measuring and comparing statistical characteristics of chaotically excited vibration response. Figure 5.18 shows crack introduced in one specimen. Our experimental study is focused on relating the crack severity vis-á-vis several characterizing parameters; thus for all specimen, we introduced cracks at the same location ($L_1 = 0.2L = 100$ mm from fixed end) with varying depth. A total of six specimens were prepared with crack depth varying from 0% to 50% of the thickness. In all the specimen, saw crack was introduced on the top surface to match with the simulation condition.

5.9.2 *Results*

The time series data obtained from the accelerometer suffered from general sensor errors (dc offset and ramp bias), causing the raw time series data to be unusable for further analysis. We used the Wavelet transformation toolbox [9] of MATLAB, to filter the raw time series data and remove the errors. This filtering technique uses a moving average of the waveform to shift its mean to 0. Figure 5.19 shows a sample sinusoidal output from sensor before and after applying the correction.

Figure 5.20 shows the corrected time series that are used to compute various statistical characterizing parameters.

Figures 5.21–5.23 provide the variation in standard deviation, skewness, and kurtosis, respectively, versus the crack depth for the time series data of Fig. 5.20. Note that these three statistical characterizing parameters show an increasing trend with increasing crack depth validating the predictions of our numerical study in Sect. 5.5.3 for SDOF case and in Sect. 5.8.1 for the continuous beam case when only tip displacement measurement is used.

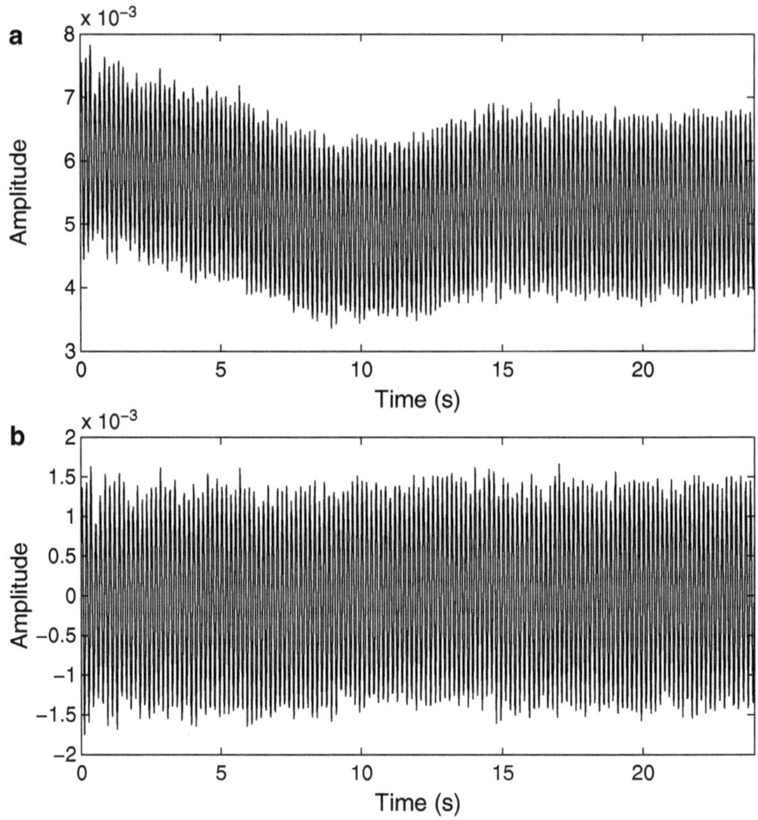

Fig. 5.19 (**a**) Sinusoidal waveform before correction and (**b**) sinusoidal waveform after correction

We note that the plots obtained from the experimental data are not as smooth as the ones resulting from numerical simulation. This may be the result of inaccuracies resulting from sample preparation or a variety of experimental errors. For example, finite bit analog to digital conversion, sampling rate, etc., lead to errors in data recording. In our experimental study, we recorded five readings for each specimen and used averages of the recorded data in performing our experimental analysis.

5.10 Conclusion

In this chapter, we modeled a beam with surface crack as a SDOF system and applied a variety of inputs, including chaotic input, as displacement at the base. The chaotic input, due to its broad frequency content, was found to be useful compared to random and sinusoidal inputs in providing meaningful trends for

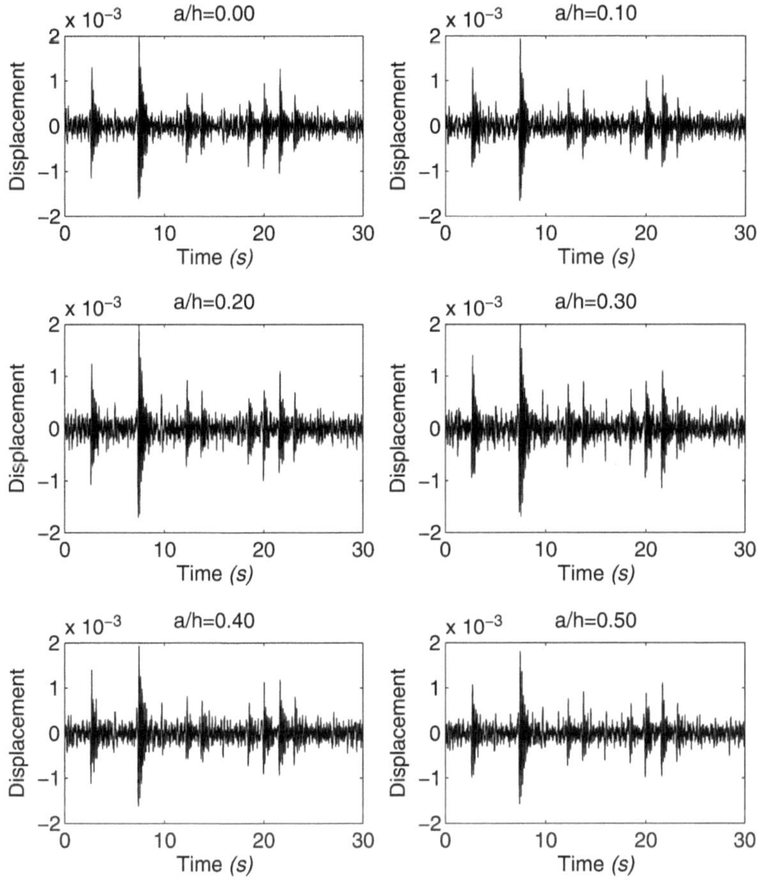

Fig. 5.20 Filtered time series for different crack depths with Duffing's input

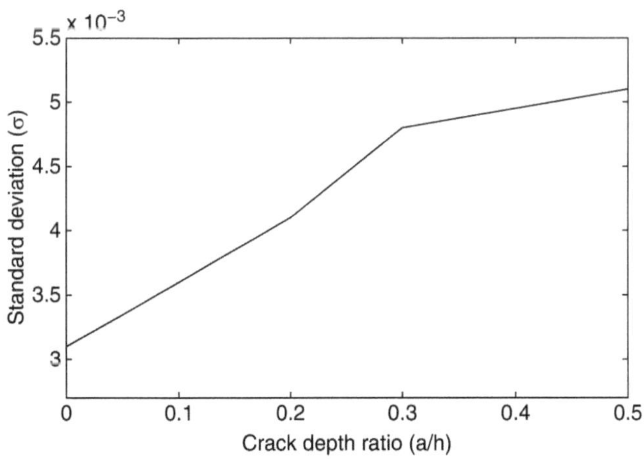

Fig. 5.21 Standard deviation (σ) for different crack depths at $L_1 = 0.2L$

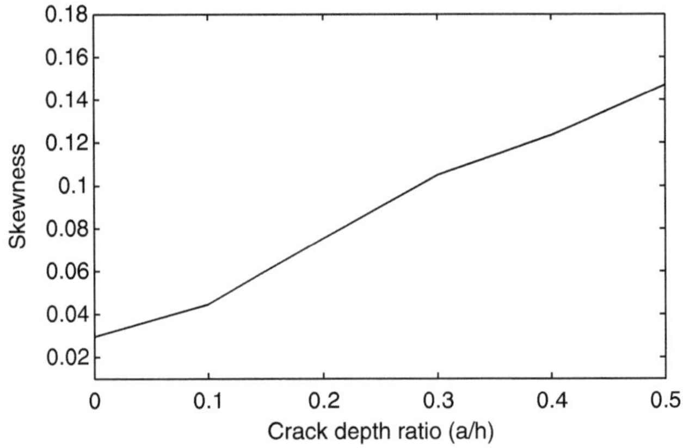

Fig. 5.22 Skewness for different crack depths at $L_1 = 0.2L$

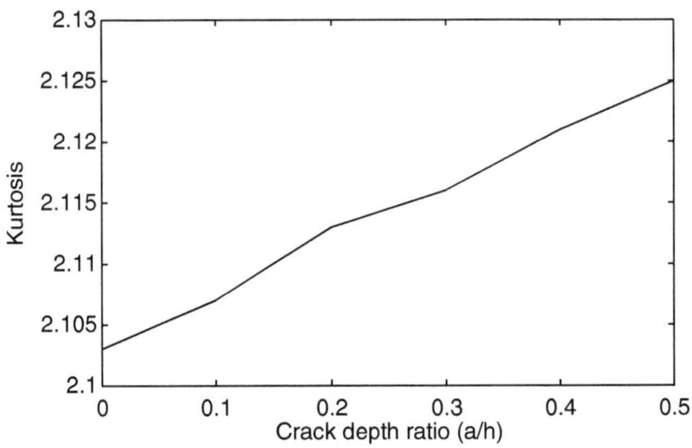

Fig. 5.23 Kurtosis for different crack depths at $L_1 = 0.2L$

statistical characteristic parameters vis-à-vis the crack severity. The results of this chapter show that for the SDOF model, crack severity can be easily and consistently predicted by using standard deviation, skewness, and kurtosis of the time series data. Next, the trend observed in the SDOF model was used for a continuous beam that was modeled and simulated using ANSYS®. In this case, the crack in the beam was modeled as a rotational spring whose stiffness depends on the depth of crack. Once again a chaotic input was applied to beam base and time series data was recorded at various locations along the beam span. For the resulting time series, statistical characteristic parameters were computed and analyzed. Standard deviation and kurtosis plots were shown to exhibit trends that can be used to predict crack location and crack depth. Finally, we used an experimental setup to validate the trends observed in statistical characteristics vis-à-vis increasing crack depth in a cantilever beam. Test specimen for experimental studies consisted of plexiglass cantilever beams that were selected to have physical properties similar to the numerical model beam. A crack in the beam was emulated by a thin, 0.1 mm saw cut on the beam surface at $L_1 = 0.2L$ from the fixed end. Analysis of the experimental time series data revealed that statistical characteristics standard deviation (σ), skewness, and kurtosis exhibit an increasing trend with increasing crack depth.

Appendix 1

In this appendix, for completeness, we include the dynamic model of a cantilever beam [11] that is used earlier sections.

Consider the bending motion of the beam of Fig. 5.24 in the absence of any external excitation. The partial differential equation for the transverse deflection $y(x,t)$ of this beam is given by

$$-\frac{\partial^2}{\partial x^2}\left[EI(x)\frac{\partial^2 y(x,t)}{\partial x^2}\right] = m(x)\frac{\partial^2 y(x,t)}{\partial t^2}, \qquad 0 < x < L, \tag{5.9}$$

where E denotes the modulus of elasticity of the beam material, $I(x)$ denotes the area moment of inertia for the beam cross section, and $m(x)$ denotes the mass per unit length of the beam. The solution $y(x,t)$ of (5.9) must satisfy the following cantilever boundary conditions

$$y(x,t) = 0, \quad \frac{\partial y(x,t)}{\partial x} = 0, \qquad \text{at } x = 0, \tag{5.10}$$

$$EI(x)\frac{\partial^2 y(x,t)}{\partial x^2} = 0, \quad -\frac{\partial}{\partial x}\left[EI(x)\frac{\partial^2 y(x,t)}{\partial x^2}\right] = 0, \qquad \text{at } x = L. \tag{5.11}$$

Using the standard assumption of separation of spatial and temporal variables, the solution of (5.9) can be written in the following form

$$y(x,t) = Y(x)\eta(t), \tag{5.12}$$

where $Y(x)$ represents the mode shape as a function of the spatial variable x and $\eta(t)$ represents the temporal variation of the amplitude.

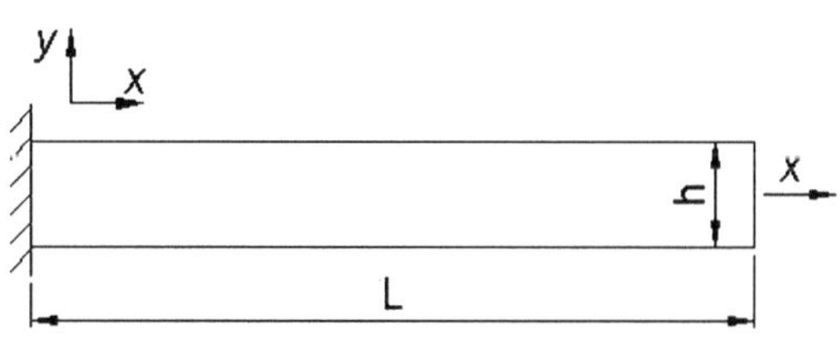

Fig. 5.24 Cantilever beam

Now we substitute (5.12) in (5.9) and separate the temporal and spatial variables to formulate the following eigenvalue problem that is used to determine the mode shapes and natural frequencies of the beam

$$\frac{d^2}{dx^2}\left[EI(x)\frac{d^2Y(x)}{dx^2}\right] = \omega^2 m(x)Y(x), \qquad 0 < x < L, \tag{5.13}$$

where ω denotes the natural frequency of the system. For a uniform beam, we use $I(x) = I$ and $m(x) = m$ and rewrite (5.13) as

$$\frac{d^4Y(x)}{dx^4} - \beta^4 Y(x) = 0, \quad 0 < x < L, \qquad \beta^4 = \frac{\omega^2 m}{EI}. \tag{5.14}$$

Moreover, the boundary conditions (5.10) and (5.11) are rewritten, respectively, as

$$Y(x) = 0, \quad \frac{dY(x)}{dx} = 0, \qquad \text{at } x = 0, \tag{5.15}$$

$$\frac{d^2Y(x)}{dx^2} = 0, \quad \frac{d^3Y(x)}{dx^3} = 0, \qquad \text{at } x = L. \tag{5.16}$$

The general solution of (5.14) is now given by

$$Y(x) = A\sin\beta x + B\cos\beta x + C\sinh\beta x + D\cosh\beta x. \tag{5.17}$$

Using the boundary conditions (5.15) and (5.16) in the general solution (5.17), we obtain

$$D = -B, \tag{5.18}$$

$$C = -A, \tag{5.19}$$

$$B = -\frac{\sin\beta L + \sinh\beta L}{\cos\beta L + \cosh\beta L}A, \tag{5.20}$$

$$\cos\beta L\cosh\beta L = -1. \tag{5.21}$$

Note that (5.21) is a transcendental equation with infinite solutions and it has to be solved numerically to determine $\beta_r L$, $r = 1, \ldots, \infty$, which are used to determine the natural frequencies ω_r, $r = 1, \ldots, \infty$, of the beam. Furthermore, the total solution is now expressed as

$$y(x,t) = \sum_{r=1}^{\infty} Y_r(x)\eta_r(t). \tag{5.22}$$

Using (5.18)–(5.21) in (5.17), the mode shapes can be expressed as

$$Y_r(x) = A_r\left[\sin\beta_r x - \sinh\beta_r x - \frac{\sin\beta_r L + \sinh\beta_r L}{\cos\beta_r L + \cosh\beta_r L}(\cos\beta_r x - \cosh\beta_r x)\right], r = 1,2,\ldots. \tag{5.23}$$

The coefficients A_r and η_r are calculated by exploiting the orthogonality of the mode shapes.

Appendix 2

In this appendix, for completeness, we include the dynamic model of a cracked cantilever beam that is used as a basis for the continuous beam analysis [17, 21].

Modeling of a crack in a beam is dependent on the crack orientation, crack location (surface or sub-surface), and mode of deformation. In this effort, we restrict consideration to a transverse surface crack of uniform depth with only bending vibration. In this case, the crack is modeled as a torsional spring whose stiffness depends on the depth of the crack as discussed in [3, 4, 17, 21]. See Fig. 5.25. The following expression can be used to compute the stiffness K_T of the torsional spring used to model the crack

$$K_T = \frac{EI}{6(1-\nu^2)h} \times \frac{1}{J(a/h)}, \tag{5.24}$$

where E denotes the modulus of elasticity of the beam material, I denotes the area moment of inertia for the beam cross-section, ν denotes the Poisson's ratio, h denotes the beam thickness, a denotes the depth of open surface crack, and $J(a/h)$ denotes the dimensionless local compliance function [21] computed using the following equation

$$\begin{aligned} J(a/h) = {} & 1.8224(a/h)^2 - 3.95(a/h)^3 + 16.375(a/h)^4 - 37.226(a/h)^5 \\ & + 76.81(a/h)^6 - 126.9(a/h)^7 + 172(a/h)^8 - 143(a/h)^9 + 66.56(a/h)^{10}. \end{aligned} \tag{5.25}$$

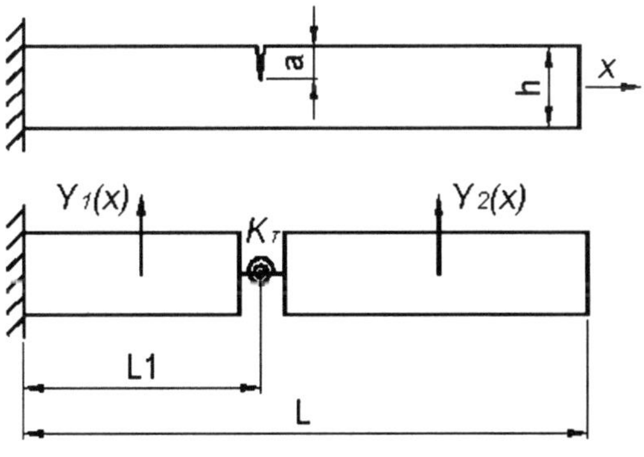

Fig. 5.25 Continuous model of a cracked beam

For the cracked beam of Fig. 5.25, the eigenvalue problems for the two parts of the beam, one to the left of the crack and the other to the right of the crack, respectively, are given by

$$\frac{d^4Y_1(x)}{dx^4} - \frac{\omega^2 m}{EI}Y_1(x) = 0, \qquad 0 \le x \le L_1, \tag{5.26}$$

$$\frac{d^4Y_2(x)}{dx^4} - \frac{\omega^2 m}{EI}Y_2(x) = 0, \qquad L_1 \le x \le L, \tag{5.27}$$

where ω denotes the natural frequency of the beam and m denotes the mass per unit length of the beam material. Following boundary conditions apply at the two ends of the cantilever beam

$$Y_1(x) = 0, \quad \frac{dY_1(x)}{dx} = 0, \qquad \text{at } x = 0,$$

$$\frac{d^2Y_2(x)}{dx^2} = 0, \quad \frac{d^3Y_2(x)}{dx^3} = 0, \quad \text{at } x = L. \tag{5.28}$$

Moreover, following conditions apply at the crack location

$$Y_1(x) = Y_2(x), \qquad \frac{d^2Y_1(x)}{dx^2} = \frac{d^2Y_2(x)}{dx^2},$$

$$\frac{d^3Y_1(x)}{dx^3} = \frac{d^3Y_2(x)}{dx^3}, \qquad \frac{EI}{K_T}\frac{d^2Y_1(x)}{dx^2} + \frac{dY_1(x)}{dx} = \frac{dY_2(x)}{dx}, \qquad \text{at } x = L_1, \tag{5.29}$$

The solution of (5.26) and (5.27) can be shown to be of the following form, respectively,

$$Y_1(x) = A_1\sin(\beta x) + B_1\cos(\beta x) + C_1\sinh(\beta x) + D_1\cosh(\beta x), \tag{5.30}$$

$$Y_2(x) = A_2\sin(\beta x) + B_2\cos(\beta x) + C_2\sinh(\beta x) + D_2\cosh(\beta x), \tag{5.31}$$

where β is as defined in (5.14). Equations (5.30) and (5.31) characterize the displacement of the beam sections to the left and right of the crack.

Using the boundary conditions (5.28) and (5.29) into (5.30) and (5.31), we can obtain the following system of equations

$$\begin{bmatrix} M_{11} & M_{12} \\ M_{21} & M_{22} \end{bmatrix} X = 0,$$

where $X, M_{11}, M_{12}, M_{21}$, and M_{22} are defined, respectively, as

$$X = [A_1 \quad B_1 \quad C_1 \quad D_1 \quad A_2 \quad B_2 \quad C_2 \quad D_2]^T,$$

$$M_{11} = \begin{bmatrix} 0 & 1 & 0 & 1 \\ 1 & 0 & 1 & 0 \\ 0 & 0 & 0 & 0 \\ 0 & 0 & 0 & 0 \end{bmatrix},$$

$$M_{12} = \begin{bmatrix} 0 & 0 & 0 & 0 \\ 0 & 0 & 0 & 0 \\ -\sin(\beta L) & -\cos(\beta L) & \sinh(\beta L) & \cosh(\beta L) \\ -\cos(\beta L) & \sin(\beta L) & \cosh(\beta L) & \sinh(\beta L) \end{bmatrix},$$

$$M_{21} = \begin{bmatrix} \sin(\beta L_1) & \cos(\beta L_1) & \sinh(\beta L_1) & \cosh(\beta L_1) \\ -\sin(\beta L_1) & -\cos(\beta L_1) & \sinh(\beta L_1) & \cosh(\beta L_1) \\ -\cos(\beta L_1) & \sin(\beta L_1) & \cosh(\beta L_1) & \sinh(\beta L_1) \\ -\frac{EI\beta}{K_T}\sin(\beta L_1) & -\frac{EI\beta}{K_T}\cos(\beta L_1) & \frac{EI\beta}{K_T}\sinh(\beta L_1) & \frac{EI\beta}{K_T}\cosh(\beta L_1) \\ +\cos(\beta L_1) & -\sin(\beta L_1) & +\cosh(\beta L_1) & +\sinh(\beta L_1) \end{bmatrix},$$

$$M_{22} = \begin{bmatrix} -\sin(\beta L_1) & -\cos(\beta L_1) & -\sinh(\beta L_1) & -\cosh(\beta L_1) \\ \sin(\beta L_1) & \cos(\beta L_1) & -\sinh(\beta L_1) & -\cosh(\beta L_1) \\ \cos(\beta L_1) & -\sin(\beta L_1) & -\cosh(\beta L_1) & -\sinh(\beta L_1) \\ -\cos(\beta L_1) & \sin(\beta L_1) & -\cosh(\beta L_1) & -\sinh(\beta L_1) \end{bmatrix}.$$

The natural frequencies of the cracked beam are determined by solving the equation $\det[M] = 0$ for values of $\beta = (\omega^2 m/EI)^{1/4}$, which in turn is solved for ω.

Acknowledgments This work is supported in part by the National Science Foundation under an RET Site grant 0807286, a GK-12 Fellows grant 0741714, and the NY Space Grant Consortium under grant 48240-7887.

References

1. Andreausa, U., Casinib, P., Vestronia, F.: Non-linear dynamics of a cracked cantilever beam under harmonic excitation. Int. J. Non. Lin. Mech. **42**, 566–575 (2007)
2. Bouraou, N., Gelman, L.: Theoretical bases of free oscillation method for acoustical non-destructive testing. Proceedings of Noise Conference, The Pennsylvania State University. 519–524 (1997)
3. Bamnios, Y., Douka, E., Trochidis, A.: Crack identification in beam structures using mechanical impedance. J. Sound Vib. **256**(2), 287–297 (2002)
4. Bamnios, G., Trochidis, A.: Dynamic behavior of a cracked cantilever beam. Appl. Acoust. **45**, 97–112 (1995)

5. Chati, M., Rand, R., Mukherjee, S.: Modal analysis of a cracked beam. J. Sound Vib. **207**, 249–270 (1997)
6. Cawley, P., Ray, R.: A comparison of natural frequency changes produced by cracks and slots. Trans. ASME. **110**, 366–370 (1998)
7. Duane, C.H., Bruce, L.L.: Mastering Matlab 7. Prentice Hall, Upper Saddle River, NJ (2005)
8. Douka, E., Bamnios, G., Trochidis, A.: A method for determining the location and depth of crack in double-cracked beams. Appl. Acoust. **65**, 997–1008 (2004)
9. Jensen, A., Anders la, C.H.: Ripples in Mathematics: The Discrete Wavelet Transform. Cambridge University Press, New York, NY, (2001)
10. Moaveni, S.: Finite Element Analysis Theory and Application with ANSYS. Prentice Hall, Upper Saddle River, NJ, (2007)
11. Meirovitch, L.: Fundamentals of Vibrations. McGraw-Hill, New York, NY, (2001)
12. Nichols, J.M., Trickey, S.T., Virgin, L.N.: Structural health monitoring through chaotic interrogation. Meccanica. **38**, 239–250 (2003)
13. Nichols, J.M., Virgin, L.N., Todd, M.D., Nichols, J.D.: On the use of attractor dimension as a feature in structural health monitoring. Mech. Syst. Signal Process. **17**(6), 1305–1320 (2003)
14. Orhan, S.: Analysis of free and forced vibration of a cracked cantilever beam. NDT&E International. **40**, 443–450 (2007)
15. Peng, Z.K., Lang, Z.Q., Billings, S.A.: Crack detection using nonlinear output frequency response functions. J. Sound Vib. **301**, 777–788 (2007)
16. Ryue, J., White P.R.: The detection of crack in beams using chaotic excitations. J. Sound Vib. **307**, 627–638 (2007)
17. Rizos, P.F., Aspragathos, N., Dimarogonas, A.D.: Identification of crack location and magnitude in a cantilever beam from the vibration modes. J. Sound Vib. **138**(3), 381–388 (1990)
18. Spiegel, M., Stephens, L.: Statistics. McGraw-Hill, New York, NY (2008)
19. Sundermeyer, J.N., Weaver, R.L.: On crack identification and characterization in a beam by non-linear vibration analysis. J. Sound Vib. **183**(5), 746–760 (1995)
20. Sprott, J.C.: Chaos and Time-Series Analysis. Oxford University Press, New York, NY, (2003)
21. Vakil-Baghmisheh, M.-T., Peimani, M., Sadeghi, M.H., Ettefagh, M.M.: Crack detection in beam like structures using genetic algorithms. Applied Soft Computing. **8**, 1150–1160 (2008)
22. Trendafilova, I., Manoach, E.: Vibration-based damage detection in plates by using time series analysis. Mech. Syst. Signal Process. **22**, 1092–1106 (2008)

Part IV
Chaos in Robotics

Chapter 6
Robotic Approaches at the Crossroads of Chaos, Fractals and Percolation Theory

Burak H. Kaygısız, Murat Karahan, Aydan M. Erkmen, and Ismet Erkmen

6.1 Focus

Chaos, fractals and percolation through a medium are inherent characteristics and behaviors in nature. Robotics nowadays tend to bridge the gap between natural behaviors and human made mechanical ones by dwelling with humanoid, bio-robots and biologically inspired systems such as swarms and mobile sensor networks. Due to complex problems within these robotic systems (such as vibration, noisy sensing, robot/irregular-environment interactions leading to turbulence and unmodelled uncertainty of different types and frequencies) robotic designs shaping their dynamics, control and guidance need not only withstand chaos related issues, but may also dip into fluidics. This chapter not only surveys autonomous robotic systems at this crossroad between chaos related designs and analyses, and deals with robots equipped with sensors designs and penetration capabilities based on percolation theory but also provides our in depth chaos analyses of robotic state spaces together with our novel percolation approaches to an active simultaneous

B.H. Kaygısız
KAREL-ARGE, Bilkent Cyberplaza B Blok Kat:3
e-mail: burak.kaygisiz@gmail.com

M. Karahan
TUBITAK-UZAY ODTU Yerleskesi
e-mail: muratkarahan@gmail.com

A.M. Erkmen (✉)
Department of Electrical and Electronics Engineering, Middle East Technical University, 06531 Ankara, Turkey
e-mail: aydan@rorqual.cc.metu.edu.tr

I. Erkmen
Electrical Engineering Department, Middle East Technical University, Turkey
e-mail: erkmen@metu.edu.tr

S. Banerjee et al. (eds.), *Applications of Chaos and Nonlinear Dynamics in Engineering – Vol. 1*, Understanding Complex Systems, DOI 10.1007/978-3-642-21922-1_6,

localization and mapping (SLAM) of mobile robots exploring highly unstructured uncertain regions such as rubbles of a natural disaster site.

This chapter begins in Sect. 6.2 by dwelling with chaotic motions in feedback controlled two-and three-degree-of-freedom robots and continues in Sect. 6.4 expending into higher degrees of freedom spanning the fields of biologically inspired robotics, from walking robots, humanoids to cooperating mobile robots ending into hyperredundant robotics namely colonies of robots and swarms. Chaos arising through bifurcations under parametric changes in different walking robotic devices, including two legged ones, is overviewed including detailed phase plane analyses of a two legged robot composed of periodic motions among chaotic, unpredictable trajectories. Hence, chaotic motion of a robot is seen in Sect. 6.2 to be typically having periodic behavior of accumulation states that are slightly disturbed by any uncertainty inherent in the states of the system dynamics.

Section 6.3 provides the detailed overview of our works in the relevant topic of chaotic analyses of robotic systems dealing with the identification of the chaotic boundaries of regular (periodic and quasiperiodic) regions in nonlinear robotic systems, using cell mapping equipped with measures of fractal dimension and rough sets. Our contribution to the literature pertaining with a proposed fractal-rough set approach considers a state space divided into cells where cell trajectories are determined using cell to cell mapping technique. All image cells in the state space, equipped with their individual fractal dimension are then classified as being members of lower approximation, upper approximation or boundary region of regular regions with the help of rough set theory. The rough set with fractal dimension as its attribute is used to model the uncertainty of the regular regions, treated as sets of cells in this method. This uncertainty is then smoothed by a reinforcement learning algorithm in order to enrich regular regions that are used for control. The main focus of this approach is to investigate, using roughness metrics, the imprecision in the chaos map boundary of the regular domain of robots. This approach is applied to robotic subsystems especially to manipulators and wheeled unmanned autonomous land systems.

Not only chaos methods are used in state space but fractals have motivated roboticists to generate scale invariant robots in their mechanical designs. Two main authors have made important contributions in this topic, which we overview in Sect. 6.4 of this chapter as well. One of them designed his own modular robotic systems and produced a few prototypes having the major characteristic of strong coupling between the modules producing stiff structures that can also accomplish reconfiguration. Because of the self reconfiguration ability of these systems, he called them "metamorphic". Similarly, another roboticist, worked on the self-reconfiguration of modular structures made up of symmetric modules. Since their hexagonal symmetric structure is similar to fractal structures, he called each of the modules as a "fractum". As it is also the case for the metamorphic robots, fractums move within the structure by "sliding" over each other.

Robots using both Chaos and percolation theories do not abound in the literature as will be clearly attested by the content of Sect. 6.5. It is a very new and very challenging area relating fluidics to chaos in mechanical systems. Percolation theory, a fluidics based perspective, has been applied to the development of biomimetic abilities of the sensors like the ones embedded in a soft-body lamprey-like robot. The sensory elements in the system are piezo-resistive and piezo-resistivity in the composite material which the sensors are made up of, is governed by both percolation theory and quantum tunneling effects. In another work, authors have modeled the temperature sensors used when constructing meteorological micro- and milirobots via percolation theory. Another big area arising from the non-linear dynamics is just stemming nowadays which is the percolation theory applied to robot navigation and path planning (Sect. 6.5). Robot navigation for search and rescue in highly irregular and unpredictable natural disaster environments needs to handle different combinations of problems including localization, coverage and motion planning issues. Exploiting percolation theory and targeting the guidance problem is the major focus of Sect. 6.5, motivated by the fact that a search and rescue (SAR) robot can navigate within and penetrate a disaster area only if the area in question possesses connected voids. Disaster area exploration methodologies that predict connected voids by segmented maps frequently use information gain based on percolation theory. As an emerging robotics application exploiting information of nonlinear dynamics, details of this percolation approach is provided in this chapter. In the relevant Sect. 6.5.2, we demonstrate our novel perspective of percolation enhanced guidance such that a search and rescue (SAR) robot can navigate within and penetrate a disaster area estimating and planning paths through connected voids while avoiding obstacles. Penetrability of a disaster area is a primary factor in the navigation of a search and rescue (SAR) robot aside its speed, since it is highly desirable that the robot, without hitting a dead end or getting stuck, keeps its mobility for its primary task of reconnaissance and mapping when searching the highly unstructured environment. The novel percolation guided prioritized exploration technique introduced in Sect. 6.5 collaborates with entropy based SLAM under a switching control dependent on priority given to either position accuracy or to map accuracy. This methodology has proven to combine the superiority of both percolator guidance and entropy based prioritization so that the active SLAM becomes speedy, with high coverage rate of the area as well as increased accuracy in localization. Our percolator guidance is based on frontier based conditioning of a posteriori occurrences of new connected voids that uses the fact that every obstacle partially seen at the frontier of the explored domain has a spatial continuity into the unexplored area. This developed modular architecture will be introduced in details and demonstrative examples will be provided and discussed. The conclusion of this chapter deals with open ended questions in the field analyzed, still partially answered.

6.2 Chaos and Robots

Chaos arising through bifurcations under parametric changes in different robotic devices, including two and three degrees of freedom is studied extensively in the literature. Paar presents a mathematical model of a robot with one degree of freedom and numerical investigation of its dynamics in a particular parameter scan leading the system through chaos [35]. A systematic study of the passive gait of a compass-like, planar, biped robot kinematically equivalent to a double pendulum is conducted in [12]. Domains of attraction of a two axis robot under different dynamic loads and conditions leading the system to chaos are given in an extensive study by Glabisz [11]. Cao studied the problems of suppressing or inducing chaotic dynamics in a simple model of robot arms and mechanical manipulators [4]. Zhang explored the dynamic behavior and chaotic phenomena of a two-degrees-of-freedom robot system through numerical simulations [59]. These are only examples from a big world of chaos arising through bifurcations in the robotic devices. Moreover specifically in this section, chaotic motions in robotics are investigated through two illustrative examples conducted on two different classes of robotic designs namely walking robots and autonomous wheeled robots. A more comprehensive work on the issue can be found in the studies of the authors Kaygisiz and Erkmen [19–23]. This section first focuses in Sect. 6.2.1 on the analysis method used while Sect. 6.2.2 provides the demonstrative examples of chaotic motion in two robotic systems bearing different dynamics.

6.2.1 Analysis Method

The global analysis of the state space is of great importance, especially when inspecting the stability of nonlinear robotic systems. One can adopt the widely used technique of generating domain of attractions using cell to cell mapping [15, 16] together with Lyapunov exponents. This method is based on partitioning an N-dimensional state space into a large number of cells. Domain of attraction is then formed as sets of cells in the cellular state space after a predefined number of integration performed by taking the center of every new cell as the initial point of the next iteration [15, 16]. More than just processing cells by this well known methodology of cell to cell mapping, we expanded the methodology by also examining the system behavior in control space through the Lyapunov exponents of the mapping in order to classify cells as regular and chaotic. Lyapunov exponents are used to quantify the expansion and contraction occurring in a dynamical system and are a generalization of eigenvectors and eigenvalues at an equilibrium point. A positive exponent corresponds to an expansion in the related eigenvector direction and a negative exponent relates to contraction. They are used to determine the stability of any type of steady state behaviour. If the cumulative of Lyapunov exponents in a system is negative, contraction outweights expansion and the corresponding trajectory is said to be bounded as in chaotic attractors. What distinguishes a strange attractor from the other types of attractor is the existence of at least one positive Lyapunov exponent. Lyapunov exponents are related to the Jacobian of the local linearized model of any nonlinear system, which is a two dimensional error propagation equation:

$$\delta \mathbf{X}_{n+1} = \mathbf{A}_n \mathbf{X}_n \tag{6.1}$$

where

$$\mathbf{A}_n = \begin{pmatrix} \frac{df}{dx_n} & \frac{df}{dy_n} \\ \frac{dg}{dx_n} & \frac{dg}{dy_n} \end{pmatrix} \tag{6.2}$$

the Jacobian matrix J_n relates $\delta\mathbf{X}_n$ to the initial condition:

$$\delta\mathbf{X}_n = \mathbf{A}_{n-1}\mathbf{A}_{n-2}\ldots\mathbf{A}_0\delta\mathbf{X}_0 = \mathbf{J}_n\delta\mathbf{X}_0 \tag{6.3}$$

and is evaluated for each iteration on the exact trajectory.

$\delta\mathbf{X}_n$ can be written in terms of eigenvalues $\mu_i(n)$ and eigenvectors $\mathbf{e}_i(n)$ of $\mathbf{J}_n$ as:

$$\delta\mathbf{X}_n = \mathbf{J}_n\delta\mathbf{X}_0 = \sum_i c_i\mu_i(n)\mathbf{e}_i(n) \tag{6.4}$$

the Lyapunov exponents are then defined as:

$$\lambda_i \approx \frac{1}{n} ln\mu_i(n), \quad n \to \infty \tag{6.5}$$

$\lambda_i > 0$ corresponds to a local expansion of small areas along $\mathbf{e}_i$ while $\lambda_i < 0$ relates to a local contraction of small areas along the corresponding eigenvector. A strange attractor should have at least one positive eigenvalue (Lyapunov exponent) to be chaotic.

Chaos which arises due to the changes in system parameters is found to have four basic onsets differing in their characteristic transition phase that leads to different routes to chaos such as: intermittency [40], period-doubling [10], the quasiperiodic-chaotic route [41] and the crisis route [13]. The common property of all routes to chaos is the change in the maximum Lyapunov exponents λ_{max}: where $\lambda_{max} < 0$ for prechaotic phase and $\lambda_{max} > 0$ at the onset of chaos. Thus, the transition of λ_{max} (largest exponent) from negative values towards zero leads the system towards chaos. Consequently, a strange attractor is determined using the maximum Lyapunov exponent of a system.

Another important feature of the strange attractor is its fractal dimension. While the dimensions of stable and periodic attractors are integer, strange attractors have fractal dimensions. The dimension of an attractor can be defined as a deterministic or probabilistic measure. There are several definitions used to find out the fractal dimension of a strange attractor such as capacity (Hausdorff), Lyapunov, information and correlation dimensions [36].

The capacity is a deterministic dimension defined as

$$d_c = \lim_{\varepsilon \to 0} \frac{log(N(\varepsilon))}{log(1/\varepsilon)} \tag{6.6}$$

where $N(\varepsilon)$ is the minimum number of cubes (volume elements) of size ε needed to cover the set. The region of state space is divided up into a grid of cubes of size ε, the equations are iterated and the number of cubes in the convergence region are counted. If for very small ε's (as ε goes to 0) the process is repeated, slope of the plot of $log(N(\varepsilon))$ versus $log(\varepsilon)$ gives the fractal dimension of the attractor [36].

Another dimension used here is the Lyapunov dimension. Let the Lyapunov exponents of a chaotic attractor be $\lambda_1 \geq \cdots \geq \lambda_n$ and k be the largest integer such that $\lambda_1 + \cdots + \lambda_k \geq 0$. Then the Lyapunov dimension is given as

$$d_L = k + \frac{\lambda_1 + \cdots + \lambda_k}{|\lambda_{k+1}|} \tag{6.7}$$

Information dimension is one of the probabilistic dimensions and utilizes the time behavior of a system such that its calculation is based on the frequency of entrance of the trajectory into a specific volume in the state space. The information dimension is

$$d_i = \lim_{\varepsilon \to 0} \frac{log(S(\varepsilon))}{log(1/\varepsilon)}, \quad S(\varepsilon) = -\sum_{i=1}^{N(\varepsilon)} P_k ln(P_k) \tag{6.8}$$

where P_k is the frequency with which a trajectory visits k^{th} volume of the state space.

Lastly, the correlation dimension is defined as

$$d_{cor} = \lim_{\varepsilon \to 0} \frac{ln\left(\sum_{i=1}^{N(\varepsilon)} P_k^2\right)}{ln(\varepsilon)} \tag{6.9}$$

and uses again the frequency of entrance P_k to a specific volume of the state space.

Every definition contains its own pros and cons. Although, the capacity definition is simple to understand, it takes the strange attractor as a static object and requires infinite time to estimate the true picture of the attractor, measured through its dimension. The Lyapunov dimension is based on Lyapunov exponents which are easily found computationally. But there is no way to estimate negative exponents in a real time system from time series measurements. As a result, neither capacity nor Lyapunov dimensions are suitable for experimental studies. Among the above definitions, the correlation and the information dimensions are more suitable for experimental studies.

In this section, chaotic analyses are based on simulation studies using cell mapping technique and Lyapunov exponents. This gives us the ability to have the strange attractors composed of cubic elements with their Lyapunov exponents. As a result, capacity measure exploiting cubes to cover the attractor region and Lyapunov dimensions computed based on the Lyapunov exponents are used throughout this section.

Lyapunov exponents, being the unique measures to determine the chaotic phase and cell to cell mapping leading to a strong tool to take the picture of the phase portraits, are employed in subsequent sections, associated with fractal dimensions in order to investigate the stability of nonlinear robotic systems.

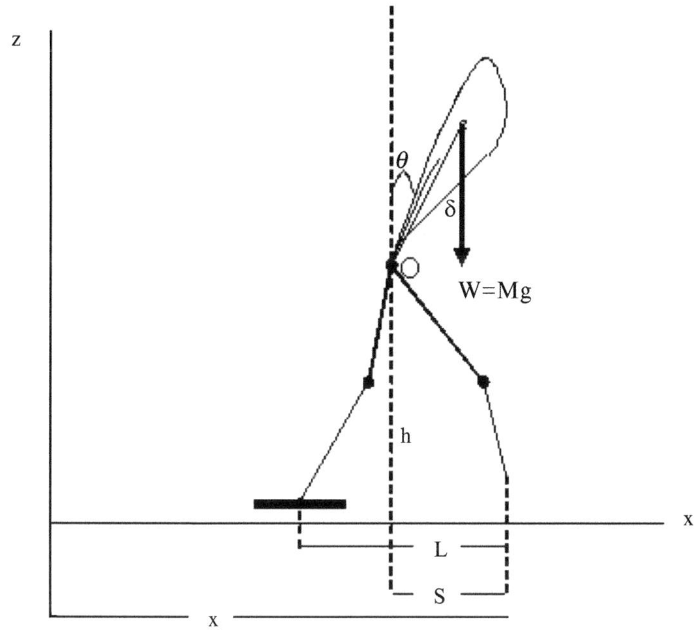

Fig. 6.1 Two legged body

6.2.2 *Illustrative Examples*

6.2.2.1 Biped Robot Locomotion

The dynamic model of a two legged robot (Fig. 6.1) is composed of a rigid body and a pair of legs whose weights are lumped within the robot body so that they can be modeled as weightless.

In the system, $W = Mg$ is the body weight, J the moment of inertia, O the hip joint, the length from center of mass c to O; L is the step length and S is the support shift giving the horizontal distance from joint O to the forward step. During the walking process, only one leg provides support at a time and the legs are lifted up to a maximum height h. The legged locomotion is considered to displace the body in a planar surface x-z with a velocity V. θ is the body tilt angle from the vertical and $\dot{\theta}$ is its rate of change. At this point, dimensionless quantities of time, step duration, and step size can be defined as in [2, 3]:

$$d\tau = \left(\frac{Mg\delta}{J + M\delta(\delta + h)}\right)^{\frac{1}{2}} dt$$

$$\tau_0 = \lambda\gamma$$

$$\lambda = \frac{L}{\delta} \tag{6.10}$$

where $T = L/V$ is the stepping period. The other dimensionless parameters are;

$$\begin{aligned}\alpha &= S/L \\ \mu_1 &= M\delta^2/[J + M\delta(\delta + h)] \\ \mu_2 &= \frac{\mu_1 h}{\delta}\end{aligned} \tag{6.11}$$

and

$$\varphi(t) = t/\tau_0 - \alpha \tag{6.12}$$

where [x] is the integer part of x. Employing these dimensionless quantities, the equation of the body motion can be expressed as follows:

$$\begin{aligned}\lambda\varphi(t) = {} & (1 - \mu_2\cos\theta + \mu_1\lambda\varphi(t)\sin\theta)\frac{d^2\theta}{dt^2} \\ & + (\mu_1\lambda\varphi(t)\cos\theta - \mu_2\sin\theta)\left(\frac{d\theta}{dt}\right)^2 - \sin\theta\end{aligned} \tag{6.13}$$

In the following analysis, system parameters are chosen to be 0.1 for μ_1 and 0.3 for μ_1 while 0.5 and 1 are assigned to α_1 and τ_0, respectively as in works by Beletskii [2,3].

We apply the methodology introduced in Sect. 6.2.1 to the controlled walking of the two legged robot just modeled. The $x1 - x2$ phase plane for different initial conditions of (0,0), (3,1) and (−3,9) are obtained respectively as in Fig. 6.2. Here, the states $x1$ and $x2$ are the aforementioned system variables θ and $\dot{\theta}$.

These initial conditions are specifically chosen to be demonstrated here in terms of robot state space behavior since they are highly characteristic for the system where the states, $x1$ and $x2$ related to the body structure of the walking robot, undergo either oscillatory behaviors with unpredictable amplitude due to uncertainty in the overall system or predictable periodic behaviors related to an initial condition of the system. The phase plane of the robot while it tries to keep its body straight up is given in Fig. 6.2a. Figure 6.2b, c gives the phase planes while the robot tries to swing its body from initial up and upside down positions with low and high angular rates, respectively. Cell to cell mapping is then applied to the obtained state space which is divided into 2D cells of dimension hxh such that a point x_i belongs to cell Z_i if

$$(Z_i - 1/2)h \leq x_i < (Z_i + 1/2)h \tag{6.14}$$

where interval size for both dimension is taken as $h_1 \times h_2 = 0.064 \times 0.128$.

The system of differential equations is discretized using the Euler integration rule with a time step of 2.5×10^{-3}. At each step, starting from cell $Z(n)$, the centre point $x(n)$ of $Z(n)$ is first taken in the discrete process as the initial state and the point $x(n+1)$ is generated as output of the Euler integration such that $Z(n+1)$, the image

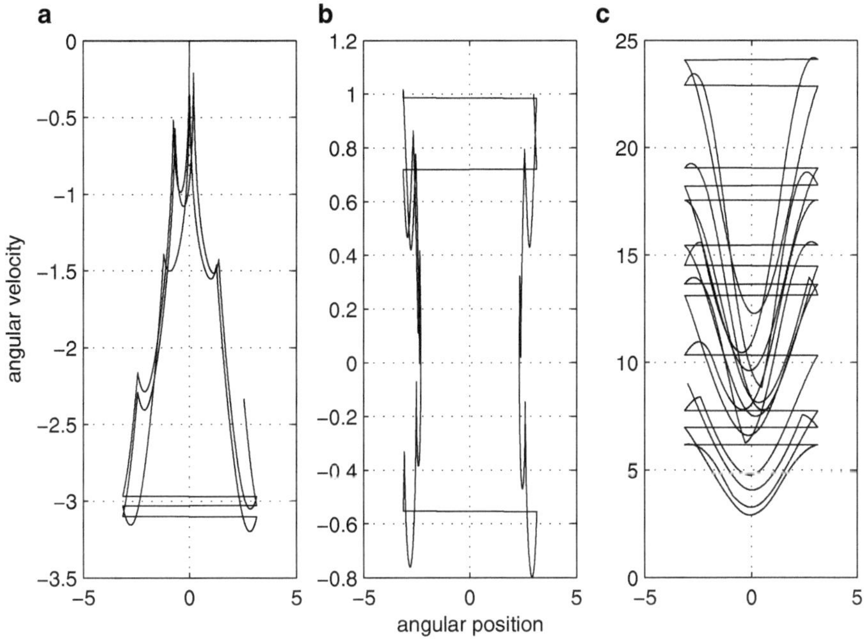

Fig. 6.2 $x1 - x2$ phase plane for different initial conditions of [0,0], [3, 1] and [−3, 9]

cell of $Z(n)$, is obtained as the cell in which $x(n+1)$ lies. The next iteration is performed by taking the centre point of $Z(n+1)$ as the initial point and this discrete cell to cell mapping is continued for 4,000 steps. At every step we also find the Jacobian matrix in order to calculate the Lyapunov exponents of the starting cell.

Using the Lyapunov exponents, we examine our illustrative system over a state space partitioned into 101×101 cells. The interval lengths h_1 and h_2 defining the sizes of each side of a cell are selected as 0.064 and 0.128 for x_1 and x_2, respectively.

The domain of attraction under these conditions is obtained as in Fig. 6.3 where the dark cells are regular while the others are chaotic according to the Lyapunov exponents. For every regular (dark) cell given in the figure, the maximum Lyapunov exponent is found to be negative. Other cells have positive maximum Lyapunov exponents.

As a result, it has been shown Fig. 6.3 that the phase plane of a two legged robot is composed of islands of regular and periodic motions among a multitude of chaotic, unpredictable trajectories. The rectangular window on the figure shows the border of chaos and regularity obtained after cell mapping.

6.2.2.2 Autonomous Land Vehicle Over Irregular Roads

Now let us consider the chaotic analysis of the state space of an autonomous offroad land vehicle (ALV). We model our vehicle as composed of a chassis

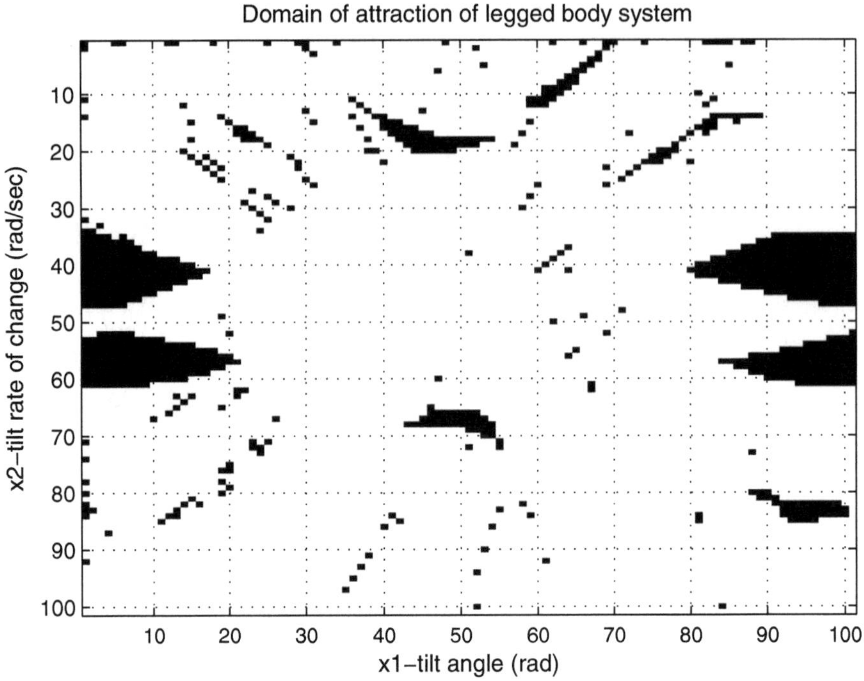

Fig. 6.3 A domain of attraction for division 101 × 101

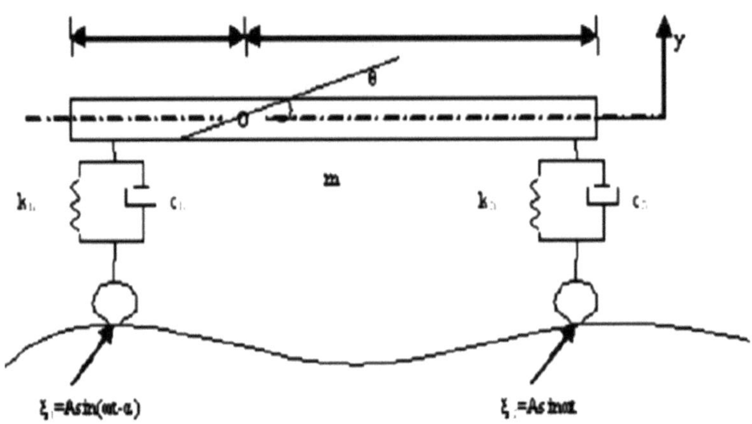

Fig. 6.4 Schematic diagram of the model

body and lumped tires represented by nonlinear springs in parallel with velocity dependent dampers (Fig. 6.4) where the ALV and terrain model parameters are given in Table 6.1. These model parameters are acquired from [17] where a detailed modelling work has been conducted using experimental data. Equations of motion are derived based on the following assumptions:

Table 6.1 System parameter values

Symbols	Parameters	Values
k_{i1}	Spring constant of back tire	496.38 kgf cm^{-i}
k_{i2}	Spring constant of front tire	553.28 kgf cm^{-i}
c_{i1}	Damping constant of back tire	4.434 kgf cm^{-i} s^{-i}
c_{i2}	Damping constant of front tire	4.434 kgf cm^{-i} s^{-i}
m	Vehicle mass	271.889 kgf cm s^{-2}
a	Distance of back tire from C.G.	84.7 cm
b	Distance of front tire from C.G.	118.5 cm
ρ	Radius of gyration	102.24 cm
L	Length between tires	203.2 cm
a	Amplitude of impressed vibration	5 cm
α	Phase angle between front and back tires	160°
ω	Forcing frequency	1 rad s^{-1}

1. The road profile is approximated by a sinusoidal shape,
2. A phase shift is introduced between the sinusoidal shape of the road portions in order to realize irregularity,
3. The vehicle is considered in the longitudinal plane only,
4. Forces and couplings due to wheel rotations are neglected,
5. Translational and angular displacements are assumed to be small enough for tire and spring motions so that nonlinear springs remain within their elastic range,
6. The vehicle is kept in cruise control with constant velocity.

ALVs generally cruise with a predetermined constant velocity. Road irregularities are generally washboards in unpaved roads with a sinusoidal shape and amplitude of nearly 50 mm [29], and such a road profile introduces a phase shift between the forces on front and back tires. The forces on tires stemming from low amplitude road irregularities cause small translational and angular displacements enough for the nonlinear springs to remain within their elastic range. Hence, the assumptions given above for modelling the interaction between our ALV and the offroad profile are general enough for our model to be as close as possible to the physical system.

The chassis displacement follows a nonlinear behavior that we modeled as:

$$m\ddot{y}+\sum_{1}^{3}k_{i1}\left(y-a\theta-\xi_1\right)^i+\sum_{1}^{3}k_{i2}\left(y-b\theta-\xi_2\right)^i$$
$$+\sum_{1}^{3}c_{i1}\left(\dot{y}-a\dot{\theta}-\dot{\xi}_1\right)^i+\sum_{1}^{3}c_{i2}\left(\dot{y}-b\dot{\theta}-\dot{\xi}_2\right)^i=0 \tag{6.15}$$

based on the mentioned assumptions [17]. The parameters included in this behavior are clearly shown in Fig. 6.4. The pitch vibration on the other hand obeys the following equation

$$m\rho^2\ddot{\theta} - \sum_1^3 ak_{i1}\left(y - a\theta - \xi_1\right)^i + \sum_1^3 bk_{i2}\left(y - b\theta - \xi_2\right)^i$$
$$-\sum_1^3 ac_{i1}\left(\dot{y} - a\dot{\theta} - \dot{\xi}_1\right)^i + \sum_1^3 bc_{i2}\left(\dot{y} - b\dot{\theta} - \dot{\xi}_2\right)^i = 0 \quad (6.16)$$

Here, θ is the body pitch angle, $\dot{\theta}$ the angular rate, y the vertical displacement, $\dot{y}$ the vertical velocity; ξ_1 and ξ_2 represent the offroad input disturbances occurring at each in tire contact position; $\dot{\xi}_1$ and $\dot{\xi}_2$ rates of changes generating vertical velocities at each tire. An average stiffness curve is adopted from the work by [6] for the radial and bias ply tires which have third order nonlinear terms. Similar terms are also used for the dampers. As a result, the spring-mass-damper system given by the above equations consists of third order nonlinear damping and stiffness coefficients.

Under certain conditions, such a system with two degrees of freedom may exhibit chaotic fluctuations [30]. In this section, we will attempt to capture this behavior by the global stability analysis conducted on our ALV system. In order to determine the system characteristics and capture a global picture of stability, cell to cell mapping technique is applied to generate the state space of our ALV dynamics given in Table 6.1, over irregular roads. y_1 and y_2 of the states vector $[y_1, y_2, y_3, y_4] = [y, \dot{y}, \theta, \dot{\theta}]$ are forced to take different initial conditions such that global analysis is conducted on a projection subspace where y_3 and y_4 are taken constant and the 2D space is directly affected by the vertical translational motion of the vehicle. The space is divided into 2D cells $h \times h$ such that a point y_i belongs to cell Z_i if

$$(Z_i - 1/2)h \leq y_i < (Z_i + 1/2)h \quad (6.17)$$

where interval size for both dimension is taken as $h = 0.096$, in our case.

As stated before, vibration absorbers with resonance frequency of 35 Hz are introduced to the physical environment and the effective vibrational amplitudes is guaranteed to be less than 35 Hz. Thus, a sampling frequency greater than 70 Hz is enough to capture the dynamical behavior based on the Nyquist theorem. The system of differential equations is discretized using the Euler integration rule with a time step of (100 Hz). This discrete cell to cell mapping is continued for 1,000 steps.

At every step, we also find the Jacobian matrix in order to calculate the cell Lyapunov exponents what we incorporate into cell to cell computations as introduced in Sect. 6.2.1 and illustrated in a gait locomotion of Sect. 6.2.2.1. Vibration sensitive and insensitive stability regions of our ALV system are obtained in Fig. 6.5 for a sinusoidal terrain roughness over a state space partitioned into 101×101 cells and using vehicle parameters given in Table 6.1. The analysis in state space is conducted over a 5 cm displacement by 101×101 cells which are squares with 0.096 cm edge size since irregularities on unpaved roads impress vibration of peak amplitude up to 5 cm as mentioned in [29].

The regular part of the domain of attraction is depicted with dark cells while the other cells are chaotic as measured by the Lyapunov exponents. The analysis using

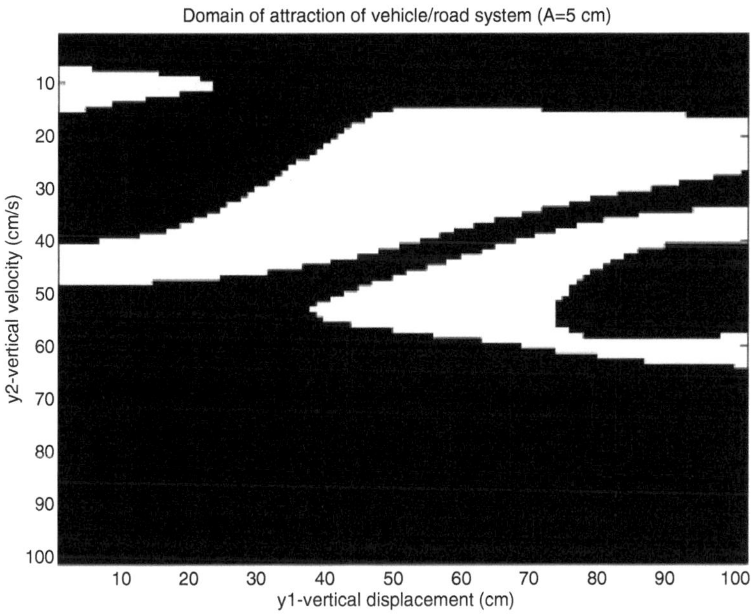

Fig. 6.5 $y_1 - y_2$ phase plane for road vibration of 5 cm

cell to cell mapping equipped with Lyapunov exponents shows us that the chassis of the robotic vehicle travelling on an irregular road produces a chaotic response depending on the initial condition since according to where the initial state falls in this state space, the state trajectories may begin in a regular region but may cross chaotic regions or may directly begin from chaos.

The state space regions obtained in Fig. 6.5 are regular sectors which continuity is frequently broken by chaotic regions thus its dimension is indicative of chaos and should have a fractal value. The dimension of the system can be found using the Lyapunov dimension definition. The Lyapunov exponents of the chaotic attractor seen in Fig. 6.5 are found to be

$$\begin{aligned} \lambda_1 &= 80.1 \\ \lambda_2 &= 73.9 \\ \lambda_2 &= -67.2 \\ \lambda_2 &= -76.5 \end{aligned} \tag{6.18}$$

Then the Lyapunov dimension of the attractor is found as follows:

$$d_{lyp} = 2 + \frac{80.1 + 73.9}{|-67.2|} = 4.28 \tag{6.19}$$

which is fractal. As given in Sect. 6.2.1, fractal dimension is an important feature of the strange attractor and while the dimensions of stable and periodic attractors are integer, strange attractors have fractal dimensions. As a result, dimension of the attractor found to be 4.28 also shows that chaotic fluctuations occur on the example robotic system that is investigated.

6.3 Detecting Onset of Chaos and Intelligent Operational Region Enlargement

6.3.1 Introduction

In analyzing nonlinear dynamical systems, one is often interested in the local behavior of the system around its equilibrium states. The local system behavior information around equilibrium point and the determination of the global regular domain are critical in order to conclude on the control strategies and to assign the relevant parameters for the system controller.

In some cases such as vehicle/road interaction, the regularity of the boundary region of a domain of operation in the state space of a dynamical system is frequently broken by chaotic regions leading to a dimension which is fractal and the precise identification of the regular region imbedded in that domain is nearly impossible. This identification uncertainty renders the use of classical approaches very inefficient.

In order to overcome the problem of fractal boundaries, instead of inefficiently attempting precise prediction, a probabilistic approach is suggested in [15,16] where domains of attractions are generated using cell to cell mapping in examining the regularity of nonlinear systems. However, such domains of attraction based on cell to cell mapping technique are far from being precise on the boundary of the region: having possible chaotic elements, the boundary region of the domain of attraction introduces an uncertainty into the regularity region. Comparing Figs. 6.5 and 6.6, one can easily see the precision problem of cell to cell mapping technique. Some of the boundary cells, which are classified as regular in Fig. 6.5 are in fact chaotic in Fig. 6.6 and vice versa. As a result, the boundary region contains an uncertainty in the identification. The extraction of regular regions in state space using cell to cell mapping technique renders the precision of the region mainly dependent on the cell size. Figure 6.5 is generated for cell size 0.096, while Fig. 6.6 is generated for the same system for cell size 0.024. To have a true picture of stability for a dynamical system, the cells should be infinitely small and the state space should be divided into infinitely many cells. This means every point in state space will then be treated as a cell and cell to cell mapping technique becomes a point to point mapping process, which takes an infinite, infeasible, amount of time.

At this point, the only approach seen as adjusting the cell size according to design purposes and taking this discretized but uncertain stability region as the global

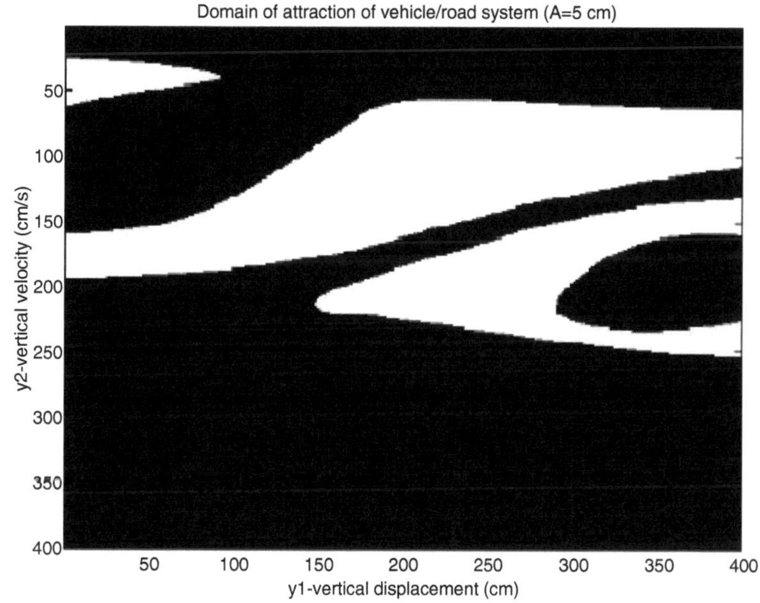

Fig. 6.6 $y_1 - y_2$ phase plane for road vibration of 5 cm (400 × 400 cells)

stability information is the work by Thompson [50]. However, until now, no study has been conducted to model the uncertainty in this discrete possibly stable region and to form a control strategy suitable for that model.

We propose a novel approach to provide a solution in this virgin field by modelling and smoothing the imprecision stemming from cell to cell mapping using an intelligent technique that integrates several phases:

1. Modelling the dynamic system and determining the states which cause uncertainty in the regular region,
2. Analyzing global stability not only with cell to cell mapping but together with Lyapunov exponents,
3. Determining the boundary region of the attraction domain and modelling the uncertainty in stability of this domain using rough set theory and the fractal dimension concept,
4. Smoothing uncertainty in this boundary region via a reinforcement learning algorithm.

In the Sect. 6.2.1, we introduced the first two phases of our approach and in Sects. 6.2.2.1 and 6.2.2.2 demonstrated the implementation of those first worksteps for two different robotic systems varying in dynamical characteristics. The last two steps of the approach given above directly attack the problem of uncertainty stemming from cell to cell mapping in an intelligent manner and is the focus of this Sect. 6.3.

In our new approach [19, 21], the fractal dimension of each cell in the regular region is analyzed and used as an attribute in a Rough Set model in order to represent metrically the uncertainty, inherent in the regular region, as a roughness texture especially of its boundary region (step 3). This boundary region is the main source of uncertainty of the regular domain generated using cell to cell mapping. Rough set theory that we use in our approach in order to model the roughness of the uncertain region in the domain of regular attractors/repellers, is a concept proposed by Zdzislaw Pawlak for modelling uncertainty and vagueness. Rough sets [37–39] is a mathematical approach modelling vagueness in uncertainty by associating a degree of uncertainty to any object of the universe so that any object without uncertainty (roughness) becomes indiscernible within a class.

A set containing indiscernible objects forms an equivalence class, called the elementary set, any union of which determines a crisp set. All other sets include uncertainty. The rough set is composed of such sets and has: (1) a boundary region where vagueness resides, with set elements that cannot be classified either as members of that set or of the complementary set; (2) a lower approximation consisting of elements that are surely members of the set; and (3) an upper approximation consisting of possible members of the set.

6.3.2 *Textured Rough Sets*

In the dynamical analysis of the chaotic systems (step 3 above) we expand rough sets by also incorporating the feature of texture measured in terms of fractal dimension.

The boundary region of a rough set is a region of uncertainty where the set elements of that region are not known to be inside or outside the set "with certainty" with respect to the attribute B:

$$BN_B(X) = \overline{B}(X) - \underline{B}(X) \tag{6.20}$$

Here $\overline{B}(X)$ is the upper approximation of a set and $\underline{B}(X)$ is the lower approximation which is formed by the classification of data points into disjoint categories using attributes. The classification represents our knowledge about data in an approximation $spaceA = (U, R)$. Here U is set known with certainty called Universe and R is an equivalence relation also denoted as indiscernibility relation on.

The upper approximation of a set X containing objects x which "possibly" belong to the set of interest with respect to the attribute B:

$$\overline{B}(X) = \{x \in U : B(X) \bigcap X \neq \phi\} \tag{6.21}$$

The lower approximation of a set X is described by objects x of the $domain(U)$, which are known "with certainty" to belong to the set of interest with respect to the attribute B:

$$\underline{B}(X) = \{x \in U : B(X) \subset X\} \tag{6.22}$$

We extend this theory by assigning a fractal dimension as an attribute of roughness to rough sets. The novelty of our work is this extension of rough sets with fractal dimension for their use in nonlinear system control. Smoothing of such a roughness in the regular region of the control space is achieved by reinforcement learning based on the chaos history of the system dynamics.

Section 6.3.3 deals basically with the third step of our approach introduced in Sect. 6.3.1, which is to identify the regular/chaotic states that make up the boundary region of the domain. This section focuses on the fractal dimensions and the rough set theory to construct a more detailed picture of the regular region by modelling the inherent uncertainty. The modelling technique is applied to the second exemplary system we used in this chapter, namely the ALV on irregular roads as well as to a two-legged locomotion system, a humanoid robot.

In Sect. 6.3.4, a reinforcement learning algorithm is employed which smoothes the roughness in the boundary region. This smoothing yields a recovery of cells by their decontamination from uncertainty that expands the lower approximation of the rough set. This expansion is generated by including the recovered cells of the boundary region into set of regular cells. Recovery is achieved through reinforcement learning according to the stability history of the cells on a cell trajectory of the dynamical system generated by cell to cell mapping (here, ALV on irregular roads as our second exemplary system).

6.3.3 Modeling Uncertainty in Stability: Boundary Region Roughness

We found that rough set theory readily applies to the problem of modeling chaos uncertainty among the regularity regions of a system where cells have to be taken as objects of our rough set. The lower approximation and the boundary region of the set consist respectively of surely regular cells and possibly regular cells, while the other cells are chaotic and not members of the regularity region. Regular cells are the ones where all points in them are regular, while chaotic cells contain only chaotic points. Possibly regular cells, which form the boundary region of the regular domain, are classified as being either regular or chaotic. Moreover, these latter cells are such that they contain both regular and chaotic points coexisting within the same cell.

To classify cells into those belonging to lower, upper and boundary regions, we extract and use homogeneity or nonhomogeneity information contained in the cells generated by the cell to cell mapping. The fractal dimension of each cell is this measure-based information that can be extracted and is suitable for the needed rough set classification. Therefore, we choose to assign the fractal dimension as an attribute of texture to the rough set as a novel expansion of the theory. We use the definition of capacity (Hausdorff dimension) introduced in Sect. 6.2.1, to compute the fractal dimension. Calculation of capacity dimension requires division of region of interest into a grid of cubes of size, as in Fig. 6.7 which is done during the cell to cell mapping process.

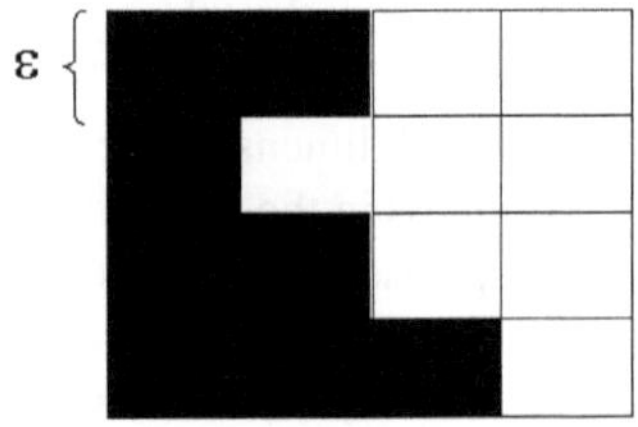

Fig. 6.7 A cell divided up into 16 subcells

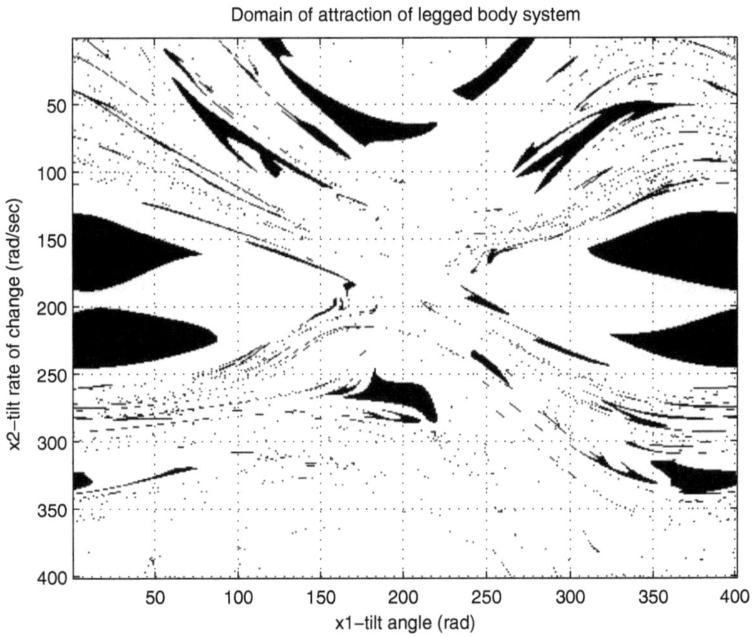

Fig. 6.8 A domain of attraction for two legged walking robot

$N(\varepsilon)$ in the computation of capacity dimension (6.6) is the number of subcells found to be nonchaotic by cell to cell mapping technique. The process is repeated by decreasing the value of subcell size, and $N(\varepsilon)$ is recorded for each case. When $logN(\varepsilon)$ is plotted versus $log(1/\varepsilon)$ for every cell, the slope of the linear plot is the fractal dimension as goes to zero in the limit. An example plot for a two legged robot system investigated in [23] is given in Fig. 6.8. The phase plane in Fig. 6.8 of a two legged robot is composed of islands of regular and periodic motions among a multitude of chaotic, unpredictable trajectories. Hence, chaotic motion of the robot is typical where periodic motion of accumulation states is slightly disturbed by any uncertainty or noise inherent in the states of the system dynamics. Consequently, fractal dimension of every cell is determined dividing each one of them into subcells to differentiate chaotic and periodic motions of the walking robot. Figure 6.9 shows

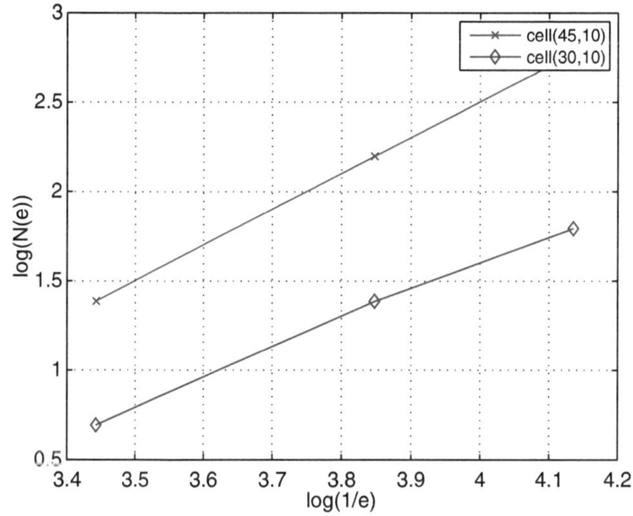

Fig. 6.9 Log-log plot to estimate the dimension of each of two cells identified in the right upper rectangle

the $logN(\varepsilon)$ versus $log(1/\varepsilon)$ plot generated for two of such cells. According to the slopes in this figure the fractal dimension attributes of the two cells in question can be found as 1.71 and 1.92.

At this point, one can realize that another attribute can be assigned to cells for their classification into rough sets, which is a fuzzy measure of possibility for the regularity of cells in a set. We form this attribute as the ratio of number of regular subcells to total number of subcells within a cell where regular subcells are the ones that are found to be regular using the cell to cell mapping technique and Lyapunov exponents. Thus,

$$R_n = \frac{Number\ of\ regular\ subcells}{Number\ of\ subcells} \tag{6.23}$$

When the subcell size ε tends to zero, this ratio R_n becomes equal to the probability of a point, in belonging to a nonchaotic, regular region. We call this attribute the regularity number R_n. It follows naturally from our definition that a regular cell has an R_n of value 1 and the R_n of a chaotic cell takes the value of 0. The regularity number complements the fractal dimension by providing a coarse overall measure of uncertainty while fractal dimension gives a measure of the roughness characteristics of uncertainty by a measure of its texture. At the limit, d_c and R_n become equivalent by tending towards equivalent information: if the fractal dimension d_c of a cell is an integer number (smooth texture), then the coarse overall measure R_n of the cell does not change with the change of number of subcells in a cell. In our previous studies [19, 21], it is observed that R_n is either 0 or 1 when the fractal dimension is an integer number. Thus, in our approach, a rough set contains regular cells of $R_n = 1$

and integer fractal dimension, while uncertain cells are those identified as possibly regular, having a fractional value as fractal dimension and an R_n number $\neq 0$ and $\neq 1$. If a cell is chaotic it does not contain any regular subcells and consequently its fractal dimension is undefined and $R_n = 0$. However in the general case of fractal dimension d_c, the overall measure R_n shows a different variability and they begin to complement each other for uncertainty measurements. This concept is given in detail in the following section.

6.3.3.1 Correlation of Regularity Number and Fractal Dimension

Examining the rough set approach, it is seen that the lower approximation for the regular cells consists of all the cells that have integer dimension. That is an important clue bringing us to a point that we can say all the integer dimensional cells or set of subcells, is stable. The relation between the number of regular cells and the cell size can be stated as follows:

$$\frac{logN(\varepsilon_1) - logN(\varepsilon_2)}{log\varepsilon_2 - log\varepsilon_1} = \frac{logN(\varepsilon) - logN(\varepsilon_1)}{log\varepsilon_1 - log\varepsilon} \tag{6.24}$$

where ε's are the subcell sizes and $N(\varepsilon)$ is the number of the subcells that are regular in a cell. Rearranging the above equation gives

$$N(\varepsilon) = N(\varepsilon_1)\left(\frac{\varepsilon_1}{\varepsilon}\right)^{log_{\frac{\varepsilon_2}{\varepsilon_1}}\frac{N(\varepsilon_1)}{N(\varepsilon_2)}} \tag{6.25}$$

Substituting the capacity dimension d_c into the above equation, we obtain the following relation between the cell dimension and regularity number after rearranging

$$N(\varepsilon) = N(\varepsilon_1)\left(\frac{\varepsilon_1}{\varepsilon}\right)^{d_c} \tag{6.26}$$

into an expression using regularity number (R_n):

$$R(\varepsilon)\frac{c}{\varepsilon^n} = R(\varepsilon_1)\frac{c}{\varepsilon_1^n}\left(\frac{\varepsilon_1}{\varepsilon}\right)^{d_c} \tag{6.27}$$

where c is the size of a region of interest, n is the number of states of the nonlinear system, ε's are the cell sizes. Change in the regularity number with the change of cell size can be defined as:

$$\frac{R(\varepsilon)}{R(\varepsilon_1)} = \left(\frac{\varepsilon}{\varepsilon_1}\right)^{n-d_c} \tag{6.28}$$

This relation thus gives the correlation between the regularity number and fractal dimension, which are the attributes of the rough set defined for state space classification. Calculating the fractal dimension of the regular region given in Fig. 6.5, using the capacity dimension definition, it is found that

$$d_c = \frac{log(118856) - log(7555)}{log(0.096) - log(0.024)} = 1.9878 \tag{6.29}$$

Employing regularity number capacity dimension relation, it is found that regularity in the region increases by 1.7% decreasing the cell size four times as in the case of Fig. 6.6. This is theoretical maximum expansion of the regular region given in Fig. 6.5.

From the relation of equation (6.28), it is seen that if the fractal dimension of a cell (or region) is integer then the regularity number does not change when the subcell (or cell) size is changed, hence there exist no uncertainty in regularity. Thus it can be stated that if the dimension is integer in a cell, finer resolutions in the state trajectory will not change the regularity region in this cell and this region can be included in the lower approximation of the rough set. Using this fact, the lower approximation of the rough set can be expanded. The expansion of the lower approximation, which is achieved by smoothing the boundary region using reinforcement learning, is based on the relation given above. The method for smoothing and its results are given in the following Sect. 6.3.4. But before investigating the smoothness control, we will demonstrate the classification of surely regular (lower approximate), possibly regular (upper approximate) and boundary cell of the cell space for the autonomous land vehicle example elaborated previously in Sect. 6.2.2.2.

6.3.3.2 Uncertainty Modelling of ALV/Road Dynamical System

For the ALV system under investigation, the control state space regular region contains 10 201 cells generated by cell to cell mapping. The state space has been transformed into textured rough sets using fractal dimension and the regularity number R_n. In this textured rough set representation of the vehicle on the irregular terrain of our university campus, we found 7364 cells with $R_n = 1$ and $d_c = 2$ which define the fully regular cells for the vehicle (map of Fig. 6.5 with cells of Fig. 6.6 removed). 2,411 cells with $R_n = 0$ and undefined d_c, form the set of fully chaotic cells where vehicle chassis vibration control is not feasible. 426 partially or possibly regular cells with $0 < R_n < 1$ and $d_c \neq 0$ but fractional valued (Fig. 6.10) are also found. These cells are not usable as is for control due to their uncertainty and vibration sensitivity. Therefore they represent lost regions for control. If we can recover them by reducing the uncertainty, they would lead to the enrichment of the region usable for control.

In the following Sect. 6.3.4, a reinforcement learning algorithm is employed which smoothes the roughness in the boundary region. This smoothing yields a recovery of cells that expands the lower approximation of the rough set.

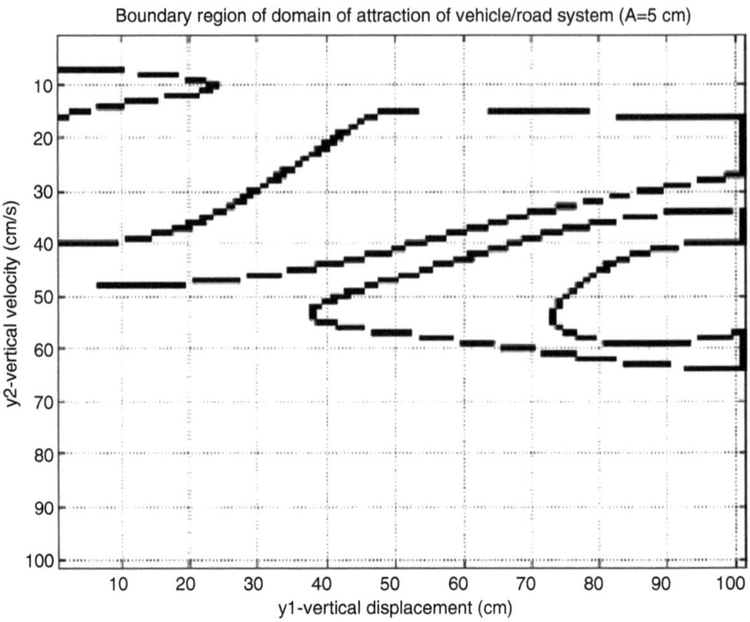

Fig. 6.10 Boundary region of the rough set (partially regular areas)

This expansion includes the recovered cells of the boundary region and is achieved through reinforcement learning according to the stability history in the state cell trajectory of the ALV chassis vibration control on the irregular roads.

6.3.4 Smoothing the Chaos Roughness in Regularity

This section aims at decreasing the uncertainty in the boundary region of the rough set and thus smoothing the chaos roughness in that region using reinforcement learning. Reinforcement learning is the technique of training an agent by reward and punishment through an environmental feedback process in order to learn the dynamical characteristics of a system navigating within this environment. Allowing systems to rearrange and improve themselves with their past space, it has been considerably popular in the field of robotics and autonomous systems [26, 42].

The most popular model free learning methods are adaptive heuristic critic introduced by Barto et al. [1], Q-learning of Watkins [55] and learning with average reward examined first by Schwartz [43].

Having a simple problem with two different states (chaoticity and regularity), we employed the linear reward-inaction algorithm of Hilgard and Bower [14], which is based on finite-state automata that adjust the probability of taking different actions according to the previous successes and failures where we consider regularity as success and chaoticity as failure. In our implementations, the cells in the boundary region are rewarded by a history of nonchaotic behavior (regularity) and the instant probability of being regular, (R_n). Reward is decremented when an event other than regularity occurs for the point in cell "*i*". Our cell recovery procedure begins by taking numerous random points from each cell and finding out if they are regular or not, R_n of each subcell is iterated by the following reward formulas and rules:

1. Pick a random point form cell "i",
2. When the point in the cell "i" is regular, the regularity measure is reinforced,

$$R_n^i(k+1) = R_n^i(k) + \alpha(1 - R_n^i(k)) \tag{6.30}$$

3. When the point in the cell "i" is chaotic, the regularity is penalized by this chaotic behaviour and decreased,

$$R_n^i(k+1) = R_n^i(k) - \alpha R_n^i(k) \tag{6.31}$$

4. Go to step 1 until Kth random point is iterated for cell "i" where K is the iteration number adjusted according to the convergence properties,
5. R_i^n converges to either 0 or 1 after Kth iterations for cell "i".

The iteration runs until all the boundary cells are iterated following the given rules. Then the cells which are assigned regularity number of 1 are recovered as regular and are added to the controllable regular set of the state space.

We will illustrate our cell recovery procedure that performs smoothing of the chaos roughness in the boundary region of the rough set, by implementing it on the vehicle chassis control implementation. In the application example, every cell is subdivided into four and 16 subcells in order to find the fractal dimension and 16-subcell structure is used to compute the regularity number. In this example, the reinforced or penalized regularity measure R_n of each subcell is found using the above formulas (6.30) and (6.31), iterated with 100 points chosen randomly within the subcell considered and this process is repeated for every subcell of the boundary region. To provide a slow convergence rate and make the algorithm fully convergent is taken as a small values, namely 0.05.

As a result, the R_n values of the subcells considered, either converged to 1 (meaning that uncertainty is fully eliminated and their fractal dimension has become integer making these cells fully regular) or to 0. The subcells with R_n values approximately 1 at the end of the reinforcement learning process are added to the lower approximation. This helps to increase the level of knowledge about the characteristics of the regular and chaotic region. Using this method, the regular region is enlarged by highly probable regular cells according to their past history of nonchaotic behavior. Thus the boundary region of the rough set in the state space is diminished enriching the region of cells to be used for control. Including subcells with $R_n > 0.95$, 15.96% of the boundary region (68 cells) has been added through the algorithm to the regular region and the region is expanded by 1% while the rest of the region is converged to the chaotic area for our ALV implementation.

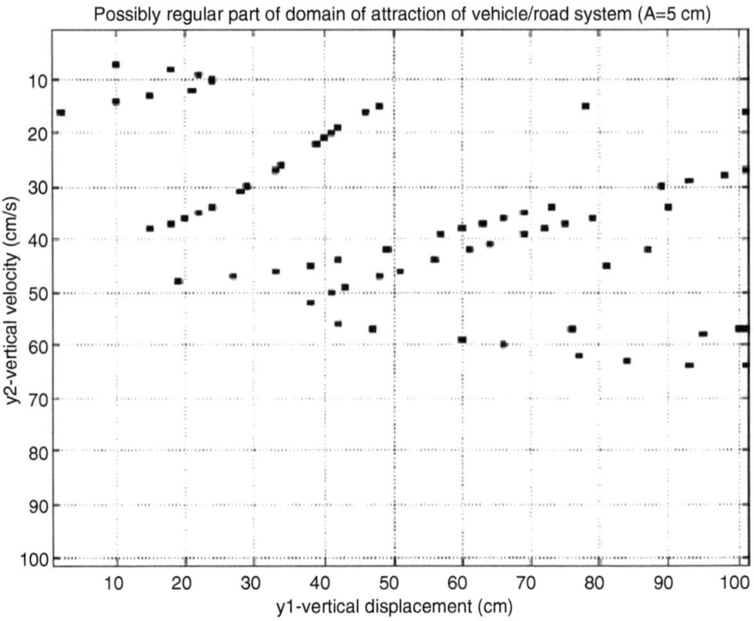

Fig. 6.11 Recovered cells of the boundary region into the lower approximation

The boundary region added to the lower approximation is given in Fig. 6.11 and represents the usable portion of the boundary region for control in the motion of the robot on irregular roads. When Fig. 6.11 is compared to that of Fig. 6.10 we can measure the degree of smoothness achieved after reinforcement learning.

It is easy to see that the smoothing algorithm using a 16-subcell structure has achieved a 1% expansion which is less than the expected theoretical maximum 1.7% found in Sect. 6.3.3.1 using relation (6.28).

6.4 Fractals and Robotic Hardware Designs

Robotic researchers have always shown interest towards reconfigurable and scalable distributed robotic structures. The need for these systems arises from the unstructured nature of the environment and the changing nature of the task in that medium. In order to adapt to such a medium with graceful degradation, the system must be hyperredundant and change its structure in an optimum way while keeping itself robust and accurate in behavior [54]. Fractal mechanisms are employed for such reconfigurable and scalable distributed structures. Deployable fractal mechanisms are highly expandible or collapsible and yet are capable of preserving their basic structural geometry in a dynamic fashion [44]. These fractal structures, mechanisms and robot manipulators are shown to have computer-controlled microsensing and dynamic capabilities. They can be visualized as motorized lego bricks with some electronics inside enabling them to communicate, control and reconfigure as a colony. With the intelligence and communication capabilities of each fractal structure, one can theoretically build any hardware such as robotic hands, bridges etc. Each fractal structure possesses a limited microcontroller with simple software which helps robots to achieve multifaceted activities in complex environments with a reduced amount of effort. Many works exist in the literature using such modular systems that pertain to the principle of independence, cooperation and simple structure. Fractal mechanism can be described as a holonic structure, the simplest physical elements of the system, which have the capabilities of independent actuation, and of cooperation with other holons in achieving global behavior for the whole colony structure. Extensive studies on the holonic models have been conducted by [7–9].

Shahinpoor designed deployable fractal mechanisms as smart structures [44]. He focused on the highly deployable mechanisms and possibly redundant, multi-axis, multi-arm, multi-finger robot manipulators whose kinematic structure is fractal. He also presented a number of such structures and mechanisms associated with simple kinematic models.

Chirikjian designed his own modular robotic system and produced a few prototypes having self-reconfiguration ability [5]. These modules, which he calls "metamorphic", show strong coupling and produce stiff structures that can also accomplish reconfiguration. These modules are designed in a hexagonal geometry due to fact that symmetry brings a great simplification to the analysis and control of self-reconfigurable systems. Chirikjian used the simulated annealing method to decide on the action of each module in the reconfiguration phase.

Similar to Chirikijan, Murata worked on the self-reconfiguration of modular structures made up of symmetric modules. Since their hexagonal symmetric structure is similar to fractal structures, he calls each of the modules a "fractum" [32]. In both of the works [5, 32], the coupling principles of the modules are the same and are based on electromagnetic fields.

These reconfigurable fractal robotic systems can adapt their structures to the changing environmental conditions and needs, such as changing the robot configuration from a legged robotic structure to a snake robot and then to a rolling robotic one [58]. Such fractal systems can be used for different missions such as fire fighting, bridge building, earthquake rescue without changing their basic fractal structures. Fractal robots possessing self-repairing capability are currently a trendy worldwide research topic [52].

6.5 Robots at Crossroads of Chaos and Percolation

Reconnaissance mission in mobile robot explorations consists of many distinct phases among which, feature extraction from the cluttered unstructured environment and the generation of available paths for continuous navigation, penetrating highly uncertain irregular media have recently been challenged by chaotic analyses. The investigation of such nonlinear approaches facing chaotic issues has led researchers to use chaotic motion planning models in developing instruments to guarantee search of the whole workspace through connected state spaces. Authors in references [28] use a Lorenz approach as a dynamical system with underlying chaotic behaviors applied to the generation of motion commands for mobile robots, implemented on the KHEPERA simulation environment. Their basic common approach is a seminal method on kinematic control of mobile robots using chaotic trajectory generation. On a similar tone, Nakamura et al. [34] have developed a method that integrates robot motion and Arnold dynamical system so as to define an open loop control system using chaotic system state variables. Moreover, robotic explorations occasionally base upon localization that needs to be accurately determined. Reconnaissance of the area explored frequently require mapping of the robot environment during the exploration task. Recently the two needs of localization and mapping met on the same robotic exploration missions generating the simultaneous localization and mapping (SLAM) techniques [31, 46]. Recently, a novel SLAM perspective considering a mobile robot exploration as an invasion of voids over occupying grid cells of the environment has been explored under the defense of obstacle cells (defender) led to the adoption of Percolation Theory in robot navigation in highly unstructured environments [54]. On a similar note, Erkmen and Topal [53] have proposed a percolator enhanced prioritized multi robot exploration so as to reach the localization information of the trapped survivors using the connected voids within the debris of the disaster area. We will elaborate on these novel approaches in Sect. 6.5.2.

6.5.1 Percolation Theory: Overview and Early Applications to Robotics

Percolation theory has been recently a particular interest to us as roboticists since it is the study of connected objects in fluid flow applications within porous media [47, 48] helping to model molecular connectivity of water, diffusion of different materials,or even city growth [27]. Percolation is also important for oil extraction since oil exists as connected cluster underneath the surface of the earth. One characteristic that has caught researchers attention is minimum path in percolation among connected clusters (spanning voids) which provides a conditional probability that is related, in oil research, to the time elapsed between injection of water until the extraction of the first bit of oil [45]. Time passed during extraction of the first bit of oil by injection of water can provide a clue for percolation probability in fluid flow, for oil recovery. In our exploration problem, spanning voids of unexplored area are not known explicitly. Furthermore we cannot obtain any percolation probability of the medium in an unknown environment. Thus, it is nonsense to measure time passed in robotic exploration problems since mobile robots do not know where to go in an unknown environment. On the other hand, gradient percolation [47, 48] such as modeling diffusion in variably porous material, is of high relevance to the reconnaissance objective. Among those, invasion percolation that models the percolating invader into a material occupied by defender molecules depending upon its porosity degrees is highly appropriate for our mobile robot exploration in unknown irregular environments that we introduce in Sect. 6.5.2. This diffusion in materials modeled by invasion percolation inspired us into using it to estimate the "diffusion" of possible paths within unknown regions based on "extrapolating" the boundaries of obstacles detected at the vicinity of frontiers of explored regions, into the unexplored sites. Invasion percolation theory finds the existence of spanning clusters [47, 56] and expands them iteratively through porosity minimization of iteratively added sites at their outer boundary. These iteratively added sites trace the percolating path of invader over defender molecules. Furthermore the primary principle of invasion percolation is to diffuse along a path of least resistance [56]. Porosity term has been used for defender resistance to traversability of invader. Consequently invasion percolator finds the site which has lowest resistance porosity among all already invaded adjacent sites and then invades it. Invasion depends on the threshold pressure of each pore which is defined as an occupation probability. Lattice representation used for site percolation is also an integrated part of invasion percolation such that each lattice site becomes equipped with invasion probability similar to the percolation probability. On the other hand, in our approach, in order to represent traversability of the invaders among the defenders, we assign porosity degrees to the lattice sites of the percolation environment.

Percolation analysis has been used to optimize swarm mobile robot planning to find the shortest path. In this work, the author has applied the phase transition phenomena of percolation theory on the determination of the routing algorithm of mobile agent colonies. In [24] the authors proposed a methodology that, from their initially given randomly selected formations, constructs connected clusters of mobile robots that do not communicate with each other. In these works, a phase transition phenomenon of percolation theory has been investigated on the formation behaviors of swarm robots. [53] also contributed percolator enhanced prioritized multi robot exploration methodology for the localization of the trapped survivors by determining the connected voids within the unstructured disaster area. However, only one paper exists to date at the crossroad of chaos and percolation. In that work the authors investigate the characteristics of a dynamic programming applied to a control problem of a chaotic system which is an underactuated robot, namely the Acrobot [49]. Percolation theory hence fluidics has been applied to the development of biomimetic abilities of the sensors like the ones embedded in a soft-body lamprey-like robot [25]. The sensory elements in the system are piezo-resistive and piezo-resistivity in the composite material which the sensors are made up of is governed by both percolation theory and quantum tunneling effects. In another work, authors have modeled the temperature sensors used when constructing meteorological micro-and milirobots via percolation theory.

6.5.2 Percolation Guided Active SLAM: The Most Recent Novel Methodology

The potential capabilities of percolation theory being the detection of connectivity of voids through an unknown environment, led us to use percolation based guidance techniques in exploration guided SLAM methodologies in order to wander in, exploit and map unknown regions more rapidly and safely. Prioritized Exploration Strategy based on percolation enhanced entropy based Fast SLAM [18] improves the performance of the classical exploration integrated SLAM approach in uncertain unstructured environments, by the enhancement provided by percolator guidance. Classical simultaneous localization and mapping (SLAM) techniques are either feature based or metric based using different Bayesian Filter approaches such as Kalman Filters, Particle Filter and their variations [31, 33, 51]. While SLAM techniques are passive methodologies, there are also active methods that guide mobile robots in their explorations [57]. Guidance for the mobile robot are generated in some works [46] to maximize the information gain that the mobile robot acquires about its environment during its navigation. Karahan's [18] guidance enriches the classical methodology of maximizing the information gain using the classical entropy based technique, by a switching mechanism that can also add percolation based guidance, that is a more adventurous way of estimating the continuity of voids in the area to be next explored by avoiding stemming obstacles at the receding frontier of the explored region. In the percolator enhanced mobile robot exploration methodology [18], the invader and defender fluids model of [56] has been modified and expanded into a model for the invading percolation of connected voids amongst defender obstacles within an unstructured environment represented on the grid based SLAM maps by adjusting the porosity degrees of its grid cells. Our expansion of the [56] model also applied for the first time to mobile sensor networks, has proven to enhance the guidance of the mobile robot navigations and it is the novelty of our active SLAM perspective as a fast, efficient yet adventurous exploration. Due to the grid cell occupation probabilities, porosity degrees are assigned to the cells [18] according to the invasion percolation model for modeling the flow of voids among obstacles on the grid based SLAM representation during mobile robot navigation. In our approach we model the spatial distributions of voids and obstacles using the displacement of invader fluids (voids) among defenders (obstacles) within the porous medium. The key motivation behind using invasion percolation in our approach is the continuity of percolating invader expanding through neighboring sites with lowest porosity degrees. Thus, in our methodology, invasion percolation ramifications continue through neighboring sites with lower degree porosity while other ramifications coming across higher porosity, die out. Since occupancy grid cell representation is also similar to the site lattice structure of [56], we have also related occupation probabilities of occupancy grid cells with the porosity degrees of the lattice sites. Our method adjusts these porosity degrees of upcoming unexplored sites so as to impose the continuity of previously detected obstacles when robots dive into the unexplored areas: the percolator modification

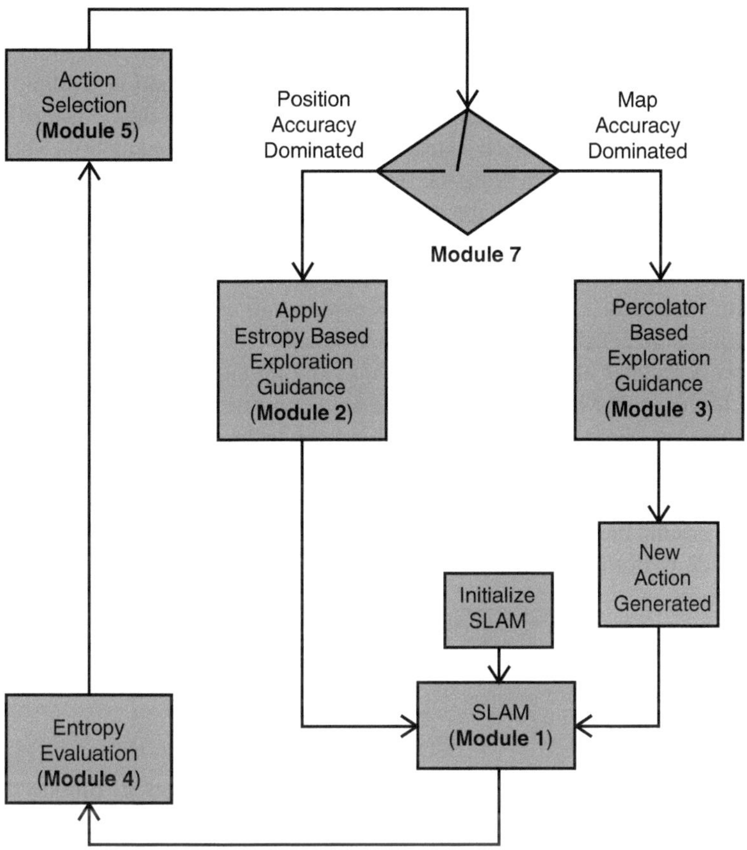

Fig. 6.12 Invasion percolator model [18]

increases porosity degrees of obstacle sites based on the continuity of obstacles previously detected and thus porosity becomes lower for unknown sites distant from such obstacle continuities, meaning possibly void in the area to be explored. Iteratively added neighbor sites having highest porosity degrees mean resisting to the invader percolation and are considered indentations of voids by obstacles leading to void ramifications within the unexplored region. Thus, we model the indentations of spatially connected spanning voids by increasing the porosity degrees at the vicinity areas of obstacles. Our robot is guided this way, by upcoming spanning voids avoiding, in the unexplored region, the possible continuities of obstacles, just detected at the boundary of the explored region. In our proposed system seen in Fig. 6.12 [18] a switching system (module 7) either enables classical Entropy Based exploitation running the Entropy based Fast SLAM (modules 2 and 1) which classically chooses the optimized trajectory from the given action set (module 5) so as to exploit unknown regions in the vicinity of the known ones for localization and mapping; or favors the percolator guidance (module 3) that we develop as a novel system, which is the percolator based exploration beyond frontiers of the explored regions, thus favoring map accuracies. If the selected action alternative (module 5) causes more entropy decrease for map information leading to the dominance of map accuracy then, the switching process of module 7 favors the use of invasion percolator guidance instead of entropy based one and module 3 becomes coupled to Fast SLAM (module 1). As a consequence of this switching process, we have been able to increase the exploration performance of the classical entropy based Fast SLAM technique using invasion percolation guidance in a switching mode, providing a percolation based enhancement to the entropy based method. Invasion Percolator mechanism provides exploration actions as a consequence of estimating connected voids beyond the frontiers of explored terrains. However invasion percolator although being an avid explorer cannot refine the localization of the mobile robot in order to preserve an acute precision. Exploitation actions are also necessary to refine the localization knowledge. Thus within the switching architecture we provide, our Invasion Percolation exploration collaborates with the entropy based exploitation that provides tuning of the localization accuracy by roaming around the actual robot position. This novel methodology [18] has proven to enhance performance in coverage and mapping as well as acute precision in localization.

6.6 Conclusion

Chaos arising through bifurcations under parametric changes in robotic devices including autonomous vehicles, legged robots has led to problematic issues in the performance of the devices such as stabilization, adaptability to dynamical en environmental changes and controllability. Towards relieving some of the problems, in the balance of this chapter we have conducted detailed phase plane analyses composed of periodic motions among chaotic, unpredictable trajectories. In this context,the analysis method that uses cell to cell mapping and fractal dimensions has been introduced. In depth analysis on two highly dynamical robotic structures have motivated the novel approach introduced by the authors that is used to increase the richness of the information in the regular region of a nonlinear robotic systems using fractal/rough set representation of that region. This information gain has been found to ease any control action involved.The fractal/rough set model developed is a new approach brought to the area of the uncertainty modelling in chaotic systems. The approach introduces the fractal dimension of the elements in the rough set as a measure of stability roughness.

Another arising area in the robotics research is the reconfigurable and scalable fractal robots. A brief survey on fractal mechanisms and their self reconfiguration ability has been given in the chapter reflecting the impact of chaos theory and fractals in the design of robotic systems: scale invariance and repeatable structures of different resolutions, together with the holonic decentralization concept have led to pioneering works in hyperredundant robots. However the field has still to mature by works on chaotic analysis of reconfiguration and formation transitions in swarms, which is the lacking piece in the research area.

Moreover,in applications using a heterogeneous hyperedundant robot network as well as single robot navigation in uncertain unstructured environments, SLAM and explorations techniques generally exhibit relative slowness and poor coverage and can be considered as major drawbacks, especially when they are used as part of time critical tasks such as search and rescue. Hence motivated by the critical exploitation issues such as time, computational burden and coverage, we introduce [18] an approach that evaluates the probability of propagation of connected voids amongst possible obstacles in an unstructured environment and uses this evaluation as a valuable guidance for robot navigation in the unstructured environments even though mobile robot possesses poor odometry measurements during navigation. Experiment results of the percolator enhanced exploration approach have shown that optimized motion commands can be chosen to pair the percolator with entropy based approach by the collaboration of a switching mechanism. Thus the percolator's coverage superiority has been integrated to high map accuracy and position awareness performances of the classical entropy based approaches using a switching control. Unstructured environments consisting of connected voids such as in search and rescue are appropriate platforms to show the efficiency of percolator guidance contributions versus the slowness and insufficiency in coverage of the entropy based methodology. However, especially in highly irregular, chaotically unstructured environments full of dead ends, deprived of connected voids, the performance enhancement of the percolator over the classical exploration methodologies is found to be hardly noticeable. If such an environment can be partitioned into regions of connected voids (regions of connectivity) the phase transition phenomena of percolation theory, used by few other authors (Sect. 6.5.1), can be applied for the determination of the routing of mobile robot colonies in passing from one region to the other.

References

1. Barto, A.G., Sutton, R.S., Anderson, C.W.: Neuronlike adaptive elements that can solve difficult learning control problems. IEEE Trans. Syst. Man Cybern. **13**(5), 834–846 (1983)
2. Beletskii, V.V.: Nonlinear effects in dynamics of controlled two-legged walking. Nonlinear Dynamics in Engineering Systems, Springer, Berlin (1990)
3. Beletskii, V.V., Pivovarov, M.L., Starostin, E.L.: Regular and chaotic motions in applied dynamics of a rigid body. American Institute of Physics, Chaos **6**(2), 155–166 (1996)
4. Cao, H., Chi, X., Chen, G.: Suppressing or inducing chaos in a model of robot arms and mechanical manipulators. J. Sound Vib. **271**(3-4), 705–724 (2004)
5. Chirikjian, G.S.: Kinematics of a metamorphic robotic system. In: Proceedings of ICRA'1994, pp. 449–455 (1994)
6. Clark, S.K.: Mechanics of pneumatic tires. Nat. Bur. Stand. (US) Monogr. 122 (1971)
7. Durna, M., Erkmen, A.M., Erkmen, I.: Holonic grasping. In: Proceedings of IROS'1998, pp. 140–145 (1998)

8. Durna, M., Erkmen, A.M., Erkmen, I.: Self localization of a holon in the reconfiguration task space of a robotic colony. In: Proceedings of ICRA'2000, pp. 1748–1754 (2000)
9. Durna, M., Erkmen, A.M., Erkmen, I.: The self reconfiguration of a holonic hand the holonic regrasp. In: Proceedings of IROS'2000, pp. 1993–1998 (2000)
10. Feigenbaum, M.J.: Quantitative universality for a class of non-linear transformations. J. Stat. Phys. **19**, 25–52 (1978)
11. Glabisz, W.: Stability of simple discrete systems under nonconservative loading with dynamic follower parameter. Comput. Struct. **69**, 105–115 (1998)
12. Goswami, A., Thuolit, B., Espiau, B.: A study of the passive gait of a compass-like biped robot: symmetry and chaos. Int. J. Robot. Res. **17**(12), 1282–1301 (1998)
13. Gregobi, C., Ott, E., Yorke, J.: Crises, sudden changes in chaotic attractors, and transient chaos. Physica **7D**, 181–200 (1983)
14. Hilgard, E.R., Bower, G.H.: Theories of Learning. Prentice Hall, New York (1975)
15. Hsu, C.S.: A theory of cell-to-cell mapping dynamical systems. ASME **47**, 931–939 (1980)
16. Hsu, C.S.: An unravelling algorithm for global analysis of dynamical systems: an application of cell-to-cell mappings. ASME **47**, 940–948 (1980)
17. Ileri, L.: Chaotic dynamic analysis of a vehicle model over road irregularities. M.Sc. Thesis. Middle East Technical University (1990)
18. Karahan, M.: Prioritized exploration strategy based on invasion percolation guidance. M.Sc. Thesis. Middle East Technical University (2010)
19. Kaygisiz, B., Erkmen, A.M., Erkmen, I.: Smoothing stability roughness of a robot arm under dynamic load using reinforcement learning. In: Proceedings of IROS'2001, pp. 2393–2397 (2001)
20. Kaygisiz, B., Erkmen, A.M., Erkmen, I.: Detection of transition to chaos during stability roughness smoothing of a robot arm. In: Proceedings of IROS'2002, pp. 1910–1915 (2002)
21. Kaygisiz, B., Erkmen, A.M., Erkmen, I.: Smoothing stability roughness of fractal boundaries using reinforcement learning. In: Proceedings of IFAC 15th World Congress (2002)
22. Kaygisiz, B., Erkmen, A.M., Erkmen, I.: Detecting onset of chaos and intelligent operational region enlargement. Int. J. Comput. Cognit. **3**, 102–108 (2005)
23. Kaygisiz, B., Erkmen, A.M., Erkmen, I.: Intelligent compensation of chaos roughness in regularity of walk for a two legged robot. Chaos, Solitons and Fractals **29**, 148–161 (2006)
24. Leung, H., Kothari, R., Minai, A.A.: Phase transition in a swarm algorithm for self-orginized construction. Phys. Rev. E **48-42**, 461111–461119 (2003)
25. Liu, W., Li, F., Stefanini, C., et al.: Biomimetic flexible/compliant sensors for a soft-body lamprey-like robot. Robot. Autonom. Syst. **58**(10), 1138–1148 (2010)
26. Mahadevan, S., Connell, J.: Automatic programming of behavior-based robots using reinforcement learning. In: Proceedings of 9th National Conference on Artificial Intelligence, pp. 768–773 (1991)
27. Makse, H.A.: Modelling urban growth patterns with correlated percolation. Phys. Rev. **58**(8), 7054–7062 (1998)
28. Martins-Filho, L.S., Machado, R.F., Rocha, R., et al.: Commanding mobile robots with chaos. ABCM Symposium Series in Mechatronics **1**, 40–46 (2004)
29. Mays, D., Faybishenko, B.A.: Washboards in Unpaved Highways as a Complex Dynamic System. Complexity **5**(6), 51–60 (2000)
30. Moon, F.C.: Chaotic Vibrations: An Introduction for Applied Scientist and Engineers. Wiley, New York (1987)
31. Montemerlo, M.: Fast SLAM: A factored solution to the simultaneous localization and mapping problem. In: Proceedings of the AAAI National Conference on Artificial Intelligence, pp. 593–598 (2002)
32. Murata, S.: Self-assembling machine. In: Proceedings of ICRA'2000, pp. 441–448 (1994)
33. Murphy, K.P.: Bayesian Map Learning in Dynamic Environments. In: Proceedings of Neural Information Processing Systems Conference, pp. 1015–1021
34. Nakamura, Y., Sekiguchi, A.: The chaotic mobile robot. IEEE Trans. Robot. Autom. **17**(6), 898–904 (2001)

35. Paar, V., Pavin, N., Paar, N., et al.: Nonlinear regular dynamics od a single degree robot model. Robotica **14**(4), 423–431 (1996)
36. Parker, T.S., Chua, L.O.: Chaos: A tutorial for engineers. Proc. IEEE **75**, 982–1001 (1987)
37. Pawlak, Z.: Rough sets: algebraic and topological approach. Institute of Computer Science Polish Academy of Sciences (1982)
38. Pawlak, Z.: Rough sets, rough functions and rough calculus. Institute of Theoretical and Applied Informatics Polish Academy of Sciences (1995)
39. Pawlak, Z.: Rough set approach to knowledge-based decision. Eur. J. Oper. Res. **99**, 48–57 (1997)
40. Pomeau, Y., Manneville, P.: Intermittent transition to the turbulence in dissipative dynamical systems. Commun. Math. Phys. **74**, 189–197 (1980)
41. Ruelle, D., Takens, F.: On the nature of turbulence. Commun. Math. Phys. **20**, 167–192 (1971)
42. Schaal, S., Atkeson, C.: Robot juggling: an implementation of memory-based learning. Control Systems Magazine **14**, 57–71 (1994)
43. Schwartz, A.: A reinforcement learning method for maximizing undiscounted rewards. In: Proceedings of the International Conference on Machine Learning **10**, 298–305 (1993)
44. Shahinpoor, M.: Deployable fractal mechanisms as smart structures. ASME-Applied Mech. Div. **167**, 119–126 (1993)
45. Soares, R.F., Corso, G., Lucena, L.S., et al.: Distribution of shortest paths at percolation threshold: Application to oil recovery with multiple wells. J. Phys. Math. Gen. **343**(1), 739–747 (2004)
46. Stachniss, C., Hahnel, D., Burgard, W.: Exploration with active loop closing for fast SLAM. In: Proceedings of IROS'2004, pp. 1505–1510 (2004)
47. Stanley, H.E., Andrade, J.S., Havlin, S., et al.: Percolation phenomena: a broad-brush introduction with some recent applications to porous media,liquid water and city growth-test of a percolation model. J. Phys. Math. Gen. **266**(1), 5–16 (1999)
48. Stauffer, D., Aharony, A.: An Introduction to Percolation Theory. Taylor and Francis (1994)
49. Ueda, R., Arai, T.: Dynamic programming for global control of the acrobat and its chaotic aspect. In: Proceedings of ICRA'2008, pp. 2416–2422 (2008)
50. Thompson, J.M.T., Soliman, M.S.: Fractal control boundaries of driven oscillators and their relevance to safe engineering design. Proc. R. Soc. Lond. A **428**, 1–13 (1990)
51. Thrun, S., Bcken, A.: Integrating grid-based and topological maps for mobile robot navigation. In: Proceedings of the Thirteenth National Conference on Artificial Intelligence, pp. 944–950
52. Tomita, K., Murata, S., Kurokawa, H., et al.: Self assembly and self repair method for a distributed mechanical system. IEEE Trans. Robot. Autom. **15-6**, 1035–1045 (1999)
53. Topal, S., Erkmen, A., Erkmen, I.: Percolation enhanced prioritised multi-robot exploration of unstructured environments. Int. J. Reas. Intell. Sys. **2**(3-4), 217–225 (2010)
54. Topal, S., Erkmen, I., Erkmen, A.: Towards the robotic "avatar": an extensive survey of the cooperation between and within networked mobile sensors. Future Internet **2**, 363–387 (2010)
55. Watkins, C.J.C.H., Dayan, P.: Q-learning, machine learning. **8**, 279–292 (1992)
56. Wilkinson, D., Willemsem, J.F.: Invasion percolation: a new form of percolation theory. J. Phys. Math. Gen. **16**(14), 3365–3376 (1983)
57. Yamauchi, B.: Frontier-based exploration using multiple robots. In: Proceedings of the International Conference on Autonomous Agents, pp. 47–53 (1998)
58. Yim, M., Shen, W.M., Salemi, B., et al.: Modular self-reconfigurable robot systems. IEEE Robot. Autom. Mag. **14**, 43–52 (2007)
59. Zhang, H., Li, Z.: Research on nonlinear dynamics of two degrees of freedom robot. Lecture Notes in AI, vol. 5314, pp. 420–426 (2008)

Part V
Chaos and Nonlinear Dynamics in Communication

Chapter 7
Chaos-Based Communication Systems: Current Trends and Challenges

José M.V. Grzybowski, Marcio Eisencraft, and Elbert E.N. Macau

7.1 Introduction

Much research has been devoted to synchronization and chaos-based communication in the last two decades or so. Since the early 1990s, many papers have addressed the subject, and much progress has been made. Considering that the research has been developed in many different – and, at certain extent, complementary – directions, it would be interesting to outline the advances and establish the bottom line to forthcoming research on the topic.

Chaos-based communication schemes, as originally conceived, use synchronization of chaotic oscillators as a backbone. It means that, under certain circumstances, the complex and highly sensitive nonlinear dynamics of coupled chaotic oscillators can synchronize, and such synchronous state can be exploited in several different manners to allow communication. Thus, to introduce this brief report, we attempt to tie synchronization and chaos-based communication together to make it clear how they are related. Ahead, the topic is briefly situated in the context of its evolution over time and the state of the art and challenges of the technique for future researches are outlined.

In the first place, concerning communication in general, what is the role of synchronization? In schemes based on coherent detection, synchronization enables *carrier recovery* and *timing recovery* at the receiver's end [43]. Carrier recovery refers to the reproduction or recovery, at the receiver's end, of the carrier signal produced in the transmitter. Once transmitter and receiver oscillators are matched,

J.M.V. Grzybowski (✉) · E.E.N. Macau
Instituto Tecnológico de Aeronáutica, São José dos Campos, Brazil
e-mail: zzmariovic@yahoo.com.br; elbert@lac.inpe.br

M. Eisencraft
Universidade Federal do ABC, Santo André, Brazil
e-mail: marcio.eisencraft@ufabc.edu.br

S. Banerjee et al. (eds.), *Applications of Chaos and Nonlinear Dynamics in Engineering – Vol. 1*, Understanding Complex Systems, DOI 10.1007/978-3-642-21922-1_7,

coherent demodulation of the modulated baseband signal is possible. On its turn, timing recovery refers to the need that both coherent and noncoherent receivers have to know the exact time and duration of each received symbol in a stream, in order to be able to assign decision times and reset the initial conditions of the correlator [43]. When mentioning *chaos synchronization* we mean a special type of carrier recovery.

In the second place, why use chaotic rather than sinusoidal carriers in communication systems? When a sinusoidal carrier is used to transmit information, the spectral power density concentrates in a narrow band of frequencies. Chaotic signals, on their turn, can occupy a large bandwidth, their autocorrelations can be impulsive and the cross-correlations between signals generated by different initial conditions present low values [15, 22, 25, 37, 38]. These characteristics have been behind the rationale for using chaotic signals as candidates for spreading information signals. When chaotic signals modulate independent narrowband sources increased bandwidths result with lower power spectral density levels in a fashion similar to what happens in Spread Spectrum (SS) systems [51]. Consequently, chaos-based and SS systems share several properties, namely (i) they are difficult to intercept by any unauthorized user; (ii) they are easily hidden from any unauthorized receiver and, in many cases, it is difficult even to detect their presence; (iii) they are resistant to jamming; and (iv) they provide a measure of immunity to distortion due to multipath propagation [51]. Figure 7.1 shows examples of periodic and chaotic carriers.

Apart from these reasons, chaos features some highly desirable characteristics from the point of view of communication, as it will be pointed in the sections ahead.

7.1.1 Context

The original idea of using chaos to transmit information was possible due to a couple of seminal publications on chaos control and synchronization in the year 1990.

Ott et al. [66] established that it is possible to control chaos to unstable periodic orbits or to fixed points by using small perturbations. They showed that, due to the property of transitivity of the chaotic motion in its attractive set, small perturbations are sufficient to render chaotic regime controlled. Subsequently, Hayes et al. showed both theoretically [33] and experimentally [34], how this idea could be used to make a chaotic signal bear information and so the idea of "communication with chaos" was introduced. Figure 7.2 shows an example of chaos synchronization using the Ott-Grebogi-Yorke (OGY) method for the Lorenz system [57].

Meanwhile, Pecora and Carroll [71] gave birth to the theoretical fundamentals of chaos synchronization by presenting a framework for determination of the stability of synchronization of a pair of identical coupled chaotic systems by means of the calculation of conditional Lyapunov exponents.

From these publications, a whole new perspective emerged in the study of nonlinear dynamics, especially chaos. The possibility of controlling and synchronizing chaos gave rise to the proposal of several practical applications, especially for communication.

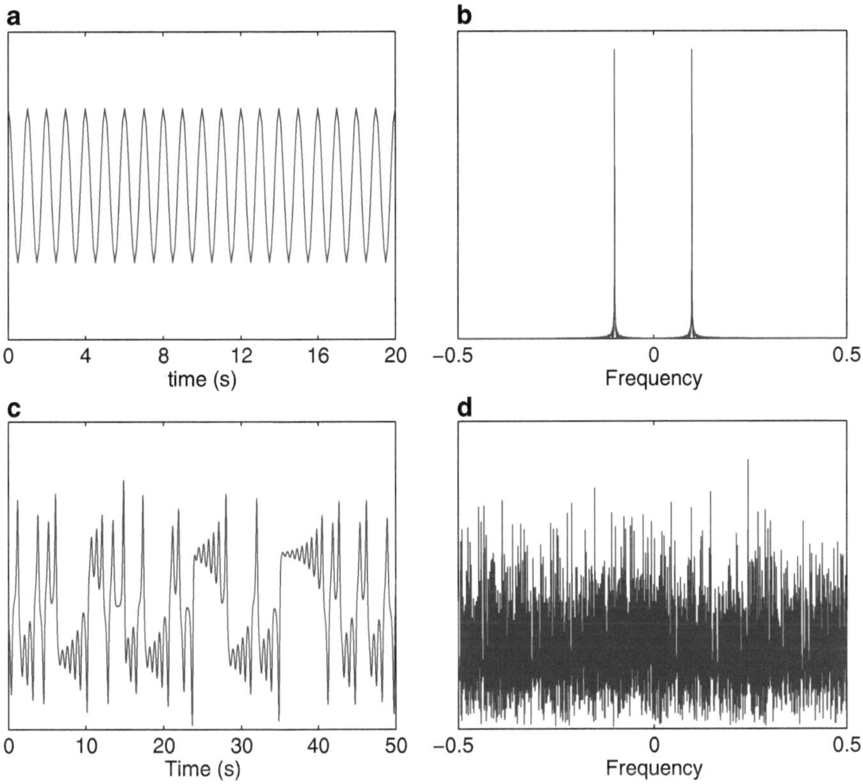

Fig. 7.1 The use of chaos as a carrier wave instead of a sinusoidal wave: the basic principle behind chaos-based communication; examples of a sinusoidal narrowband carrier in (**a**) time and (**b**) frequency domain and of a chaotic broadband carrier in (**c**) time and (**d**) frequency domain

Given that chaos became an important research topic, it did not take long until the first full chaos-based communication scheme was proposed and demonstrated experimentally by Cuomo and Oppenheim [13]. The idea was simple: to exploit the stable master-slave synchronous behavior between a pair of chaotic systems to encode, transmit and decode information in the chaotic evolution. The coupling signal consisted of a state variable of the master system, added with a low-power message. When adequately used as an input in the slave system, the chaotic waveform not only enabled systems to realize synchronization, but it also allowed the recovery of the information hidden in the chaotic dynamics. This system is known as *chaotic masking* and its block diagram is shown in Fig. 7.3.

Although the original idea itself was simple, it promoted insight into a remarkable underlying phenomenon: the slave system would synchronize only with the chaotic part of the input signal, regardless that some amount of noise and a low-power message were also present. Later on, this property became known as

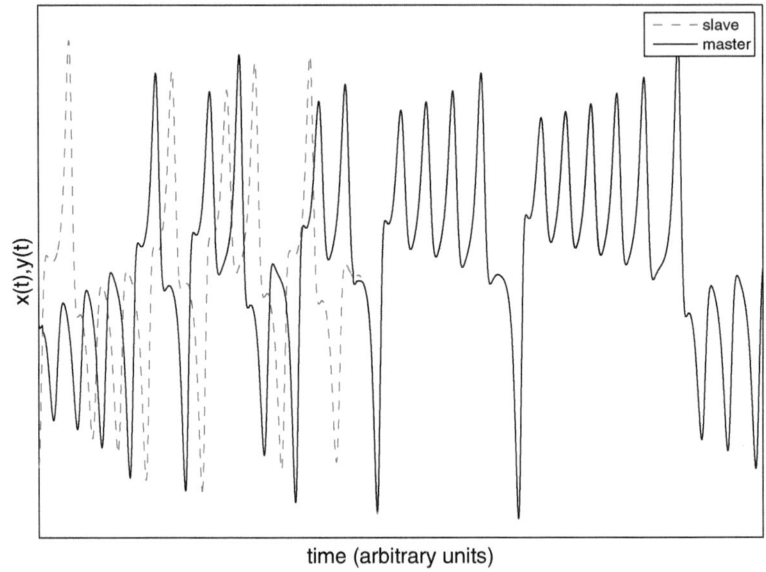

Fig. 7.2 Chaos under control: Lorenz systems get in synchrony by means of the OGY method [66]

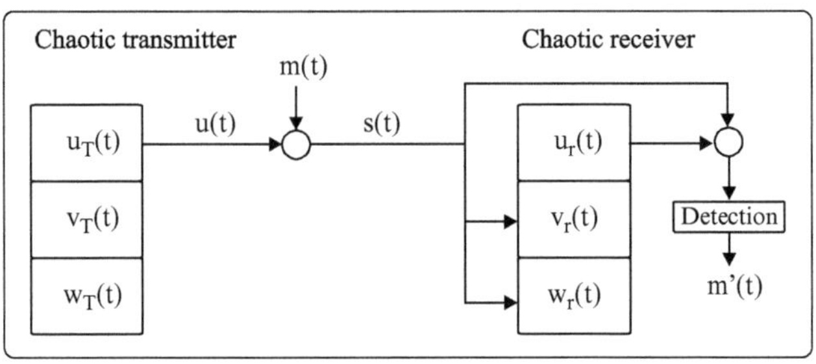

Fig. 7.3 Chaos communication design based on the Cuomo and Oppenheim [13] original scheme

chaos-pass filtering, that is, the property of synchronous systems to discard the non-chaotic part of the signal, which allows the message to be separated from the chaotic carrier signal [64, 70].

In the same year, Wu and Chua [87] proposed the *chaotic modulation*, which consists of the injection of the information signal in the equations of the chaotic system. The injection functions as a perturbation that alters the dynamics of the transmitter such that the chaotic signal itself contains information. At the receiver's end, by means of synchronization, the receiver is capable of regenerating the corresponding unperturbed chaotic signal and information retrieval follows by comparing the received perturbed signal and the generated unperturbed signal.

About ten years had passed since the first publications on chaos synchronization when another counterintuitive realization was made in delay-coupled chaotic systems subject to delayed self-feedback: a slave system can anticipate the trajectory of its master, what generates an emergent behavior that was named *anticipating synchronization*. According to Voss [83], such phenomena is nontrivial and of universal nature, since its underlying mechanism consists of the interplay between dissipation and memory. As such, anticipating synchronization can be exploited for communication purposes between remote oscillators or even for fast prediction, that is, state prediction without use of computation [83].

Many other possibilities of applications of chaos, ranging from digital and analog modulation to pseudorandom sequences generation and watermarking were proposed [37,38,77,81]. Chaos has also been shown in connection to devices used in signal processing such as nonlinear adaptive filters and phase-locked loop networks [23, 31, 63, 80].

As studies evolved and more elaborated scenarios were set to study chaos synchronization, time delays were introduced due to the fact that they are almost surely present in any real systems and, as a consequence, theoretical results with technological significance have to consider their influence. Besides, the realization that time delays in the feedback loop allow the emergence of chaos in laser emitters and also have potential to highly enhance chaos complexity has led to intensive study of delay differential equations, in order to precisely model and understand the emergence of chaos in such systems [24, 30].

As a result, the synchronization problem gained more elements, which allowed the emergence of unknown phenomena. A particularly interesting one resulted from the study of mutually delay-coupled identical chaotic oscillators, which could realize zero-lag synchronization despite of coupling delay. Such phenomenon is named *isochronal synchronization*. It was successfully tested for simultaneous chaos-based communication using identical mutually delay-coupled chaotic systems [40, 82, 84].

Since the first publications on chaos communication, much effort has been deployed towards the development of techniques and physical devices that can support and sustain efficient and cost-effective chaos-based communication systems. The efficiency aspect is mainly related to (i) the intensive exploitation of the chaotic dynamics to encode information, (ii) the low-power signal needed to control chaotic dynamics and (iii) the possibility of source coding, channel coding and encryption to be performed all at once in the same process [7]. The cost-effective aspect is related to (i) the simplicity of chaos-based communication setups [7], (ii) the low power-consuming devices needed to generate and control high-power chaotic signals and (iii) the use of all nonlinear operation excursion of electronic and optical components, which avoids the rather complicated and energy-consuming measures to avoid nonlinearities in the generation of the sinusoidal wave signals that serve as carriers in traditional communication schemes.

Nevertheless, desirable characteristics of chaos are not limited to the ones cited so far. Increased robustness against noise, parameter mismatch and multipath fading have been reported [7, 8, 56, 88], and consist of important points in favor of chaos-based communication schemes. Besides, recent results estimate that some chaos-based optical communication schemes can bear 15% more users in multi-user communication schemes for the same Bit Error Rate (BER) [55].

On the other hand, analytical upper limits in the BER performance for chaos-based communication schemes were found to be considerably lower than the performance of their conventional counterparts [37, 42, 46, 86]. This somewhat comes as a practical drawback since it diminishes the efficiency of information transmission.

As a result of such good and bad aspects, chaos-based communication is still a controversial subject. The main question is whether the overall performance of chaos-based communication schemes can surpass the performance of their conventional counterparts. This more concrete analysis would come as a result of the fair development of the various topics involving the practical implementation of chaos communication. So far, although it is not possible to answer that question in a broader sense, recent results indicate that there are concrete specific applications in which chaos-based communication can perform quite well, indeed [61, 69, 73].

As an imminently interdisciplinary matter, chaos-based communication has been investigated by physicists, engineers and mathematicians, among others. The comprehensive theory of chaos-based communication involves topics such as Nonlinear Dynamics and Chaos, Chaos Control and Synchronization, Lyapunov Stability of Motion Theory, Optics, Electronics, Telecommunication Engineering, among others. As a result of the contributions of different fields, the topic has evolved towards different and mutually complementary directions. Of course, in some of them the results have advanced more than in the others, and the consequence is that the state-of-the-art of chaos-based communication is hard to define if not in terms of its constituent individual matters. It is true that chaos-communication has been successfully tested in real environments, and a special reference should be made to a field experiment in Greece [4], probably the most relevant one up to date. But it is also true that not even in such experiment chaos-based communication was taken to the edge of the full possibilities it theoretically allows.

7.1.2 Advantages and Disadvantages of Chaos Communication

Is it worth investing so much time and energy in the research of chaos-based communication systems once conventional systems are managing perfectly? That seems to be the most fundamental and obvious question, as we watch the rise of faster and faster communication systems based on conventional techniques. The answer is not obvious, though. Chaos-based communication systems have unveiled a whole set of new possibilities, while at the same time they have been proven not to be able to reach the same BER performance as conventional systems [42, 86]. As such, the answer for this question may be obtained by pondering the advantages and disadvantages concerning the specific application that is desired.

At this point of evolution of hardware devices, such as lasers and electronic components, there are cases when the use of conventional sinusoidal waveforms to carry information can itself be seen as drawback. While optical devices are known to feature chaotic regime for a wide range of its parameters, the exclusive use of sinusoidal waves may consist of a waste of energy, due to the need for rather complicated measures to maintain linearity in inherently nonlinear systems. Besides, as lasers and optical fiber technologies are fairly developed, the use of their capacity to the fullest implies the efficient use of the wide range of signals that can be generated through their use. Indeed, the fact that both laser and electronic devices have much larger parameter range operation in chaotic regime than in periodic regime per se magnifies the capacity of generating broadband carrier signals.

While in conventional SS communication systems the broadband signals are generated using pseudorandom sequences to spread signals in frequency, in chaos-based schemes the bare fact that a chaos-generating device is running is enough to generate broadband signals [53, 79]. Typically, such broadband signal features a dense set of unstable periodic orbits within a range of frequencies with flat power spectral density. In practice, it means that chaotic systems make use of the input energy to generate broadband signals.

It is well known that arbitrarily nearby trajectories of chaotic systems exponentially diverge in the time evolution, due to the hallmark property of exponential sensitivity to initial conditions [2]. As a consequence, given two chaotic signals started in arbitrarily close initial conditions, such signals become statistically uncorrelated over time, which enables the use of multiplexing techniques. In the case of frequency division multiplexing, it is known that even the band-pass filtered chaotic signal will remain chaotic due to the existence of a dense set of unstable periodic orbits in the chaotic attractive set [36].

Concerning the overall implementation, conventional communication systems are known to require the intensive use of modulators, source encoders, channel encoders and filters. On the other hand, it is claimed that chaos-based communication schemes can be implemented by using only one subsystem at each end of the communication link yet providing all the basic processes needed for digital information transmission [7].

Furthermore, chaos-based communications are intended to make use of nonlinearities and complexity instead of suppressing them. In fact, by using such complexity adequately, as a result of a good understanding of the nonlinear dynamics, efficiency of chaos-based communication schemes can possibly supplant patterns of communication efficiency based on conventional schemes, yet saving energy and using the nonlinear characteristics of components to their fullest.

Note also that, however source coding and compaction techniques are available, they have rather limited resources when compared to eminent chaos-based coding techniques. This is due to the fact that in chaos-based coding, the redundancy needs not be sent over the channel; rather, they are built in the deterministic nature of the chaotic dynamics [7].

Concerning applications of communication such as mobile or indoor radio, an effect arises out of the interaction of signals traveling different paths and arising at the receiver. Such effect, known as *multipath fading*, consists of the destructive interference among signals and can result in severe signal degradation [51]. Theoretically, such effect can be reduced in chaos-based communication schemes, due to the inherent lower cross-correlation existing between any two time-shifted segments of the chaotic waveform, comparatively to that of periodic signals [53].

Apart from the theoretical advantages of chaos-based communication schemes, hardware equipments are somewhat converging to the use of nonlinear features of optical, optoelectronic and electronic components at low noise rates. Besides, according to [3], some of them could be easily included in the existing optical network infrastructure, such as the built-on-chip chaotic emitters.

Digital communication systems using chaotic carriers have been demonstrated to possess inferior upper limits for BER performance in Additive White Gaussian Noise (AWGN) channel than their corresponding versions based on conventional digital techniques [86]. This means that, under the same conditions of signal-to-noise ratio, conventional techniques have lower BER than their chaotic counterparts. Although it is true that the BER performance can be enhanced through the use of optimized coding and decoding or estimation techniques [32, 58], it is also true that conventional systems often present better BER performance even in such case. The alternative might be to direct chaos-based communication schemes to applications in which such disadvantage is not critical, such as in ultra-wideband radio, low data rate personal networks and optical laser systems [54].

From the last paragraphs, it is possible to enjoy some of the potentials of chaos for communications. In Sect. 7.2 we attempt to briefly review some state-of-the-art research areas. Afterwards, in Sect. 7.3 we comment on some challenges chaotic systems must surpass to become practical. Finally, on Sect. 7.4 we draft some conclusions.

7.2 State-of-the-Art of Chaos-Based Communication

In this section, a glimpse of some recent research topics on chaos-based communications is provided. We clearly do not intend to cover all the field. Rather, we succinctly describe areas that are closely related to practical real-world applications, for historical or technical reasons.

7.2.1 Source Coding Based on the Chaotic Dynamics

Chaos-based communication schemes present great potential for source coding and information transmission due to the inherent determinism of chaotic dynamics. Consider a chaotic map $x_{n+1} = f(x_n)$ that is able to generate a sequence $X_0 = \{x_0^0, x_1^0, \ldots, x_n^0\}$ of length $n+1$ out of a given initial condition x_0^0. Such chaotic sequence can be made to correspond to a given data stream $M_0 = \{m_0, m_1, \ldots m_n\}$ by creating a topological correspondence between the units of the alphabet and the disconnected regions of the invariant chaotic set Thus, a given data stream M_i corresponds to a unique chaotic sequence $X_i = \{x_0^i, x_1^i, x_2^i, \ldots\}$, as shown in [7].

Once the quantized initial condition x_0 is transmitted over a communication channel, the sequence X_0 can be generated at the receiver's end. As a consequence, the data stream M_0 can be retrieved. It is worth noting that only the initial condition x_0^0 of the chaotic sequence X_0 travels through the communication channel. As such, this principle allows the transmission of large amount of data as a much lower amount actually travels through the communication channel [7]. In other words, for each set of n+1 symbols produced by the information source, only one value must travel through the communication channel, namely, the initial condition of the chaotic sequence.

Note that the chaotic signal as it is used in this scheme is not only the carrier of information; rather, it also contains information in itself as it allows the receiver to recreate a whole data stream $M_0M_1 \ldots M_l$ from pieces of information $x_0^0x_0^1 \ldots x_0^l$. As such, the chaotic dynamics can be exploited for source coding.

7.2.2 Controlling Chaos and Modulating Information with Small Perturbations: The OGY Method

What does chaos control has to do with chaos communication? Chaos control strategies are responsible for generating a desired trajectory within the chaotic attractive set, such that information is encoded in the chaotic evolution.

In this context, the OGY method has been successfully used for more than two decades in the problems of controlling and synchronizing chaos. According to Grebogi and Lai [27], the key ingredient for chaos control is the observation that, within the chaotic set, there are infinite unstable periodic orbits which are recurrently visited by the trajectory as the chaotic process evolves in time. Some of these periodic orbits may correspond to a desired trajectory on which we may want the system to settle. Also according to [27], another key ingredient is that chaotic systems are subject to sensitive dependence on initial conditions, which also means that we can easily alter the trajectory by using small perturbations. The OGY method consists of using small perturbations to 'capture' and stabilize a chaotic trajectory on a given final state, i.e., a periodic orbit or fixed point, whenever such trajectory passes sufficiently close to this desired final state [66].

The use of small perturbations to control chaos is intimately related to the efficiency of chaos-generating devices, since the process of modulating a digital message into the chaotic evolution can be managed in practice with a small amount of energy. Although it was the first well-documented effective strategy for chaos control and synchronization, the OGY method remains widely used due to its energy-saving characteristic yet preserving effectiveness of the overall process.

The point here is that knowledge about the dynamics allows the use of little energy to control the chaotic dynamics and to encode information in its evolution. This fact has been widely explored in the development of chaos-based communication schemes through the use of the OGY method [9, 33]. For instance, Fig. 7.4 shows Poincaré sections to which the symbols 0 and 1 can be assigned, such that the chaotic evolution generates a binary bit stream, as shown in Fig. 7.5.

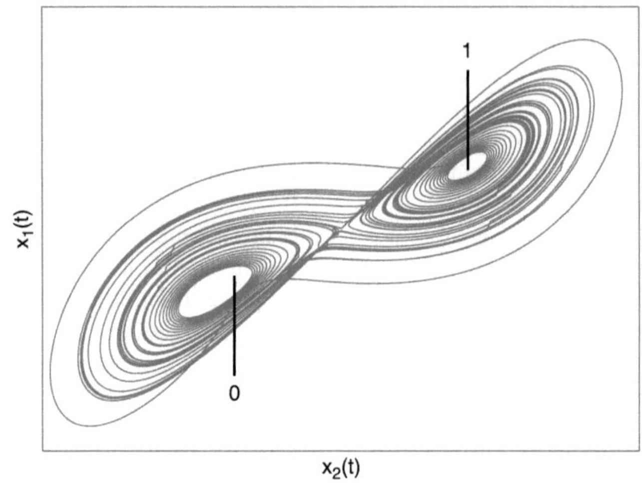

Fig. 7.4 Bits 0 and 1 are modulated in the chaotic evolution as the chaotic trajectory crosses a predetermined threshold [33]

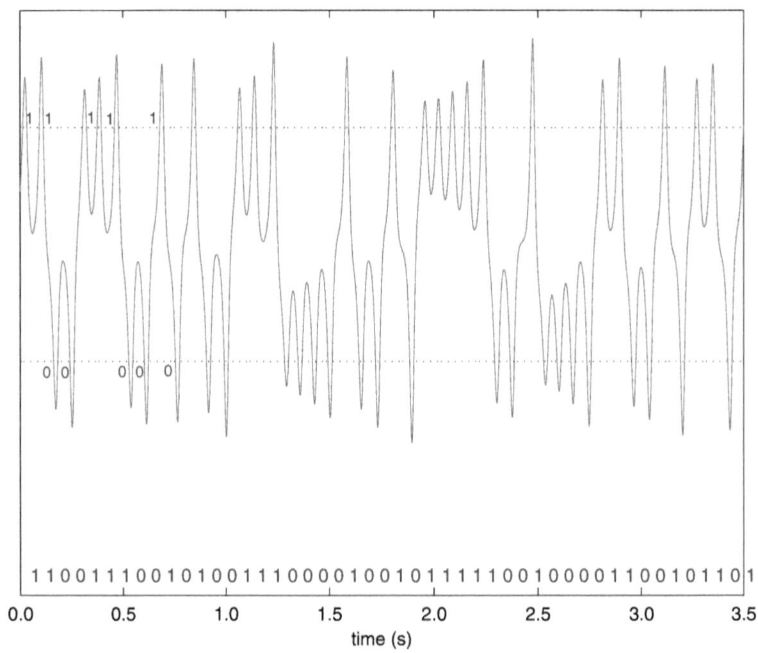

Fig. 7.5 Modulation process using the chaotic evolution [33]

7.2.3 Chaos-Based Communication in Delay Channels

Chaos communication can be effective in scenarios involving channel delays. Simultaneous bidirectional communication between delay-coupled oscillators has been presented in numerical simulations using lasers [82] and in experimental setups using chaotic electronic circuits [84]. In both cases, it was achieved in a scenario involving two coupled chaotic systems linked together by a delay channel. As one considers real situations, where time delays are almost surely present, chaos-based communication may be exploited as a means to overcome the inconveniences caused by channel delay yet maintaining a conceptually simple framework. Nevertheless, as the concept of simultaneous bidirectional chaos-based communication relies on synchronization, one needs to determine what synchronization means in the context of delay-coupled oscillators. The aim of this section is to briefly address the phenomenon of isochronal synchronization, which is the backbone of the simultaneous bidirectional chaos-based communication framework, and to outline how such synchronous state can be explored in simultaneous bidirectional communication schemes.

Recent results have shown that mutually coupled chaotic systems are capable of realizing zero-lag synchronization even in an environment with channel delay. Such form of synchronization is named isochronal synchronization and it has appeared in numerical simulations [41, 50, 85, 89], in experimental setups [6, 41, 84] and, more recently, in analytical results based on the Lyapunov-Krasovskii stability theory [29].

As the problem of communication in real environments is considered, time delays in the information transmission must be taken into account. The effect of such transmission time can be experienced in different forms, according to the unidirectional or bidirectional nature of the communication scheme. In unidirectional communication schemes in the form transmitter-receiver, such effect is limited to time delay in the information reception. As an example, suppose that a waveform $x(t)$ representing a symbol is sent by the master or transmitter system over a channel at the instant t and it is subject to delay τ due to the inherent characteristics of such channel. In this case, the transmitted waveform $x(t)$ is received at the receiver end at the instant $t+\tau$, and the principle behind decoding the symbol is the achronal synchronization between transmitter and receiver systems, that is, $x(t) = y(t+\tau)$, in which the receiver follows the transmitter with a time lag of τ [84] and recovers the transmitted symbol due to the chaos-pass filtering property of the synchronized oscillators.

On its turn, considering bidirectionally delay-coupled oscillators, isochronal synchronization allows the conception of simultaneous bidirectional communication schemes due to the fact that $x(t) = y(t)$ despite of the delay time introduced by the channel. Isochronal synchronization allows not only the realization of simultaneous bidirectional communication, but also the encryption and decryption of information and negotiation of secret keys [40, 84]. Although mutual driving of chaotic systems subject to coupling delay are sensitive to the magnitude of time delay, parameter

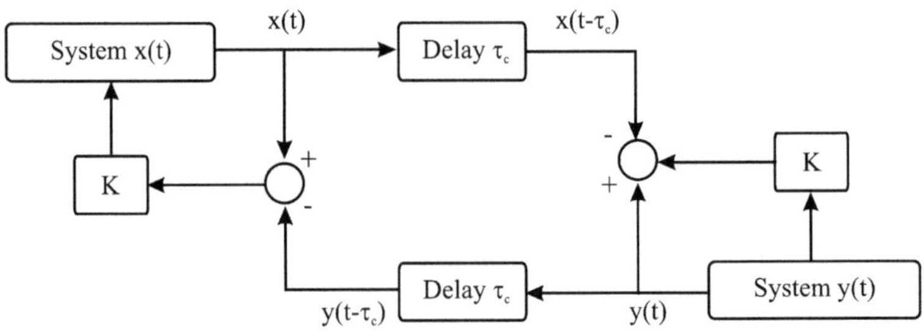

Fig. 7.6 Scheme of mutually driven systems [84]

mismatches and noise, successful realizations of chaos-based communication in simple experimental setups reinforce the possibility of fully exploiting chaotic dynamics to transmit information in such real environments presenting delay.

The idea of simultaneous bidirectional transmission and reception is based on the fact that the systems can synchronize with zero lag in the presence of channel delay and thus information can be injected and retrieved properly at both ends of the communication link, as if no delay were present. As simultaneous transmission occurs, there are two possibilities of symbol encoding at a given moment: (i) either both systems are encoding the same symbol or (ii) each system is encoding a different symbol. That being considered, two different circumstances concerning the maintenance of the synchronous state arise. In the first case, as mutually coupled systems are coding the same symbol at a given instant, the synchronization is maintained due to the fact that both systems are subject to the same perturbation caused by the information injection. In this case, an eavesdropper would have no clue of which symbol is being encoded [40, 84]. In the second case, as the systems encode different symbols at a given instant the message is treated as additive noise at the receiving circuit and it is filtered due to chaos-pass filtering of the synchronized systems. It is claimed that as the systems encode the same symbol at the same time, an eavesdropper could not possibly infer the symbol being encoded, and thus simultaneous bidirectional communication can be used in practice to negotiate secret keys, as both sides of the link can agree on a secret key consisting on the first N symbols that coincide, in such way that a key with dimension as large as desired can be negotiated in a public channel [84]. This framework is advantageous since it eliminates the need for a private channel to exchange key information and thus greatly simplifies the communication process. The block diagram of this system is shown in Fig. 7.6. Figure 7.7 illustrates the outcome of simultaneous data transmission: $m_x(t)$ and $m_y(t)$ are bit streams transmitted by systems $x(t)$ and $y(t)$, respectively. The dynamics of the synchronization error allows the identification of spikes that correspond to each system encoding a different bit. It follows that the received bit streams can be recovered after a simple XOR operation, as it is shown in the figure.

It is worth noting that the modulation of different symbols in a given instant causes bursts in the synchronous state. As resynchronization occurs, the next symbol

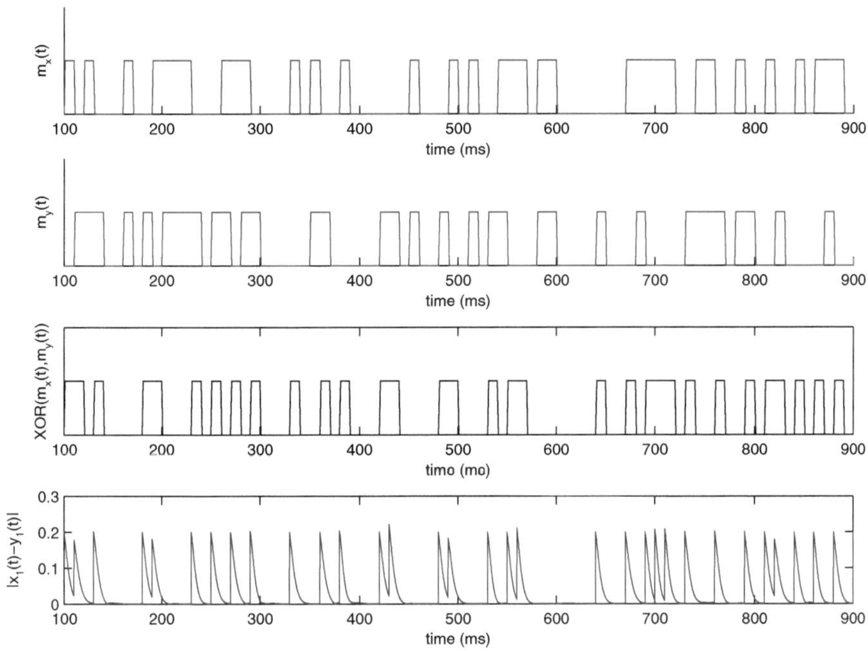

Fig. 7.7 Simultaneous bidirectional chaos-based communication scheme [84]

in the symbol sequence can be encoded at both ends and, as a consequence, the encoding rate is given by the inverse of the resynchronization time. For instance, in the case of semiconductor lasers, resynchronization takes around 0.3 ns, what allows a maximum encoding rate of 3Gbps per system [82].

Considering that source coding efficiency and data compression rates using chaotic dynamics can overpass conventional source coding and data compressing techniques [7], chaos-based simultaneous bidirectional communication based on isochronal synchronization tend to be of increasing research interest due to its great information bearing capacity yet maintaining simple framework and cost-effective implementation.

7.2.4 Chaos-Based Communication in Bandlimited and Noisy Channels

Although chaos-based communication systems based on chaos synchronization work well in ideal environments, the presence of noise and distortion in the channel, which are almost surely present, for instance, in wireless channels, brings unsatisfactory results in terms of BER when compared to conventional communication systems [37,53,58,86].

In the last years, there have been many researches whose objective is to approximate the performance of chaos communication systems to that of conventional ones, considering realistic environments. Here we briefly review some of these current techniques that may allow chaotic signals to be used in practical applications in the near future.

7.2.4.1 Bandlimited Channels

The papers by Pecora and Carroll [71], Cuomo and Oppenheim [13] and Wu and Chua [87] have inspired numerical and theoretical studies on the feasibility of master-slave communication systems based on chaotic synchronization. However, these schemes do not usually present satisfactory performance when the bandwidth limitations imposed by the communication channel are taken into account [21, 60]. This is a matter that in practice can not be neglected. In fact, because of the nonlinear nature of the nodes composing the network, if any spectral component is amiss in the transmission, then all spectral components can be affected. Consequently, the message sent by the master can not be faithfully recovered at the slave.

To overcome this problem, Rulkov and Tsimring [74] and Eisencraft and Gerken [18] independently proposed a method for synchronizing master and slave, described by chaotic differential equations, under bandwidth limitations. The idea is to employ an identical filter on both nodes in order to confine the spectral content of the transmitted signal to the available bandwidth. Afterwards, these results were extended to difference equations [21].

Figures 7.8–7.10 show the results obtained from numerical simulations by using the two-dimensional Hénon map [35]. Figure 7.8 exhibit the performance of the scheme described in [21] for an ideal channel. Notice that after a transient, the original message sent by the master is fully recovered in the slave. In Figs. 7.9 and 7.10, the channel is considered a Finite Impulse Response (FIR) filter [65] of order 50 and cut-off frequency $\omega_c = 0.8\pi$. In Fig. 7.9, no filter is used in the master and in the slave, in order to confine the spectrum of the transmitted message; hence, such a message can not be recovered. In Fig. 7.10, FIR filters of order 30 and $\omega_s = 0.4\pi$ are employed in both nodes and the message is fully rebuilt after a transient.

Therefore, within this framework, master-slave systems based on chaotic synchronization can satisfactorily work even when bandwidth limitations imposed by the communication channel are considered.

7.2.4.2 Additive Channel Noise

As it has been stated before, one of the main problems of chaos-based communication schemes is their poor performance under AWGN. If, on the one hand, the use of signals engendered by chaotic phenomena is a promising alternative towards

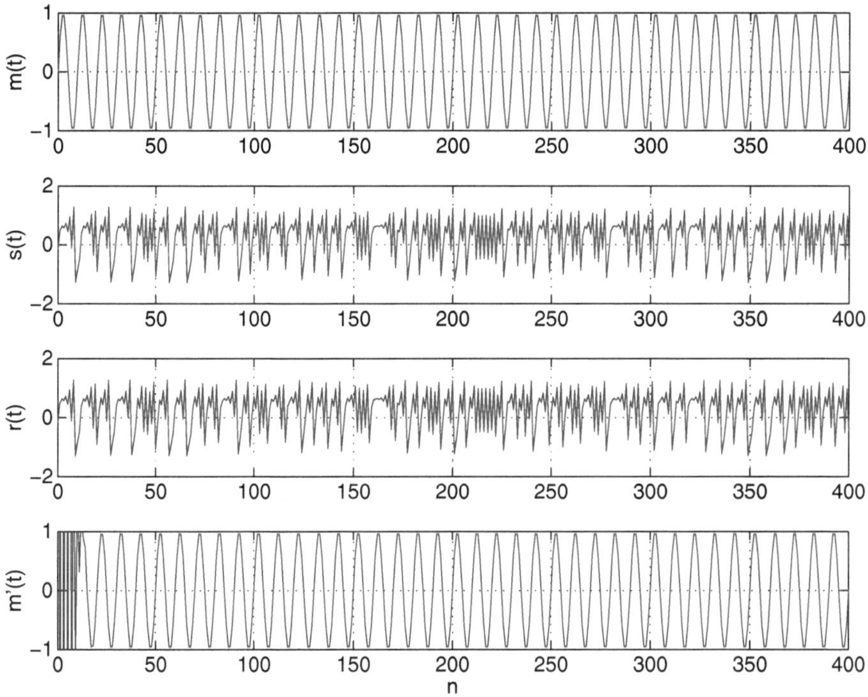

Fig. 7.8 Message $m(t)$, transmitted $s(t)$ and received $r(t)$ signals and recovered message $m'(t)$ in the ideal channel case [21]

efficient, secure and low power-consuming communication systems, on the other hand, certain intrinsic features of these signals – e.g. aperiodicity, sensitivity to initial conditions and broadband spectrum – pose significant difficulties to the performance of essential signal processing steps, such as denoising. A straightforward way to illustrate this point is to notice that the spectral similarities between chaotic signals and noise renders unsuitable the classical linear filtering approach based on direct frequency response shaping.

However, there have been many papers on recovering chaotic time-series from noisy environments. Among them, we can mention several different approaches: estimation theory, e.g. [14, 16, 17, 20, 59, 67, 68]; local polynomial approximation [47,48]; singular value decomposition and local geometric projection [11,26,49,75] and, more recently, blind signal separation [12, 76].

Based on these methods, many chaos-based communication schemes that use denoising techniques before or during demodulation have been proposed lately [20, 39]. Although their performance in terms of BER in AWGN are better than the chaos-based systems that do not use them, there is still a long path to them become comparable with conventional system in this aspect.

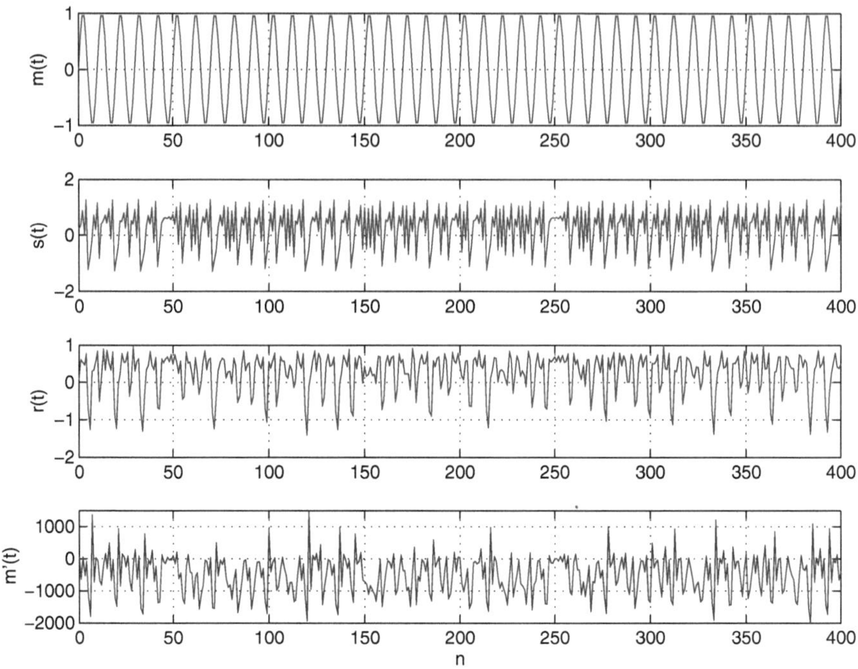

Fig. 7.9 Message $m(t)$, transmitted $s(t)$ and received $r(t)$ signals and recovered message $m'(t)$. The channel is a low-pass filter with normalized cut-off frequency $\omega_c = 0.8\pi$ [21]. The message could not be recovered

7.2.4.3 Limits of BER Performance

To qualitatively explain why digital chaos-based communications systems have BER performance issues, we focus now in one of the simplest ones, the Chaos Shift Keying (CSK) and its variants based on noncoherent or differential demodulation [37,43–45,53,77]. In one of the most relevant experiments on chaos communication up to date, the Optical Communications Laboratory of the Athens University in Greece, implemented an 120 km optical fiber link in metropolitan Athens and managed to transmit at gigabit rates using CSK [4, 78].

CSK is a digital modulation where each symbol to be transmitted is encoded as coefficients of a linear combination of signals generated by different chaotic attractors [43].

CSK can be basically implemented in two ways: the *coherent* and the *non-coherent* CSK. In the first case, the chaotic signals are regenerated at the receiver via chaos synchronization and the transmitted symbol is decoded using correlators. In the second case, each symbol is transmitted by chaotic signals with different mean energy and the receiver can decode the transmitted symbol using a threshold decision [37].

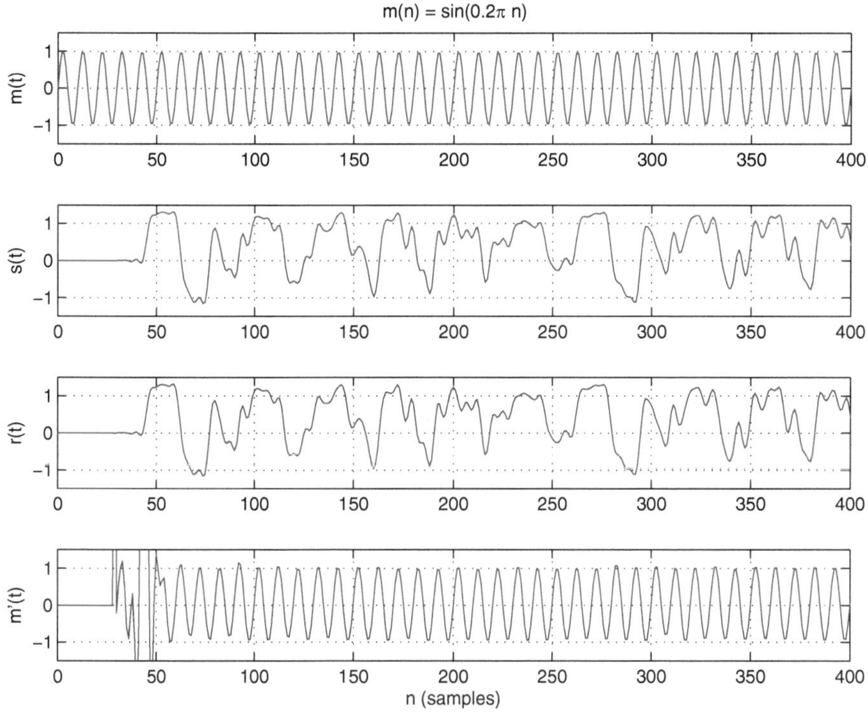

Fig. 7.10 Message $m(t)$, transmitted $s(t)$ and received $r(t)$ signals and recovered message $m'(t)$. for the same channel as in Fig. 7.9 but now low-pass filters are included in master and slave systems [21]. The message is fully rebuilt after a transient

As in conventional digital communication schemes, coherent demodulation is optimal if the basis function can be perfectly recovered at the receiver [51]. However, as chaos synchronization is sensitive to noise it turns out that non-coherent CSK has better performance in terms of BER in AWGN channel [86]. Nevertheless, an important drawback of non-coherent CSK is that the decision threshold depends on the power of channel noise that has to be estimated before demodulation.

An alternative to eliminate the problem of dependence of the decision threshold on the noise power in the channel is to use differential schemes, as the Differential CSK (DCSK) [44]. DCSK is a variant of CSK with two maps whose basis sequences consist of repeated segments of chaotic waveforms. To transmit a "1" two identical segments of length $N/2$ integer are sent. To transmit a "0" the second segment is multiplied by -1. The decision on the transmitted bit is based on the correlation between these two segments and the decision threshold is zero, independently of the channel noise.

A typical binary DCSK signal $x(n)$ corresponding to the symbol sequence $\{1,1,0,1,0,0,1,0\}$ using the tent map [22] as the chaotic generator is shown in Fig. 7.11.

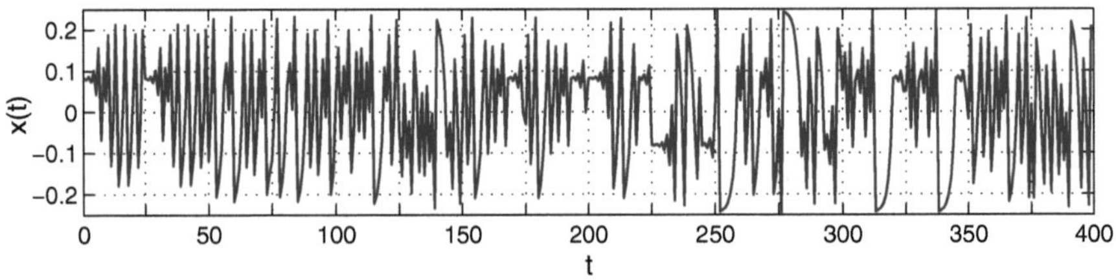

Fig. 7.11 DCSK signal transmitted for the data sequence $\{1,1,0,1,0,0,1,0\}$ with $N = 50$ samples per symbol [44]

For a conventional modulation scheme using only one periodic basis function $s(n)$ composed of N samples per symbol as an integer multiple of its period, the energy per symbol E_s is constant for each distinct symbol. In contrast, chaotic signals are by definition aperiodic. Thus, when using a chaotic basis, $s(n)$ is different at each interval and the energy can be different for each transmitted symbol.

Compared to conventional systems, the fact that the energy per symbol is not constant is a major disadvantage of noncoherent CSK and DCSK. The practical consequence of such drawback is that errors in reception can occur even in ideal noiseless channels, which is undesirable in practice.

An alternative solution is to modify the modulation scheme so that the transmitted energy for each symbol is kept constant. That is the aim of Frequency Modulated DCSK (FM-DCSK) [45].

The FM-DCSK transmitter generates a DCSK signal with constant energy per symbol. The idea is to take advantage of the fact that the power of a frequency modulated signal is independent of the signal, as long as it is slowly-varying compared to the carrier [51]. Thus, the chaotic signal is fed into a frequency modulator. If the output of this modulator is used in implementing DCSK, then the output of the correlator at the receiver will be a constant in the absence of noise and the problem of energy variability disappears.

Despite the fact that FM-DCSK eliminates the threshold dependence, the energy variability and the synchronization problems, it still suffers from the fact that it does not use any information on the chaos generating system in the demodulation process. In conventional systems, as the carrier is perfectly known by the receiver, optimal receivers as correlators or matched filters can be used with superior performance.

Table 7.1 summarizes the problems encountered in the digital modulations described. The column **Threshold** concerns the problem of dependence of the decision threshold on the noise power in the channel. The column **Energy** represents the problem of variability of energy per symbol. The column **Sync.** means the need for recovery of basis chaotic functions at the receiver and the last column, **Map Info** when signalized means that the system does not use properties of the chaotic attractor in the estimation of the transmitted symbol. Among the modulations cited, FM-DCSK has the best results because it does not depend on chaotic synchronization, its decision level threshold is independent of noise and the mean energy per symbol is constant.

Table 7.1 Problems of chaotic modulations studied in this section

System	Threshold	Energy	Sync.	Map Info
Coherent CSK		X	X	
Noncoherent CSK	X	X		X
DCSK		X		X
FM-DCSK				X

The analyzed non-coherent and differential receivers have a common feature: they do not use any characteristic of the dynamics of the systems that generate the chaotic signals to process the demodulation. These techniques are limited to estimating characteristics of the received signal and to comparing them with an adequate decision threshold.

So, to obtain better BER performance it is necessary to use the transmitter dynamics information on the receiver in a way robust to noise. Many works on this research field have been published lately as [14,16,17,20,59,67,68]. This research is fundamental for chaos-based communication to have a chance on noisy non-optical communications channels.

7.2.5 *Experimental Realization of Chaos-Based Communication: Photonic Integrated Devices*

Optical networks are regarded as a highly suitable environment for the proper functioning of chaos-based communication, due to (i) the great potential of devices that generate chaotic light (semiconductor lasers, laser diodes) and (ii) the low level of distortion and noise to which chaotic light is submitted when traveling through optical fiber links. As a consequence, optical links appear as a subject of intensive and fruitful study in the topic nowadays.

Fruitful, indeed, as it has been reported in [3], where compact, fully controllable and stably operating monolithic photonic integrated circuits (PICs) were successfully used for peer-to-peer optical link communication over 100km of optical fiber links, with bit rates of 2.5Gb/s and BER below 10^{-12}. Such devices are said to generate high-complexity broadband chaos, which are desirable characteristics from the standpoint of chaos-based communication. The major argument in favor of chaos in [3] is that chaotic carriers offer an extra layer of encryption that help secure data from unauthorized users, such that eventual eavesdroppers would be unable to recover the information embedded in the chaotic signal due to the technical difficulties of tuning their PICs accordingly. The main element responsible for the proper functioning of this setup seems to be the precise thermoelectric cooling of the devices, which provides stability and controllability of important synchronization properties, such as wavelength, optical power, spectral distribution and phase matching conditions [3].

More importantly, according to Argyris et al. [3], PICs are available to be directly and efficiently incorporated to the existing optical network.

7.3 Challenges of Chaos-Based Communication

In this section, we point out some challenges regarding the future applications of chaos in telecommunications.

7.3.1 Chaos-Based Versus Conventional Communication

The fact that conventional communication is managing perfectly and still improving incredibly fast, as optical techniques, such as *Dense Wavelength Multiplexing* (DWM), providing potential for efficient and reliable transmission at rates of the order of terabits per second [1], makes out of conventional techniques and devices something of a more and more well-established communication standard. Concerning chaos-based communication, there are still many points that must be improved and tested before they can be effectively used beyond experimental setups with scientific purpose.

It seems, however, that rather than mutually exclude each other, both techniques can coexist. This is due to the fact that the interesting aspects of each of them find their way in different applications, depending on the demands of the scenarios involved. For example, while conventional mobile radio communication experiences the undesirable effects of multipath fading and narrow band interference, chaos-based experimental setups have shown that chaos synchronization manages poorly in noisy channels. On the other hand, while conventional communication is quite suitable for ultra-high-speed optical communication using DWM, for example, it is not likely that data encryption at the software level will possibly be able to keep up at such high transmission rate. On its turn, chaos-based communication can provide an extra layer of high-speed real-time encryption at the physical level, which may allow the reduction of software encryption strength while maintaining the original level of secrecy, thus allowing the secure transmission of data at very high rates.

Finally, it seems that at this point, the challenge consists of directing the adequate technique to the applications it is best suited for, taking into account the desired aspects and demands of the final application and at the same time ensure that conventional and chaos-based appliances can interact properly.

7.3.2 Chaos-Based Communication in the Existing Infrastructure

It is not likely that the operating communication systems will be simply substituted, no matter how developed any other technique might be. Some relevant subjects related to the development of new techniques and technologies are (i) at what extent it can use and (ii) how easily it can be incorporated in the existing infrastructure and (iii) how compatible it is with the standard technology.

Bearing this in mind, some recent papers have dealt with this subject. Lau et al. [52] and Chaps. 7 and 8 of [53] have studied the coexistence of CSK and DCSK with conventional narrowband digital communication systems and with conventional spread spectrum systems. They have obtained theoretical and numerical BER results which allow the evaluation of the performance of the resulting hybrid system for different (i) power ratios and spreading factors, (ii) chaotic signal powers, (iii) conventional spread-spectrum signal powers, and (iv) noise power spectral densities. Although the results themselves are not much conclusive, they are an important step towards deeper understanding of the advantages and drawbacks of the coexistence of the communication techniques.

It is also worth citing the results presented in [69] as examples of the use of on-the-shelf pieces of equipment to create practical chaos-based applications. This paper presents the design and validation by means of suitably improved randomness tests of two different implementations of high-performance true-random number generators which use a discrete-time chaotic circuit as their entropy source. The proposed system has been developed from a standard pipeline analog-to-digital converter design, easily available, modified to operate as a set of piecewise-linear chaotic maps.

7.3.3 *Further Development of Efficient Multi-User Schemes*

For any practical communication system, the multiple-access capability is essential. Early studies on chaos-based communication systems were focused on a single-user case. In the past few years, more effort has been employed in the investigation of systems with multiple-access capability, which is a key feature of spread-spectrum communication systems [79]. However, the study of the multi-user capacities of chaos-based communications in real environments is still in its early developments.

A method based on multiplexing chaotic signals has been proposed by Carroll and Pecora [10] and some chaos-based approaches for generating spreading codes have been applied to conventional code division multiple access systems [62, 72]. A patent on the subject was granted to Yang and Chua (US Patent 6331974).

Multiple access using DCSK and FM-DCSK has also been proposed, see [79] and references therein. To minimize the co-channel interference [51], Kolumban et al. [46] proposed a multi-user FM-DCSK scheme, in which a chaotic signal is combined with two Walsh functions to form the basis functions representing the symbols $+1$ and -1 for each user.

As another possibility, Eisencraft and Kato [19] have numerically demonstrated the possibility of generating band limited chaotic signal and to employ conventional frequency-division multiplexing for multi-user capability of chaos-based communication system.

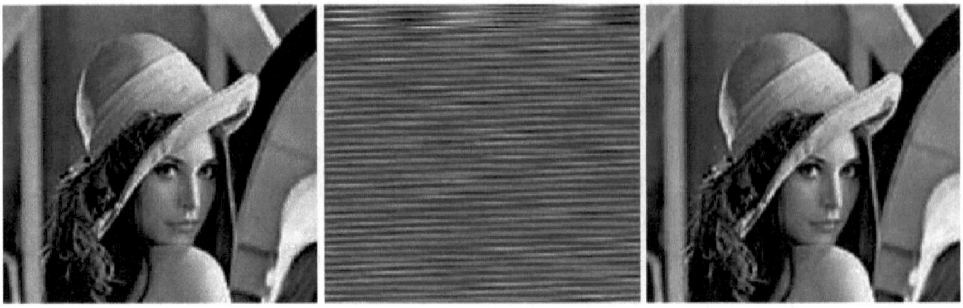

Fig. 7.12 Lenna picture encrypted and decrypted using chaos [28]

7.3.4 Encryption at the Physical Level and Legal Issues

One of the remarkable characteristics of chaos-based communication is the possibility of implementing cryptography at the physical level, since chaos has many desirable characteristics from the point of view of cryptography, namely, complexity, ergodicity, transitivity, determinism and sensitivity to initial conditions and perturbations [40]. Although the cryptographic predicates of chaos are rather controversial, especially concerning security issues, the widespread use of chaos might touch restrictions in several countries that have strict policies concerning the private use of encryption.

The massive use of real-time high-speed encryption at the physical level would be barely a by-product of chaos-based communication and, if on the one hand security of chaotic encryption is questionable, on the other hand it does make it harder to control and access information. Such fact can surely become a barrier to widespread use of chaos-based communication schemes, since it can touch long-standing legal issues concerning information secrecy in many countries.

Despite this questions, chaos-based cryptographic systems and their analysis have been areas with large number of contributing papers lately, mainly on image cryptography and watermarking, where conventional solutions are not established yet, see e.g. [5] and references therein. For example, Fig. 7.12 shows the Lenna picture encrypted and decrypted using chaos [28].

7.4 Final Remarks

This chapter is aimed to outline the main aspects involved in chaos-based communication in terms of the development of the technique and physical devices up to date. Topics are not intended to be treated in depth. Rather, newcomers to the topic are provided a suitable starting point to their investigation. At the same time, expert researchers may profit from a global overview that may allow the identification of their own work while acquiring a general idea of what has been done in different directions of research.

Based on the topics presented in this communication and in the cited references, final remarks are given regarding current applications of chaos-based communications:

- Optical Communications: coherent CSK was successfully implemented with performance similar to conventional systems. As optical channels approach a noiseless one, sensitivity of chaos synchronization to noise is not a relevant issue and chaos can provide an extra security level;
- Wireless Communications: real wireless channels present noise, multipath and delay, so that many of the proposed chaos-based communication systems simply do not work. However, many researches are in course to investigate solutions to the problems that arise due to these non-ideal and inherent characteristics of the channel;
- Satellite Formation Flying: in this application, many smaller satellites cooperate to perform tasks that would not be possible with a single monolithic satellite, such as distributed measurement for synthetic aperture radar. Due to the distributed nature of the mission, synchronization among the clocks of the satellites is necessary for coordination, for instance, when data acquired by different satellites must be matched up. Due to the distances, time-delays are generated when information is transmitted between satellites, and isochronal synchronization has potential to be exploited in this scenario;
- Cryptography: this is perhaps the topic receiving most part of the publications within chaos applications. The characteristics of chaos have been exploited in hundreds of methods proposed in the literature. Nowadays, the applications are concentrated in image encryption and watermarking, where conventional solutions are not established.

Since conventional communication techniques are managing quite well, the deployment of so much time and energy in the development of chaos-based techniques gives rise to the question of whether it is worth the try. However, given the whole new theoretical possibilities that chaos brings about and the increasingly demanding scenarios where reliable, high-speed and effective communication is demanded, it is a general consensus that researching chaos communication is a far-sighted measure. Either in the case the outcome will be successful or not in terms of widespread use of such technique, chaos-related subjects have provided many insights into different important areas of scientific interest, such as Medicine, Social Sciences, Physics and Engineering. This, per se, can be regarded as a substantial scientific contribution.

Acknowledgements J.M.V. Grzybowski thanks FAPESP, grant nr. 2008/11684-0; M. Eisencraft and E.E.N. Macau thanks CNPq for the support on this research.

References

1. Aburakawa, Y., Otsu, T.: Dense wavelength division multiplexed optical wireless link towards terabit transmission. In: Microwave Photonics, 2003. MWP 2003 Proceedings. International Topical Meeting on, pp. 135–138, DOI 10.1109/MWP.2003.1422846 (2003)
2. Alligood, K.T., Sauer, T.D., Yorke, J.A.: (1997) Chaos: An Introduction to Dynamical Systems. Textbooks In Mathematical Sciences, Springer, New York (2003)
3. Argyris, A., Grivas, E., Hamacher, M., Bogris, A., Syvridis, D.: Chaos-on-a-chip secures data transmission in optical fiber links. Opt. Express **18**(5), 5188–5198, DOI 10.1364/OE.18.005188 (2010)
4. Argyris, A., Syvridis, D., Larger, L., Annovazzi-Lodi, V., Colet, P., Fischer, I., Garcia-Ojalvo, J., Mirasso, C., Pesquera, L., Shore, K.: Chaos-based communications at high bit rates using commercial fibre-optic links. Nature **438**(7066), 343–346, DOI {10.1038/nature04275} (2005)
5. Arroyo, D., Alvarez, G., Fernandez, V.: A basic framework for the cryptanalysis of digital chaos-based cryptography. In: Systems, Signals and Devices, 2009. SSD '09. 6th International Multi-Conference on, pp. 1 –6, DOI 10.1109/SSD.2009.4956652 (2009)
6. Avila, J.F.M., Leite, J.R.R.: Time delays in the synchronization of chaotic coupled lasers with feedback. Opt. Express **17**(24), 21,442–21,451, DOI 10.1364/OE.17.021442 (2009)
7. Baptista, M.S., Macau, E.E., Grebogi, C., Lai, Y.C., Rosa, E.: Integrated chaotic communication scheme. Phys. Rev. E **62**(4), 4835–4845, DOI 10.1103/PhysRevE.62.4835 (2000)
8. Bollt, E., Lai, Y.C., Grebogi, C.: Coding, channel capacity, and noise resistance in communicating with chaos. Phys. Rev. Lett. **79**(19), 3787–3790, DOI 10.1103/PhysRevLett.79.3787 (1997)
9. Carroll, M., Williams, C.: Symbolic dynamics method for chaotic communications. In: MILCOM 2002. Proceedings, vol. 1, pp. 232–236, DOI 10.1109/MILCOM.2002.1180445 (2002)
10. Carroll, T.L., Pecora, L.M.: Using multiple attractor chaotic systems for communication. Chaos: An Interdisciplinary Journal of Nonlinear Science **9**(2), 445–451, DOI 10.1063/1.166425 (1999)
11. Cawley, R., Hsu, G.H.: Local-geometric-projection method for noise reduction in chaotic maps and flows. Phys. Rev. A **46**(6), 3057–3082, DOI 10.1103/PhysRevA.46.3057 (1992)
12. Chen, H., Feng, J., Fang, Y.: Blind extraction of chaotic signals by using the fast independent component analysis algorithm. Chin. Phys. Lett. **25**, 405–408 (2008)
13. Cuomo, K.M., Oppenheim, A.V.: Circuit implementation of synchronized chaos with applications to communications. Phys. Rev. Lett. **71**(1), 65–68, DOI 10.1103/PhysRevLett.71.65 (1993)
14. Dedieu, H., Kisel, A.: Communications with chaotic time series: probabilistic methods for noise reduction. Int. J. Circ. Theor. Appl. **27**(6), 577–587 (1999)
15. Djurovic, I., Rubezic, V.: Chaos detection in chaotic systems with large number of components in spectral domain. Signal Process. **88**(9), 2357–2362, DOI 10.1016/j.sigpro.2008.03.003 (2008)
16. Eisencraft, M., do Amaral, M.A.: Estimation of nonuniform invariant density chaotic signals with applications in communications. In: Second IFAC meeting related to analysis and control of chaotic systems, London, England, pp. 1–6 (2009)
17. Eisencraft, M., Baccalá, L.A.: The Cramer-Rao bound for initial conditions estimation of chaotic orbits. Chaos, Solitons Fractals **38**(1), 132–139, DOI 10.1016/j.chaos.2006.10.067 (2008)
18. Eisencraft, M., Gerken, M.: Comunicação utilizando sinais caóticos: influência de ruído e limitação em banda. In: Anais do XVIII Simpósio Brasileiro de Telecomunicações, Gramado, Brasil, pp. 1–6, (in Portuguese) (2001)
19. Eisencraft, M., Kato, D.M.: Spectral properties of chaotic signals with applications in communications. Nonlinear Anal. Theor. Meth. Appl. **71**(12), e2592–e2599, DOI 10.1016/j.na.2009.05.071 (2009)

20. Eisencraft, M., do Amaral, M.A., Lima, C.A.M.: Estimation of chaotic signals with applications in communications. In: Proc. 15th IFAC Symposium on System Identification, Saint-Malo, France, pp. 1–6 (2009)
21. Eisencraft, M., Fanganiello, R., Baccala, L.: Synchronization of discrete-time chaotic systems in bandlimited channels. Math. Probl. Eng., DOI 10.1155/2009/207971 (2009)
22. Eisencraft, M., Kato, D., Monteiro, L.: Spectral properties of chaotic signals generated by the skew tent map. Signal Process. **90**(1), 385–390, DOI 10.1016/j.sigpro.2009.06.018 (2010)
23. Endo, T., Chua, L.: Chaos from phase-locked loops. IEEE Trans. Circ. Syst. **35**(8), 987–1003, DOI 10.1109/31.1845 (1988)
24. Erneux, T.: Applied delay differential equations, 1st edn. Springer (2009)
25. Faleiros, A.C., Perrella, W.J., Rabello, T.N., Santos, A.S., Soma, N.Y.: Chaotic signal generation and transmission. In: [77] (2005)
26. Grassberger, P., Procaccia, I.: Measuring the strangeness of strange attractors. Phys. D Nonlinear Phenom. **9**(1-2), 189–208, DOI 10.1016/0167-2789(83)90298-1 (1983)
27. Grebogi, C., Lai, Y.C.: Controlling chaotic dynamical systems. Syst. Contr. Lett. **31**(5), 307–312, DOI 10.1016/S0167-6911(97)00046-7 (1997)
28. Grzybowski, J.M.V., Rafikov, M.: Sincronização do sistema caótico unificado via controle ótimo linear feedback e aplicação em comunicação segura. Tendências em Matemática Aplicada e Computacional **9**(1), 105–114, (in Portuguese) (2008)
29. Grzybowski, J.M.V., Macau, E.E.N., Yoneyama, T.: Isochronal synchronization of time delay and delay-coupled chaotic systems. Journal of Physics A: Mathematical and Theoretical **44** (2011) 175103
30. Gu, K., Kharitonov, V., Chen, J.: Stability of time-delay systems, 1st edn. Birkhäuser Boston (2003)
31. Harb, B.A., Harb, A.M.: Chaos and bifurcation in a third-order phase locked loop. Chaos, Solitons Fractals **19**(3), 667–672, DOI 10.1016/S0960-0779(03)00197-8 (2004)
32. Hasler, M., Schimming, T.: Optimal and suboptimal chaos receivers. Proc. IEEE **90**(5), 733–746, DOI 10.1109/JPROC.2002.1015004 (2002)
33. Hayes, S., Grebogi, C., Ott, E.: Communicating with chaos. Phys. Rev. Lett. **70**(20), 3031–3034 (1993)
34. Hayes, S., Grebogi, C., Ott, E., Mark, A.: Experimental control of chaos for communication. Phys. Rev. Lett. **73**(13), 1781–1784 (1994)
35. Hénon, M.: A two-dimensional mapping with a strange attractor. Comm. Math. Phys. **50**(1), 69–77 (1976)
36. Itoh, M., Chua, L.: Multiplexing techniques via chaos. In: Circuits and Systems, 1997. ISCAS '97., Proceedings of 1997 IEEE International Symposium on, vol. 2, pp. 905–908, DOI 10.1109/ISCAS.1997.621860 (1997)
37. Kennedy, M., Setti, G., Rovatti, R. (eds.): Chaotic Electronics in Telecommunications. CRC Press, Boca Raton, FL, USA (2000)
38. Kennedy, M.P., Kolumbán, G.: Digital communications using chaos. Signal Process. **80**(7), 1307–1320, DOI 10.1016/S0165-1684(00)00038-4 (2000)
39. Kisel, A., Dedieu, H., Schimming, T.: Maximum likelihood approaches for noncoherent communications with chaotic carriers. IEEE Trans. Circ. Syst. I. Fund. Theor. Appl. **48**(5), 533–542, DOI 10.1109/81.922456 (2001)
40. Klein, E., Gross, N., Kopelowitz, E., Rosenbluh, M., Khaykovich, L., Kinzel, W., Kanter, I.: Public-channel cryptography based on mutual chaos pass filters. Phys. Rev. E **74**(4), 046,201, DOI 10.1103/PhysRevE.74.046201 (2006)
41. Klein, E., Gross, N., Rosenbluh, M., Kinzel, W., Khaykovich, L., Kanter, I.: Stable isochronal synchronization of mutually coupled chaotic lasers. Phys. Rev. E **73**(6), 066,214, DOI 10.1103/PhysRevE.73.066214 (2006)
42. Kolumban, G., Krébesz, T.: Chaotic communications with autocorrelation receiver: Modeling, theory and performance limits. In: Kocarev, L., Galias, Z., Lian, S. (eds.) Intelligent Computing Based on Chaos, Studies in Computational Intelligence, vol. 184, pp. 121–143, Springer Berlin/Heidelberg, 10.1007/978-3-540-95972-4_6 (2009)

43. Kolumban, G., Kennedy, M., Chua, L.: The role of synchronization in digital communications using chaos. I . fundamentals of digital communications. IEEE Trans. Circ. Syst. I. Fund. Theor. Appl. **44**(10), 927–936, DOI 10.1109/81.633882 (1997)
44. Kolumban, G., Kennedy, M., Chua, L.: The role of synchronization in digital communications using chaos. II. chaotic modulation and chaotic synchronization. IEEE Trans. Circ. Syst. I. Fund. Theor. Appl. **45**(11), 1129–1140, DOI 10.1109/81.735435 (1998)
45. Kolumban, G., Kennedy, M., Kis, G., Jako, Z.: FM-DCSK: a novel method for chaotic communications. In: Proceedings of the 1998 IEEE International Symposium on Circuits and Systems, 1998. ISCAS '98. (1998)
46. Kolumban, G., Kennedy, M., Jako, Z., Kis, G.: Chaotic communications with correlator receivers: theory and performance limits. Proc. IEEE **90**(5), 711–732, DOI 10.1109/JPROC. 2002.1015003 (2002)
47. Kostelich, E.J., Schreiber, T.: Noise reduction in chaotic time-series data: A survey of common methods. Phys. Rev. E **48**(3), 1752–1763, DOI 10.1103/PhysRevE.48.1752 (1993)
48. Kostelich, E.J., Yorke, J.A.: Noise reduction: Finding the simplest dynamical system consistent with the data. Phys. D Nonlinear Phenom. **41**(2), 183–196, DOI 10.1016/0167-2789(90) 90121-5 (1990)
49. Landa, P., Rosenblum, M.: Time series analysis for system identification and diagnostics. Phys. D Nonlinear Phenom. **48**(1), 232–254, DOI 10.1016/0167-2789(91)90059-I (1991)
50. Landsman, A.S., Schwartz, I.B.: Complete chaotic synchronization in mutually coupled time-delay systems. Phys. Rev. E **75**(2), 026,201, DOI 10.1103/PhysRevE.75.026201 (2007)
51. Lathi, B.P.: Modern Digital and Analog Communication Systems, 4th edn. Oxford University Press, New York, NY, USA (2009)
52. Lau, F., Tse, C., Ye, M., Hau, S.: Coexistence of chaos-based and conventional digital communication systems of equal bit rate. IEEE Transactions on Circuits and Systems I: Regular Papers, **51**(2), 391–408, DOI 10.1109/TCSI.2003.822398 (2004)
53. Lau, F.C.M., Tse, C.K.: Chaos-based digital communication systems. Springer, Berlin (2003)
54. Lawrance, A.: Recent theory and new applications in chaos communications. In: Proceedings of 2010 IEEE International Symposium on Circuits and Systems (ISCAS) (2010)
55. Liu, J., Chen, H., Tang, S.: Optical-communication systems based on chaos in semiconductor lasers. IEEE Trans. Circ. Syst. I. Fund. Theor. Appl. **48**(12), 1475–1483, DOI 10.1109/TCSI. 2001.972854 (2001)
56. López-Gutirrez, R., Posadas-Castillo, C., López-Mancilla, D., Cruz-Hernndez, C.: Communicating via robust synchronization of chaotic lasers. Chaos, Solitons, Fractals **42**(1), 277–285, DOI 10.1016/j.chaos.2008.11.019 (2009)
57. Lorenz, E.N.: Deterministic nonperiodic flow. J. Atmos. Sci. **20**(2), 130–141, DOI 10.1175/ 1520-0469(1963)020⟨0130:DNF⟩2.0.CO;2 (1963)
58. Luengo, D., Santamaria, I.: Secure communications using OFDM with chaotic modulation in the subcarriers. In: 2005 IEEE 61st Vehicular Technology Conference, 2005. VTC 2005-Spring, vol. 2, pp. 1022–1026, DOI 10.1109/VETECS.2005.1543461 (2005)
59. Luengo, D., Santamaría, I., Vielva, L.: Asymptotically optimal maximum-likelihood estimator of a class of chaotic signals using the Viterbi algorithm. In: 13th European Signal Processing Conference (EUSIPCO 2005), Antalya, Turkey, pp. 1–4 (2005)
60. Macau, E.E.N., Marinho, C.M.P.: Communication with chaos over band-limited channels. Acta Astronautica, DOI 10.1016/S0094-5765(03)80007-3, The New Face of Space Selected Proceedings of the 53rd International Astronautical Federation Congress. **53**(4-10), 465–475 (2003)
61. Marinho, C.M., Macau, E.E., Yoneyama, T.: Chaos over chaos: A new approach for satellite communication. Acta Astronautica, DOI 10.1016/j.actaastro.2005.03.019, Infinite Possibilities Global Realities, Selected Proceedings of the 55th International Astronautical Federation Congress, Vancouver, Canada, 4–8 October 2004. **57**(2-8), 230–238 (2005)
62. Mazzini, G., Setti, G., Rovatti, R.: Chaotic complex spreading sequences for asynchronous DS-CDMA. i. system modeling and results. IEEE Trans. Circ. Syst. I. Fund. Theor. Appl. **44**(10), 937–947, DOI 10.1109/81.633883 (1997)

63. Monteiro, L., Lisboa, A., Eisencraft, M.: Route to chaos in a third-order phase-locked loop network. Signal Process. **89**(8), 1678–1682, DOI 10.1016/j.sigpro.2009.03.006 (2009)
64. Murakami, A., Shore, K.A.: Chaos-pass filtering in injection-locked semiconductor lasers. Phys. Rev. A **72**(5), 053,810, DOI 10.1103/PhysRevA.72.053810 (2005)
65. Oppenheim, A.V., Schafer, R.W.: Discrete-Time Signal Processing. Prentice Hall, Upper Saddle River, NJ, USA (2009)
66. Ott, E., Grebogi, C., Yorke, J.A.: Controlling chaos. Phys. Rev. Lett. **64**(11), 1196–1199, DOI 10.1103/PhysRevLett.64.1196 (1990)
67. Pantaleon, C., Luengo, D., Santamaria, I.: Optimal estimation of chaotic signals generated by piecewise-linear maps. IEEE Signal Process. Lett. **7**(8), 235–237, DOI 10.1109/97.855451 (2000)
68. Papadopoulos, H., Wornell, G.: Optimal detection of a class of chaotic signals. IEEE International Conference on Acoustics, Speech, and Signal Processing, ICASSP-93, vol. 3, pp. 117–120, DOI 10.1109/ICASSP.1993.319449 (1993)
69. Pareschi, F., Setti, G., Rovatti, R.: Implementation and testing of high-speed CMOS true random number generators based on chaotic systems. IEEE Transactions on Circuits and Systems I: Regular Papers, **57**(12), 3124–3137, DOI 10.1109/TCSI.2010.2052515 (2010)
70. Paul, J., Lee, M.W., Shore, K.A.: Effect of chaos pass filtering on message decoding quality using chaotic external-cavity laser diodes. Opt. Lett. **29**(21), 2497–2499, DOI 10.1364/OL.29.002497 (2004)
71. Pecora, L.M., Carroll, T.L.: Synchronization in chaotic systems. Phys. Rev. Lett. **64**(8), 821–824, DOI 10.1103/PhysRevLett.64.821 (1990)
72. Rovatti, R., Setti, G., Mazzini, G.: Chaotic complex spreading sequences for asynchronous DS-CDMA. Part II. Some theoretical performance bounds. IEEE Trans. Circ. Syst. I. Fund. Theor. Appl. **45**(4), 496–506, DOI 10.1109/81.669073 (1998)
73. Rovatti, R., Mazzini, G., Setti, G.: On the ultimate limits of chaos-based asynchronous DS-CDMA-I: basic definitions and results. IEEE Transactions on Circuits and Systems I: Regular Papers, **51**(7), 1336–1347, DOI 10.1109/TCSI.2004.830700 (2004)
74. Rulkov, N.F., Tsimring, L.S.: Synchronization methods for communication with chaos over band-limited channels. Int. J. Circ. Theor. Appl. **27**, 555–567 (1999)
75. Sauer, T.: A noise reduction method for signals from nonlinear systems. Phys. D Nonlinear Phenom. **58**(1-4), 193–201, DOI 10.1016/0167-2789(92)90108-Y (1992)
76. Soriano, D.C., Suyama, R., Attux, R.: Blind extraction of chaotic sources from white gaussian noise based on a measure of determinism. In: Adali, T., Jutten, C., Romano, J.M.T., Barros, A. (eds.) Independent Component Analysis and Signal Separation. Lecture Notes in Computer Science, vol. 5441, pp. 122–129. Springer Berlin/Heidelberg (2009)
77. Stavroulakis, P. (ed.): Chaos Applications in Telecommunications. CRC Press, Boca Raton, FL, USA (2005)
78. Syvridis, D.: Optical Chaos Encoded Communications: Solutions for Today and Tomorrow. In: 2009 IEEE LEOS Annual Meeting Conference Proceedings, Vols. 1 and 2, IEEE Photon Soc., IEEE, IEEE Lasers and Electro-Optics Society (LEOS) Annual Meeting, pp. 759–760 (2009)
79. Tam, W.M., Lau, F.C.M., Tse, C.K.: Digital Communications with Chaos: Multiple Access Techniques and Performance. Elsevier, NY, USA (2006)
80. Tavazoei, M.S., Haeri, M.: Chaos in the APFM nonlinear adaptive filter. Signal Process. **89**(5), 697–702, DOI 10.1016/j.sigpro.2008.10.032 (2009)
81. Tsekeridou, S., Solachidis, V., Nikolaidis, N., Nikolaidis, A., Tefas, A., Pitas, I.: Statistical analysis of a watermarking system based on Bernoulli chaotic sequences. Signal Process. **81**(6), 1273–1293, DOI 10.1016/S0165-1684(01)00044-5 (2001)
82. Vicente, R., Mirasso, C.R., Fischer, I.: Simultaneous bidirectional message transmission in a chaos-based communication scheme. Opt. Lett. **32**(4), 403–405, DOI 10.1364/OL.32.000403 (2007)
83. Voss, H.U.: Anticipating chaotic synchronization. Phys. Rev. E **61**(5), 5115–5119, DOI 10.1103/PhysRevE.61.5115 (2000)

84. Wagemakers, A., Buldú, J.M., Sanjuán, M.A.F.: Experimental demonstration of bidirectional chaotic communication by means of isochronal synchronization. Europhys. Lett. **81**(4), 40,005 (2008)
85. Wagemakers, A., Buldu, J.M., Sanjuán, M.A.F. Isochronous synchronization in mutually coupled chaotic circuits. Chaos **17**(2), 023128 DOI {10.1063/1.2737820} (2007)
86. Williams, C.: Chaotic communications over radio channels. IEEE Trans. Circ. Syst. I. Fund. Theor. Appl. **48**(12), 1394–1404, DOI 10.1109/TCSI.2001.972846 (2001)
87. Wu, C.W., Chua, L.O.: A simple way to synchronize chaotic systems with applications to secure communication systems. Int. J. Bifurcat. Chaos **3**(6), 1619–1627 (1993)
88. Xia, Y., Tse, C., Lau, F.: Performance of differential chaos-shift-keying digital communication systems over a multipath fading channel with delay spread. IEEE Transactions on Circuits and Systems II: Express Briefs, **51**(12), 680–684, DOI 10.1109/TCSII.2004.838329 (2004)
89. Zhou, B.B., Roy, R.: Isochronal synchrony and bidirectional communication with delay-coupled nonlinear oscillators. Phys. Rev. E **75**(2), 026,205, DOI 10.1103/PhysRevE.75.026205 (2007)

Chapter 8
Chaotic Synchronization and Its Applications in Secure Communications

Rafael Martínez-Guerra, Juan L. Mata-Machuca, Ricardo Aguilar-López, and Andrés Rodríguez-Bollain

8.1 Introduction

Synchronization in chaotic systems has been investigated since its introduction in [58]. This research area has received a great deal of attention among scientist in many fields due to its potential applications mainly in secure communications [2,6,11,21,30,52,74]. However, a certain number of drawbacks have been revealed in the practical implementation of most chaos-based secure communications algorithms [7,17,59].

During the last years (almost two decades), many different approaches related to chaos synchronization have been proposed. For instance, we mention the works [3, 10, 31, 47, 53, 56] in which the authors propose the employment of state observers, where the main applications pertain to the synchronization of nonlinear oscillators; in references [19,71] use feedback controllers, which allow to achieve the synchronization between nonlinear oscillators, with different structure and order; in [29, 70] use nonlinear backstepping control; in papers [25, 26] consider synchronization time delayed systems; in works [22, 41] consider directional and bidirectional linear coupling; papers [14, 15] use nonlinear control; in [1, 54] use adaptive control; in [18, 20] employ adaptive observers; [5] considers an adaptive sliding mode observer and so on.

R. Martínez-Guerra (✉) · J.L. Mata-Machuca · A. Rodríguez-Bollain
Departamento de Control Automático, CINVESTAV-IPN, Av. IPN 2508, San Pedro Zacatenco, C.P. 07360, Distrito Federal, México
e-mail: rguerra@ctrl.cinvestav.mx; jmata@ctrl.cinvestav.mx; grodriguez@ctrl.cinvestav.mx

R. Aguilar-López
Departamento de Biotecnologa y Bioingeniería, CINVESTAV-IPN, Av. IPN 2508, San Pedro Zacatenco, C.P. 07360, Distrito Federal, México
e-mail: raguilar@cinvestav.mx

S. Banerjee et al. (eds.), *Applications of Chaos and Nonlinear Dynamics in Engineering – Vol. 1*, Understanding Complex Systems, DOI 10.1007/978-3-642-21922-1_8,

As we have mentioned there are many applications to chaotic communications. The techniques can be divided into three categories:

1. *Chaos masking*, the information signal is added directly to the transmitter.
2. *Chaos modulation*, it is based on the master-slave synchronization, where the information signal is injected into the transmitter as a nonlinear filter.
3. *Chaos shift keying*, the information signal is supposed to be binary, and it is mapped into the transmitter and the receiver. In these three cases, the information signal can be recovered by a receiver if the transmitter and the receiver are synchronized.

From the above, in this book chapter, there are considered several cases for chaotic synchronization and some applications, in particular:

- *Observer-based synchronization*. Of particular interest is the connection between the observers for nonlinear systems and the chaos synchronization, which is also known as master-slave configuration. Thus, chaos synchronization problem can be posed as an observer design procedure, where the coupling signal is viewed as output and the slave system is regarded as observer.
- *Control of chaotic Liouvillian systems*. Application to a class of chemical reacting system via sliding-mode observer based feedback.
- *Synchronization of chaotic Liouvillian oscillators*. Application to a Colpitts chaotic oscillator by means of observers: exponential observer of polynomial type, and asymptotic observer of reduced order.
- *Application to secure communications*. The general idea for transmitting information via chaotic systems is that, an information signal is embedded in the transmitter system which produces a chaotic signal, the information signal is recovered when the transmitter and the receiver are identical. In this work, a novel design approach for chaotic communication is proposed, where the receiver is a pure sliding-mode observer.

8.2 Observer-Based Synchronization

Now, we mention a brief note about observers theory. The design of observers for nonlinear systems is a challenging problem, that has received a considerable amount of attention. Since the observers developed by Kalman [35] and Luenberger [44], several years ago for linear systems, different state observation techniques have been propose to handle the systems nonlinearities. A first category of techniques consists in applying linear algorithms to the system linearized around the estimated trajectory. These are known as the extended Kalman and Luenberger observers. Alternatively, the nonlinear dynamics are split into a linear part and a nonlinear one. The observer gains then are chosen large enough so that the linear part dominates over the nonlinear one. Such observers are known as, high gain observers [23, 62]. And many other approaches such as [36, 40, 73].

In this work the synchronization method is based on a *master–slave* configuration [58]. The main characteristic is that the coupling signal is unidirectional, that is, the signal is transmitted from the master system (transmitter) to the slave system (receiver), the receiver is requested to recover the unknown (or full) state trajectories of the transmitter. By this fact, the terminology *transmitter–receiver* is also used. Thus, chaos synchronization problem can be regarded as observer design procedure, where the coupling signal is viewed as output and the slave system is the observer [16, 55, 57].

Let us consider the following nonlinear system,

$$\begin{aligned} \dot{x} &= f\,(x,u) \\ y &= C\,x, \qquad x_0 = x(t_0) \end{aligned} \tag{8.1}$$

where $x \in \mathbb{R}^n$, is the state vector; $u \in \mathbb{R}^p$, is the input vector, $p \leq n$; $f(\cdot) : \mathbb{R}^n \times \mathbb{R}^p \to \mathbb{R}^n$ is locally Lipschitz on x and uniformly bounded on u; $y \in \mathbb{R}$ is the output of the system. To show the relation between the observers for nonlinear systems and chaos synchronization we recall the observer definition.

Definition 8.1 (Observer). An observer system for (8.1) is a system with state $\hat{x}$ such that $\|x - \hat{x}\| \to 0$ as $t \to \infty$.

In the context of master-slave synchronization, $\hat{x}$ can be viewed as the state variable of the slave system. Hence, the master-slave synchronization problem can be solved by designing an observer for (8.1).

8.2.1 Some Definitions

As we can note, there exists several methods to solve the synchronization problem since the control theory perspective, in this work, we study the synchronization in master-slave configuration [58] by means of state observers based on the differential algebraic approach [48]. In order to solve the synchronization problem as an observation problem we introduce the following observability property.

Definition 8.2 (Algebraic Observability Condition–AOC). A state variable $x \in \mathbb{R}$ is said to be algebraically observable if it is algebraic over $\mathbb{R}\langle u,y\rangle$,[1] that is, x satisfies a differential algebraic polynomial in terms of $\{u,y\}$ and some of their time derivatives, i.e.,

$$P(x,u,\dot{u},\ldots,y,\dot{y},\ldots) = 0 \tag{8.2}$$

with coefficients in $\mathbb{R}\langle u,y\rangle$.

---

[1] $\mathbb{R}\langle u,y\rangle$ denotes the differential field generated by the field $\mathbb{R}$, the input u, the measurable output y, and the time derivatives of u and y.

Example 8.1. Consider the nonlinear system

$$\begin{aligned}\dot{x}_1 &= -x_1 x_2\\ \dot{x}_2 &= -x_2^2 + x_1 + u\\ y &= x_2\end{aligned} \tag{8.3}$$

System (8.3) is algebraically observable since $x_1 - \dot{y} - y^2 + u = 0$ and $x_2 - y = 0$. That is to say, x_1 and x_2 can be reconstructed knowing $\{u, y\}$ and their time derivatives.

Example 8.2. Let us consider the following nonlinear system

$$\begin{aligned}\dot{x}_1 &= x_2 + x_3^2\\ \dot{x}_2 &= x_3\\ \dot{x}_3 &= u\end{aligned} \tag{8.4}$$

If we define $y = x_2$, then

$$\begin{aligned}x_2 &= y\\ x_3 &= \dot{y}\\ \dot{x}_1 &= y + \dot{y}^2\end{aligned} \tag{8.5}$$

The above system is not algebraically observable since x_1 cannot be expressed as a differential algebraic polynomial in terms of $\{u, y\}$.

Motivated by this fact, we present the next definition.

Definition 8.3 (Liouvillian System). A dynamical system is said to be Liouvillian if the elements (for example, state variables or parameters) can be obtained by an adjunction of integrals or exponentials of integrals of elements of $\mathbb{R}$.

Example 8.3. We consider the nonlinear system as in Example 8.2. From (8.5) we can observe that, although x_1 does not satisfy the AOC we can obtain it by means of the integral

$$x_1 = \int (y + \dot{y}^2)$$

Therefore the nonlinear system (8.4) is Liouvillian.

For further information we recommend to see [8, 49].

8.3 Control of a Class of Chaotic Liouvillian Chemical Systems

The aim of this section is the synthesis of a robust control law for the control of a class of nonlinear systems named Liouvillian (Definition 8.3). The control design is based on sliding-mode uncertainty estimator, developed under the framework of algebraic-differential concepts. The estimation convergence is done via Lyapunov-type analysis and the closed-loop system stability is shown via the regulation error dynamics. Robustness of the proposed control scheme is proven in the face of noise output measurements and model uncertainties. The performance of the proposed control law is illustrated with numerical simulations, when a class of chaotic chemical system is used as application example.

Non-linear approaches to design control laws have been tested successfully in theoretical research. In particular, the Input/Output linearizing technique shows attractive characteristics for the control of the non-linear systems.

To motivate the control problem, consider the following non-linear Liouvillian system, which represents the mathematical model of a continuous stirred tank reactor (CSTR):

$$\begin{aligned}\dot{x}_1 &= \theta(x_{1e} - x_1) - E\, R(x_1, x_2)\\ \dot{x}_2 &= \theta(x_{2e} - x_2) + \Delta H\, R(x_1, x_2) + \gamma(u - x_2)\end{aligned} \tag{8.6}$$

where x_1 is a n-dimensional vector of chemical species, $R(x_1, x_2)$ is a m-dimensional vector of reaction kinetics, ΔH is a m-dimensional vector of reaction enthalpies, E is the stoichiometric matrix, x_2 is the reactor temperature, u is the cooling jacket temperature, $1/\theta$ and γ are the residence time and the heat-transfer global coefficient, respectively. If the reactor temperature x_2 is the controlled output, in compact form, the Liouvillian system (8.6) can be rewritten as follows:

$$\begin{aligned}\dot{x}_1 &= f_1(x_1, x_2)\\ \dot{x}_2 &= f_2(x_1, x_2) + B(x_2)u\\ y &= h(x) = x_2\end{aligned} \tag{8.7}$$

The *zero-dynamics* are given by the n-dimensional dynamics of the chemical species concentration at a constant temperature, which are assumed to be locally stable [24]. The study of relative-degree one systems is very important for many control applications, since the dynamics of a wide class of chemical reactors can be described in this form. Such systems are mathematically modeled as *affine* systems with respect to the control input. Systems that present relative-degree one display some interesting features, such as the *equivalent dissipativeness* by means of state or output feedback. In general, it is easier to stabilize dissipative systems than *non-dissipative* ones [65].

In what follows, non-linear systems of the form (8.7) will be considered. In order to stabilize the system defined by (8.7) via regulation of x_2, the following nominal I/O linearizing feedback control is proposed:

$$u = B^{-1}(x_2)\left[-\tau_g^{-1} e_y - f_2(x_1, x_2)\right] \tag{8.8}$$

where $\tau_g > 0$ is a prescribed time-constant. As usual, $e_y = y - y_{sp}$ and y_{sp} are tracking error and set point, respectively. The controller defined by (8.8) guarantees asymptotic stability of non-linear system (8.7) with no uncertainties and perfect measurements [33]. Moreover, it imposes a linear behavior to the system I/O dynamics by canceling the nonlinearities.

8.3.1 Feedback Controller Design

As it can be noticed, the synthesis of the ideal control law requires accurate knowledge of the mathematical model of the process to be realizable. However, a perfect model is difficult or even impossible to be obtained in practice and, consequently, for uncertain systems a conventional I/O linearizing controller design is not adequate.

Let us assume that trajectories x_1 and x_2 are bounded for all $t > 0$ (i.e., the system is bounded input to bounded output state). The basis of the non-ideal controller design is the nominal control law (8.8). In order to design the practical robust control law, let us propose the following non-linear dynamic system representation:

$$\begin{aligned} \dot{x}_1 &= f_1(x_1,x_2) \\ \dot{x}_2 &= f_2(x_1,x_2) + (\bar{B}(x_2) + \Delta B(x_2))\,u \\ y &= h(x) = x_2 \end{aligned} \tag{8.9}$$

The functions $f_2(x_1,x_2)$ and $\Delta B(x_2)$ are model uncertainties related to the non-linear system, and $\bar{B}(x_2)$ is a nominal value of the control input coefficient. In the most general case, the functions $f_2(x_1,x_2)$ and $\Delta B(x_2)$ are assumed to be unknown. Now, the following function is introduced, which corresponds to the I/O modeling error:

$$\zeta(x,u) = f_2(x_1,x_2) + \Delta B(x_2)u \tag{8.10}$$

By using (8.10) into (8.9), a new representation of the system is obtained:

$$\begin{aligned} \dot{x}_1 &= f_1(x_1,x_2) \\ \dot{x}_2 &= \zeta(x,u) + \bar{B}(x_2)u \\ y &= h(x) = x_2 \end{aligned} \tag{8.11}$$

Since the uncertainty term, $\zeta(x,u)$, is an unknown function of the states and the control input, the ideal control law for the regulation of x_2 is not causal and therefore it can not be implemented in practice. Nevertheless, there is another way to develop an input-output linearizing controller that is robust against uncertainties. The procedure described below provides a method to estimate the uncertainty term, $\zeta(x,u)$. Estimators or observers for states and uncertainties can play a key role during the early detection of hazardous and unsafe operating conditions. Following this spirit, several researches have been focused in the proposition of estimation methodologies for states and uncertainties for monitoring and control purposes [39,63].

In order to estimate the uncertain term $\zeta(x,u)$ let us consider the following dynamic subsystem:

$$\begin{aligned}\dot{x}_2 &= \zeta + \bar{B}(x_2)u \\ \dot{\zeta} &= \Phi(x,u) \\ y &= h(x) = x_2\end{aligned} \tag{8.12}$$

From this subsystem, the following algebraic-differential equation can be obtained:

$$\dot{y} - \zeta(x,u) - \bar{B}(y)u = 0 \tag{8.13}$$

Remark 8.1. From Definition 8.2 it follows that ζ is algebraically observable, however, it is necessary the estimation of $\dot{y}$.

The corresponding Input-Output representation of (8.13) can be rewritten in new coordinates as follows:

$$\eta_i = \frac{\mathrm{d}^{i-1}y}{\mathrm{d}t^{i-1}} \qquad (i = 1,2) \tag{8.14}$$

This implies that

$$\begin{aligned}\dot{\eta}_1 &= \eta_2 \\ \dot{\eta}_2 &= \Phi(\eta_1,\eta_2,\dot{u}) \\ y &= \eta_1\end{aligned} \tag{8.15}$$

It should be noted that a partial change of coordinate enables us to estimate $\eta_1 = y$ and $\dot{\eta}_1 = \dot{y} = \eta_2$ (or, equivalently, x_2 and $\dot{x}_2$).

Now, considering the noise case presence:

$$y = \eta_1 + \delta$$

where δ is an additive bounded noise. Our aim is to design an observer to obtain η_2 (the uncertainty term in the transformed space). However, as it can be seen from the nature of the system given by (8.12), a standard structure of an observer, based on a copy of the system plus measurement error correction is not realizable in this case since the term Φ is a priori unknown.

The following dynamic system is a *sliding-mode asymptotic type observer* of the system (8.15) to estimate the variables η_1 and η_2, respectively:

$$\begin{aligned}\dot{\hat{\eta}}_1 &= \hat{\eta}_2 + m\tau^{-1}\mathrm{sign}(y-\hat{y}), \qquad m > 0, \\ \dot{\hat{\eta}}_2 &= m^2\tau^{-2}\mathrm{sign}(y-\hat{y})\end{aligned} \tag{8.16}$$

where

$$y = \eta_1 + \delta,\ \hat{y} = \hat{\eta}_1,\ \text{sign} := \begin{cases} 1 & \text{if } (y-\hat{y}) > 0 \\ -1 & \text{if } (y-\hat{y}) < 0 \\ \text{undefined} & \text{if } (y-\hat{y}) = 0 \end{cases}$$

Now, returning back to the original state space, in view of (8.10), the heat of reaction can be evaluated as:

$$\zeta = -\hat{\eta}_2 - \theta(x_{2e} - \hat{\eta}_1) - \gamma(u - \hat{\eta}_1)$$

According to the variable change given by (8.11), the variable η_1 is the thermodynamic reactor temperature (system output). From the above equation for $\hat{\zeta}$, if temperature measurements are noisy, the noise would be transmitted to the estimation of the heat of reaction that may lead to poor performance in the estimation procedure. That is because it is necessary to filter the temperature measurements. This is the main reason that the structure of the proposed observer (8.16) makes sense.

Convergence analysis. Let us define the following estimation errors:

$$e_1 = \eta_1 - \hat{\eta}_1 \tag{8.17}$$

$$e_2 = \frac{\eta_2 - \hat{\eta}_2}{m} \tag{8.18}$$

By (8.15) and (8.16), it follows that the estimation errors $e = (e_1, e_2)^T$ verify the following ordinary differential equation:

$$\dot{e} = A_{\bar{\mu}}e - Ksign(Ce + \delta) + \Delta s \tag{8.19}$$

where $\bar{\mu} > 0$ is a regularizing parameter, $A_{\bar{\mu}} = \begin{bmatrix} -\bar{\mu} & m \\ 0 & -\bar{\mu} \end{bmatrix}$, $K = m\tau^{-1}\begin{bmatrix} 1 \\ m\tau^{-1} \end{bmatrix}$, $C = \begin{bmatrix} 1 & 0 \end{bmatrix}$ and $\Delta s = \begin{bmatrix} \bar{\mu}e_1 \\ \frac{1}{m}\Phi + \bar{\mu}e_2 \end{bmatrix}$ is an uncertainty term (or unmodelled dynamics term).

Assumption 8.1. There exist nonnegative constants L_{0s}, L_{1s}, such that the following generalized quasi-Lipschitz condition holds

$$\|\Delta s\| \leq L_{0s} + (L_{1s} + \|A_{\bar{\mu}}\|)\|e\|. \tag{8.20}$$

Assumption 8.2. The additive output noise δ, is bounded, namely

$$|\delta| \leq \delta^+ < \infty, \tag{8.21}$$

Assumption 8.3. There exists a positive definite matrix $Q_0 = Q_0^T > 0$, such that the following matrix Riccati equation

$$PA_{\bar{\mu}} + A_{\bar{\mu}}^T P + PRP + Q = 0 \tag{8.22}$$

with

$$R := \Lambda_s^{-1} + 2\|\Lambda_s\| L_{1s} I,\ \Lambda_s = \Lambda_s^T > 0,$$

$$Q = Q_0 + 2(L_{1s} + \|A_{\bar{\mu}}\|^2) I$$

has a positive definite solution $P = P^T > 0$.

Remark 8.2. Assumption 8.1 only limits the maximum slope present in the uncertainty term Δs which depends on the Lipschitz properties of η_2. Assumption 8.2 is a standard assumption that allows us to avoid involving the statistic behavior of the noise signal. The expression (8.22) from Assumption 8.3 has a positive definite solution if the matrix $A_{\bar{\mu}}$ is stable, which is true for any $\bar{\mu} > 0$. Since $P > 0$, there exists $k > 0$ such that $K = kP^{-1}C^T$, then Assumption 8.3 provides an additional degree of freedom to choose the gain k which can be used to establish the size of the region defined by $\tilde{\mu}$.

Theorem 8.1. *If Assumptions 8.1 to 8.3 are satisfied, then*

$$[V - V^*]_+ \to 0 \tag{8.23}$$

where

$$V = V(e) = \|e\|_P^2 := e^T Pe,$$

$$V^* = \frac{2\|\Lambda_s\| L_{0s}^2 + 4k\delta^+}{\lambda_{\min}(P^{-1/2} Q^T Q P^{-1/2})},$$

and the function $[\cdot]_+$ *is defined as follows*

$$[x]_+ = \begin{cases} x \text{ if } x \geq 0 \\ 0 \text{ if } x < 0 \end{cases}. \tag{8.24}$$

Proof. Let $V(e)$ be the following Lyapunov candidate function

$$V(e) \triangleq e^T Pe = \|e\|_P^2 \tag{8.25}$$

where $0 < P = P^T \in R^{r\times r}$ is the solution of the Riccati equation (8.22). By taking the time derivative of (8.25) and taking into account (8.19) it yields

$$\dot V(e) = 2e^T P\dot e = 2e^T P\left[A_{\bar\mu} e - K\text{sign}(Ce+\delta) + \Delta s\right],$$

according to Assumption 8.3, $K = kP^{-1}C^T$, then the previous equation can be written as

$$\dot V(e) = 2e^T PA_{\bar\mu} e - 2ke^T C^T \text{sign}(Ce+\delta) + 2e^T P\Delta s.$$

By using the following matrix inequality

$$X^T Y + Y^T X \le X^T \Lambda_s X + Y^T \Lambda_s^{-1} Y$$

which is valid for any $X, Y \in R^{r\times m}$, $0 < \Lambda_s = \Lambda_s^T \in R^{r\times r}$ [60], then it follows that

$$\dot V(e) \le e^T (PA_{\bar\mu} + A_{\bar\mu}^T P)e - 2ke^T C^T \text{sign}(Ce+\delta) + e^T P\Lambda_s^{-1} Pe + (\Delta s)^T \Lambda_s \Delta_s,$$

from Assumption 8.1 the following is obtained

$$(\Delta s)^T \Lambda_s \Delta_s \le \|\Delta s\|^2 \|\Lambda_s\| \le \left[L_{0s} + (L_{1s} + \|A_{\bar\mu}\|)\|e\|\right]^2 \|\Lambda_s\|,$$

then

$$\begin{aligned}\dot V(e) \le\ & e^T (PA_{\bar\mu} + A_{\bar\mu}^T P + P\Lambda_s^{-1} P + Q)e - e^T Qe \\ & -2ke^T C^T \text{sign}(Ce+\delta) + 2\left[L_{0s}^2 + (L_{1s} + \|A_{\bar\mu}\|)^2 \|e\|^2\right] \|\Lambda_s\|,\end{aligned}$$

from the definition of matrix R in Assumption 8.3, the previous expression can be rewritten as

$$\dot V(e) \le e^T (PA_{\bar\mu} + A_{\bar\mu}^T P + PRP + Q)e - e^T Qe - 2ke^T C^T \text{sign}(Ce+\delta) + 2L_{0s}^2 \|\Lambda_s\|,$$

and taking into account (8.22), it follows that

$$\dot V(e) \le -e^T Qe - 2ke^T C^T \text{sign}(Ce+\delta) + 2L_{0s}^2 \|\Lambda_s\|. \tag{8.26}$$

In order to eliminate the discontinuity contained in the function sign$(\cdot)$ in (8.26) the following inequality, valid for any $x, y \in \mathbb{R}$, is considered

$$\begin{aligned} x\text{sign}(x+y) &= (x+y)\text{sign}(x+y) - y\text{sign}(x+y) \\ &\ge |x+y| - |y| \end{aligned}$$

and furthermore $|x+y| \ge |x| - |y|$, then

$$x\text{sign}(x+y) \ge |x| - 2|y|. \tag{8.27}$$

Now using (8.27) in (8.26) the following inequality is obtained

$$\dot{V}(e) \leq -e^T Qe - 2k|Ce| + 2L_{0s}^2 \|\Lambda_s\| + 4k\delta^+, \tag{8.28}$$

which can be rewritten as

$$\dot{V}(e) \leq -\|e\|_Q^2 + 2L_{0s}^2 \|\Lambda_s\| + 4k\delta^+,$$

that is to say,

$$\dot{V}(e) \leq -\alpha_Q V(e) + \beta, \tag{8.29}$$

where

$$\alpha \triangleq \lambda_{\min}\left(P^{-1/2}Q^T QP^{-1/2}\right) > 0,$$

$$\beta = 2L_{0s}^2 \|\Lambda_s\| + 4k\delta^+.$$

Now, considering the following differential equation related to (8.29)

$$\dot{V}(e) = -\alpha V + \beta, \tag{8.30}$$

which is linear and stable and such that $V \to V^*$ as $t \to \infty$, where V^* is the single equilibrium point of equation (8.30)

$$V^* = \frac{\beta}{\alpha} \geq 0,$$

it follows that the function

$$G_t \triangleq [V - V^*]_+^2$$

where $[\cdot]_+$ is defined as in (8.24), according to (8.29) satisfies (for any $V \neq V^*$)

$$\dot{G}_t \leq -2[V - V^*]_+ [-\alpha V + \beta] \leq 0$$

subtracting $-\alpha V^* + \beta = 0$, it yields

$$\dot{G}_t \leq -2\alpha(V - V^*)[V - V^*]_+ \leq 0$$

that is to say,

$$\dot{G}_t \leq -2\alpha G_t \leq 0.$$

Integrating the last inequality it follows that

$$G_t - G_0 \leq -2\alpha \int_0^t G_\tau d\tau,$$

in other words

$$2\alpha\int_0^t G_\tau d\tau \le G_0 - G_t \le G_0$$

then

$$\lim_{t\to\infty} 2\alpha\int_0^t G_\tau d\tau \le G_0.$$

From Barbalat Lemma [38], it follows that $G_t \to 0$, which is equivalent to say that $[V - V^*]_+ \to 0$. □

Remark 8.3. Theorem 8.1 states that the weighted estimation error norm $V(e)$ asymptotically converges to the zone bounded by β/α. In other words, it is ultimately bounded.

The final expression for the input-output non-ideal linearizing controller with uncertainty estimation can be obtained, introducing the estimate of the uncertain term in (8.8), to generate:

$$u = B^{-1}(x_2)\left[-\tau_g^{-1}e_y - \hat{\zeta}\right] \tag{8.31}$$

Since the proposed controller uses estimated values of the uncertainty, it cannot cancel the system nonlinearities completely. Thus, the system trajectory remains inside a neighborhood close to the set point. Practical stability is achieved as long as the uncertainty estimation error is bounded. The restraint of the boundedness of the heat of reaction (uncertain term) is common for a wide class of chemical reactions and is consequence of characteristics of the mathematical modeling commonly employed; chemical reactions are usually Lipschitz with respect to temperature. It is not hard to see that global Lipschitz property of $\Delta H\ R(x_1,x_2)$ is found if the functionality $R(x_1,x_2)$ with respect to temperature is of Arrhenius-type [49].

Notice that it is not hard to implement in standard technology (e.g., PLCs) the practical controller given by Eqs. (8.16) to (8.18). In fact, the implementation only requires output measurements and the on-line solution of the dynamical system (8.16). Moreover, the implementation effort is equivalent to other control strategies, such as PI and predictive control. As a matter of fact, standard (industrial) predictive control is more complex than the proposed one, since the former requires implementation of a non-linear optimization method.

In order to analyze the closed-loop stability of the reactor temperature trajectories in the reactor, the closed-loop dynamic equation of the energy balance should be used.

$$e_y = g\, e_y + \left(\zeta - \hat{\zeta}\right) \tag{8.32}$$

If $\zeta \to \hat{\zeta}$ then $\zeta - \hat{\zeta} \to 0$, the ideal control law is recovered together with its stability properties; otherwise, the estimation error is limited as $\|\zeta - \hat{\zeta}\| \le \alpha\sqrt{\Omega} = \Pi$, accordingly with the above development.

Assumption 8.4. If $\lambda_1, \lambda_2, \ldots, \lambda_k$ are the distinct eigenvalues of the matrix A, where λ_j has multiplicity n_j and $n_1 + n_2 + \ldots + n_k = n$ and ρ is any number larger than the real part of $\lambda_1, \lambda_2, \ldots, \lambda_k$ that is $\rho > max(\mathscr{R}e(\lambda_j))$, then there exists a constant $j > 0$ that satisfies:

$$\|\exp(mAt)e_y\| \leq j\exp(-m\rho t)\|e_y\|$$

Solving (8.17), the error can be expressed as:

$$e_y = \exp(mAt)e_{y0} + \int_0^t \exp\left[mA(t-s)\right](\zeta - \hat{\zeta})\mathrm{d}s \tag{8.33}$$

Considering the Assumptions 8.1 and 8.2, it is possible to find a bound for (8.33),

$$\|e_y\| \leq j\exp(-m\rho t)\left[\|e_{y0}\| - \frac{j\Pi}{m^2\rho}\right] + \frac{j\Pi}{m^2\rho} \tag{8.34}$$

Taking the limit when $t \to \infty$:

$$\|e_y\| \leq \frac{j\Pi}{m^2\rho} \tag{8.35}$$

The above inequality implies that the closed-loop error can be made as small as desired, if the observer parameter m is chosen large enough.

8.3.2 Application Example

The chemical reactor proposed as application example has been studied previously in [27, 34], the reactors model shows periodic or even chaotic dynamic behavior depending of the set of parameters employed [27]. The reactor temperature is regulated by means of water flowing through a cooling jacket. A stream with a reactive A enters into the continuous reactor and it is converted to an intermediate product B, which reacts to transforming to the final product C, such that

$$A \to B, \quad rate_1 = k_1 a$$
$$B \to C, \quad rate_2 = k_2 b$$

A description of symbols introduced in this section is given in Table 8.1. Both two reactions are first order process with exothermic chemical reactions and the kinetic constant is modeled by the classical Arrhenius model to include the temperature dependence, as follows:

$$k_i = A_i \exp\left(-\frac{E_{ai}}{RT}\right), \quad \text{for} \quad i = 1, 2.$$

Table 8.1 Notation

Symbol	Description	Symbol	Description
a	Reactant A concentration	α	Dimensionless concentration of reactant A
a_0	Concentration of A in feed	A_c	Effective jacket heat transfer area
b	Reactant B concentration	C_p	Specific heat of reacting mixture
b_0	Concentration of B in feed	E_i	Activation energy for reaction i
k_i	Rate constant for reaction i	U	Overall heat transfer coefficient
t_N	Newtonian cooling time	β	Dimensionless concentration of reactant B
t_{res}	Mean residence time	θ_c	Dimensionless temperature of cooling water
T	Reactor temperature	θ_{sp}	Set point of reactor dimensionless temperature
T_c	Cooling water temperature	θ	Dimensionless temperature
V	Reactor volume	ρ	Density of reacting mixture
ΔH_i	Heat of i-th reaction	τ	Dimensionless time
ϕ	Arrhenius constants ratio	τ_N	Dimensionless Newtonian cooling time
t	Time	τ_{ch}	Dimensionless chemical time

Via the standard mass and energy balances the following reactor's governing equations are presented:

$$\begin{aligned}
\frac{da}{dt} &= \frac{1}{t_{res}}(a_0 - a) - k_1 a \\
\frac{db}{dt} &= \frac{1}{t_{res}}(b_0 - b) - k_1 a - k_2 b \\
C_p\rho\frac{dT}{dt} &= \frac{1}{t_{res}}C_p\rho(T_0 - T) + (-\Delta H_1)k_1 a + (-\Delta H_2)k_2 b - \frac{1}{V}UA_c(T - T_c)
\end{aligned} \tag{8.36}$$

In accordance with [27] the following condition of the reactor's model are imposed; There is not inflow of the compounds B and C, two both reactions have the same reaction heats, the same activation energy and the inflow reactor temperature is the same as the cooling jacket device. Based on the model structure proposed above, the following system representation is done:

$$\begin{bmatrix} \dot{x}_1 \\ \dot{x}_2 \\ \dot{x}_3 \end{bmatrix} = \begin{bmatrix} \zeta_1(x) \\ \zeta_2(x) \\ \zeta_3(x) \end{bmatrix} + \begin{bmatrix} l_1 & 0 & 0 \\ 0 & l_2 & 0 \\ 0 & 0 & l_3 \end{bmatrix} \begin{bmatrix} u_1 \\ u_2 \\ u_3 \end{bmatrix} \tag{8.37}$$

where:

$$l_1 = \frac{1}{t_{res}}(a_0 - a), \qquad l_2 = \frac{1}{t_{res}}(b_0 - b), \qquad l_3 = \frac{1}{V}UA_c,$$

$$\begin{bmatrix} \zeta_1(x) \\ \zeta_2(x) \\ \zeta_3(x) \end{bmatrix} = \begin{bmatrix} k_1 a \\ k_1 a - k_2 b \\ \frac{1}{t_{res}} C_p \rho (T_0 - T) + (-\Delta H_1) k_1 a + (-\Delta H_2) k_2 b - \frac{1}{V} U A_c T \end{bmatrix},$$

$$x = \begin{bmatrix} x_1 \\ x_2 \\ x_3 \end{bmatrix} = \begin{bmatrix} a \\ b \\ T \end{bmatrix}, \quad \text{and} \quad \begin{bmatrix} u_1 \\ u_2 \\ u_3 \end{bmatrix} = \begin{bmatrix} 0 \\ 0 \\ T_c \end{bmatrix}$$

Now, applying the mean residence time and the reactive A concentration as the time a concentration scales, it is obtained the following set of dimensionless mass and energy balance equations

$$\begin{aligned} \frac{d\alpha}{d\tau} &= 1 - \alpha - \frac{1}{\tau_{ch}} \alpha \exp(\theta) \\ \frac{d\beta}{d\tau} &= \frac{1}{\tau_{ch}} \alpha \exp(\theta) - \frac{1}{\tau_{ch}} \phi \beta \exp(\theta) - \beta \\ \frac{d\theta}{d\tau} &= \frac{1}{\tau_{ch}} \theta_c \alpha \exp(\theta) + \frac{1}{\tau_{ch}} \theta_c \phi \beta \exp(\theta) - (1 + \tau_N^{-1})\theta \end{aligned} \tag{8.38}$$

The corresponding dimensionless concentrations and temperature are the follows

$$\alpha = \frac{a}{a_0}; \ \theta = \frac{E_a (T - T_c)}{R T_c^2}; \ \beta = \frac{b}{a_0}; \ \tau = \frac{t}{t_{res}}.$$

The parameters related with the named *chemical* time, dimensionless temperature of the cooling jacket and the Newtonian cooling time, are respectively

$$\tau_{ch} = \frac{1}{k_1 t_{res}}; \ \theta_c = -\frac{\Delta H_1 a_0 E_{a1}}{C_p \rho R T_c^2}; \ \tau_N = \frac{t_N}{t_{res}} = \frac{C_p \rho V}{U A_c t_{res}}, \ \phi = \frac{A_2}{A_1}.$$

An important characteristic of this reactor's model is its minimum phase behavior, i.e. the corresponding internal dynamic when the temperature of the reactor is regulated is stable.

As mentioned above, given that the controller regulates only x_3 the analysis of the inner dynamics is related with closed-loop behavior of x_1 and x_2 while x_3 is kept constant. Therefore the system (8.38) is reduced to:

$$\begin{aligned} \dot{x}_1^* &= 1 - (1 + \delta_1)_1^* \\ \dot{x}_2^* &= \delta_1 x_1^* - (1 + \delta_2)_2^* \end{aligned} \tag{8.39}$$

where:

$$\delta_1 = \frac{\exp(\theta_{sp})}{\tau_{ch}}, \qquad \delta_2 = \frac{\phi \exp(\theta_{sp})}{\tau_{ch}}$$

Now, solving (8.39) we have

$$x_1^* = \left[x_{10}^* - \frac{1}{1+\delta_1}\right]\exp(-\{1+\delta_1\}\tau) + \frac{1}{1+\delta_1}$$
$$x_2^* = [x_{20}^* - (\delta_3+\delta_4)]\exp(-\{1+\delta_1\}\tau) + \delta_3\exp(-\{1+\delta_1\}\tau) + \delta_4 \tag{8.40}$$

with

$$\delta_3 = \left(\frac{1}{\phi-1}\right)\left(x_{10}^* - \frac{1}{1+\exp(\theta_{sp})/\tau_{ch}}\right),$$
$$\delta_4 = \frac{\exp(\theta_{sp})/\tau_{ch}}{[1+\exp(\theta_{sp})/\tau_{ch}]\,[1+\phi\exp(\theta_{sp})/\tau_{ch}]}$$

From (8.40) the reactor's inner dynamic is asymptotically stable such that:

$$\lim_{\tau\to\infty} x_1^* = \frac{1}{1+\exp(\theta_{sp})/\tau_{ch}} = \alpha_{eq}$$
$$\lim_{\tau\to\infty} x_2^* = \frac{\exp(\theta_{sp})/\tau_{ch}}{[1+\exp(\theta_{sp})/\tau_{ch}]\,[1+\phi\exp(\theta_{sp})/\tau_{ch}]} = \beta_{eq}$$

Numerical simulations for the closed-loop system were performed in order to show the properties of the control scheme proposed. The set of parameters of the chemical reactor are chosen as in [34], and initial conditions of system (8.37) are $x_1 = 0.45$, $x_2 = 0.1$, and $x_3 = 0.9$. For comparison purposes, an ideal I/O linearizing control, standard sliding-mode and a high order sliding-mode controllers are implemented too. The temperature set point is $\theta_{sp} = 3$, and the nominal value of the control input is $u_0 = 17.5$, the controller is tuned-on at $t = 15$, the order of the high order sliding-mode controller is considered as $p = 3$. The temperature measurements are corrupted with a $\pm 5\%$ around the current temperature value.

Figure 8.1 shows the open-loop behavior of the corresponding space portrait; note that oscillatory behavior of the corresponding trajectories makes the open-loop mode inadequate for industrial operation.

Closed loop performance of temperature trajectories show that the ideal I/O linearizing controller shows the better performance, so that it cancels the nonlinearities, imposing a desired linear behavior, with a satisfactory performance. The proposed controller tries to compensate the nonlinear terms via the integral high order sliding-mode contribution, besides it is able to reach the set point value required (Fig. 8.2), exhibiting smaller oscillations around the regulated point ($\theta_{sp} = 3$) than the other sliding-mode controllers. As predicted by the theoretical frame presented; sliding-mode and high order sliding-mode controllers can suppress nonlinear oscillations; however both controllers exhibit a considerable off-set from the corresponding set point. The estimation of the uncertain term is depicted in Fig. 8.3.

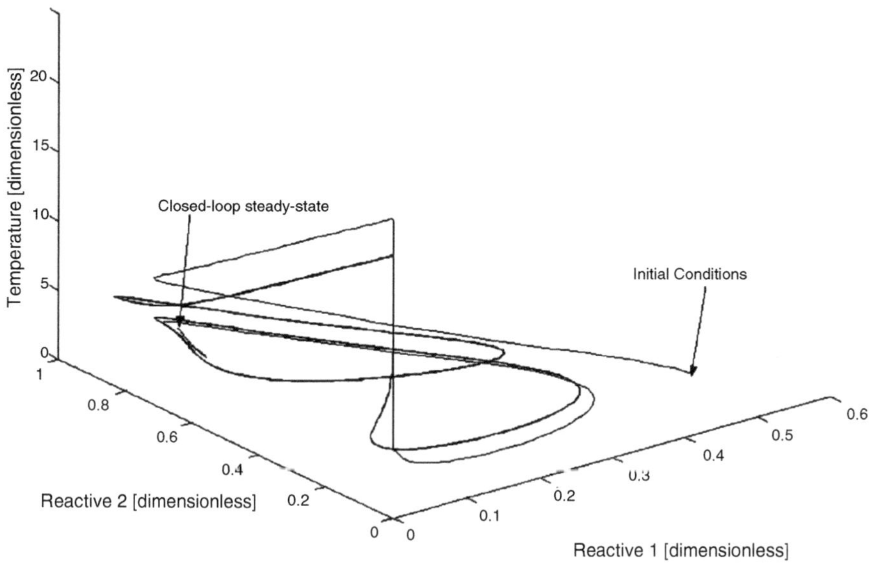

Fig. 8.1 Open-loop space portrait

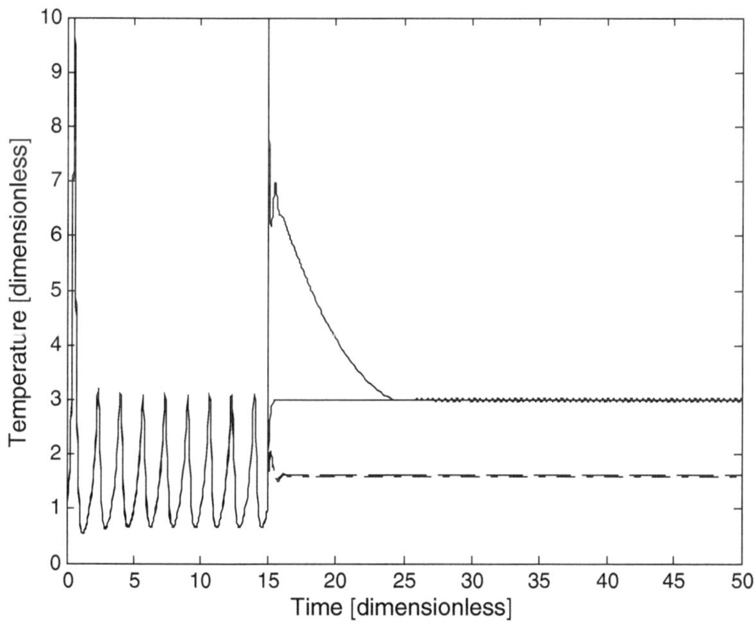

Fig. 8.2 Temperature synchronization. (*Dashed lines*: Proposed controller; *Dotted lines*: Ideal I/O linearizing controller); *Solid lines*: High-order sliding-mode controller; *Dashed dotted lines*: Sliding mode controller

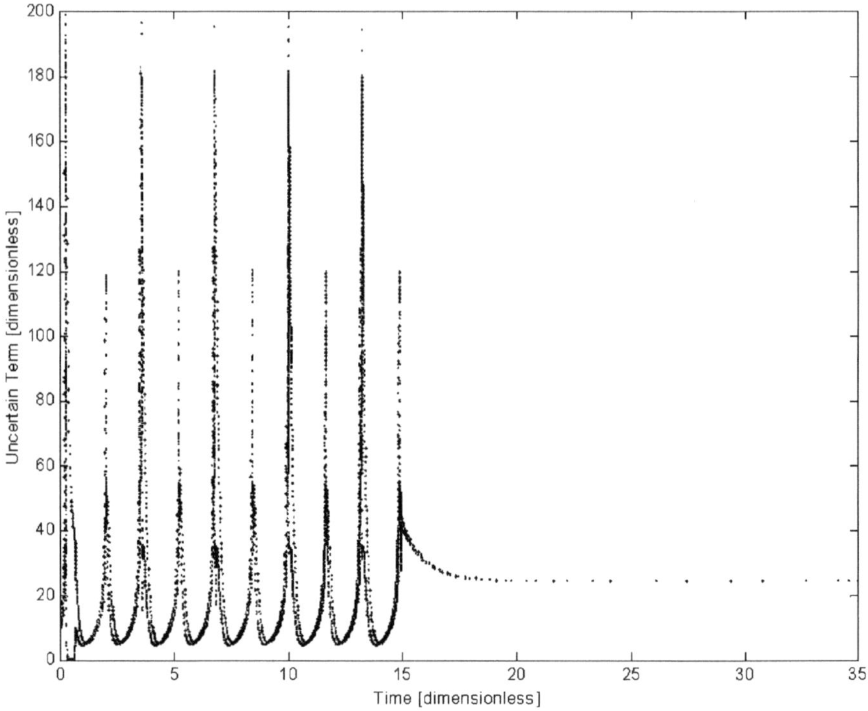

Fig. 8.3 Estimation of the uncertainty

Another important difference is that effort performed by the manipulate variable (Fig. 8.4) is very different for each controller. As it can be noticed, the I/O linearizing controller posses the best performance, sliding-mode controller exhibits the second smoothest behavior, followed by high-order sliding-mode control, which exhibits more demanding effort at the start up of the regulation task. Finally, the proposed methodology presents the more demanding control action, where small oscillations are present.

Comparing performance and control effort it is possible to note that high-order sliding-mode control is not very efficient because the regulated variable, temperature, exhibits the largest off-set, even when the effort is higher than in the case of the sliding-mode controller; nevertheless, performance of the sliding-mode controller is not satisfactory because the set point is not reached. At the expense of higher control effort, the I/O linearizing controller and the proposed controller are able to reach the set point; this is not a disadvantage because this controller is able to force to the temperature trajectory through the desired value. This is a great advantage, especially because this controller is not considering the model of the process.

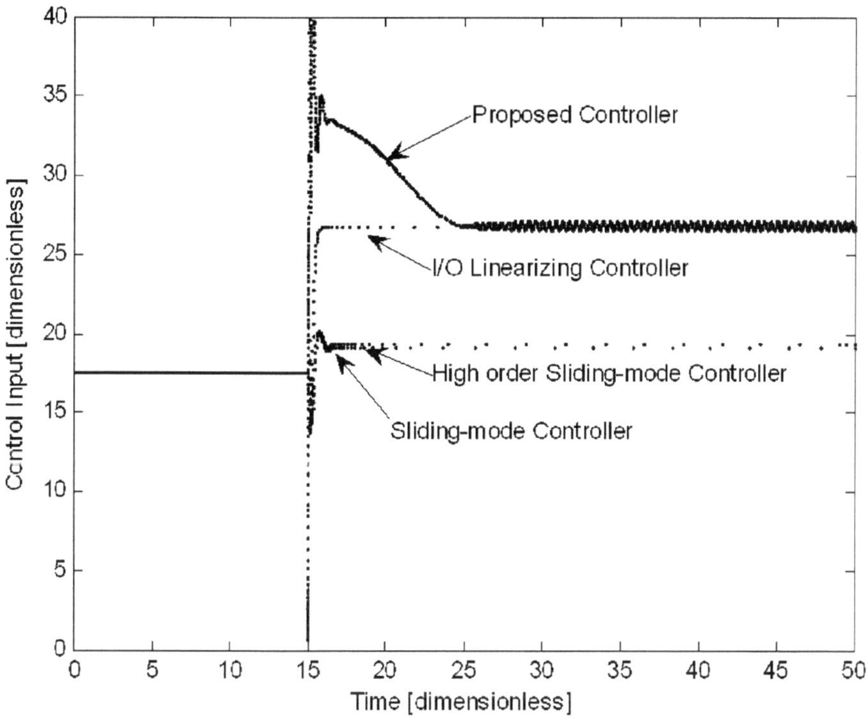

Fig. 8.4 Practical effort of the controllers

It is important to mention that the value of the control gains, has to be chosen very carefully; other numerical simulations (not presented here) showed that, sometimes (smaller values of control gains), it is not possible to stabilize the oscillatory behavior of the chemical reactor; whereas for large values of these parameters, it is possible to lead to unacceptable control efforts or, even worse, to provoke additional closed-loop instabilities.

8.4 Synchronization of Chaotic Liouvillian Oscillators

This section deals with the synchronization and parameter estimation of the Colpitts oscillator considered as a *Chaotic Liouvillian System* (CLS) in a real-time implementation. Even though the algebraic approach has been applied to the synchronization problem for almost one decade [1,47], there are not reported works containing real time applications based on this theoretical framework. A polynomial observer is used for synchronizing the Colpitts oscillator. A comparison with a reduced order observer is given to asses the performance of the proposed observer.

8.4.1 Exponential Polynomial Observer

In this section we solve the synchronization problem by using a polynomial observer based upon the Lyapunov method [38]. To this end, we first compute the dynamics of the synchronization error (difference between the master and the slave systems). Next, by means of a simple quadratic Lyapunov function, we prove the exponential convergence. The system (8.1) can be expressed in the following form,

$$\begin{aligned} \dot{x} &= Ax + \Psi(x,u) \\ y &= Cx \qquad x_0 = x(t_0) \end{aligned} \tag{8.41}$$

where $\Psi(x,u)$ is a nonlinear vector that satisfies the Lipschitz condition with constant φ, that is:

$$\|\Psi(x,u) - \Psi(\widehat{x},u)\| \leq \varphi \|x - \widehat{x}\| \tag{8.42}$$

The observer structure. The observer for system (8.41) has the next form

$$\dot{\widehat{x}} = A\widehat{x} + \Psi(\widehat{x},u) + \sum_{i=1}^{m} K_i \left(y - C\widehat{x}\right)^{2i-1} \tag{8.43}$$

where $\hat{x} \in \mathbb{R}^n$, and $K_i \in \mathbb{R}^n$, for $1 \leq i \leq m$.

Remark 8.4. The meaning of m can be understood as follows. As it is well known, an Extended Luenberger observer can be seen as a first order Taylor series around the observed state, therefore to improve the estimation performance high order terms are included in the observer structure. In other words, the rate of convergence can be increased by injecting additional terms with increasing powers of the output error.

Let us consider the following assumptions:

Assumption 8.5. For $\overline{A} := A - K_1C$, there exist a unique symmetric positive definite matrix $P \in \mathbb{R}^{n\times n}$ which satisfies the following linear matrix inequality (LMI)

$$\begin{bmatrix} -\overline{A}^T P - P\overline{A} - I & \varphi P \\ \varphi P & I \end{bmatrix} > 0, \qquad \text{where } \varphi \text{ is the Lipschitz constant.}$$

Assumption 8.6. Let us define $M_i := PK_iC$, then

$$\lambda_{\min}\left(M_i + M_i^T\right) \geq 0, \quad \text{for } i \in \{2,\ldots,m\}.$$

Remark 8.5. By using Schur complement (see Chap. 11 in [60]) the LMI in Assumption 8.5 can be represented as an algebraic Riccati equation:

$$\overline{A}^T P + P\overline{A} + \varphi^2 PP + I < 0,$$

or for some $\varepsilon > 0$

$$\overline{A}^T P + P\overline{A} + \varphi^2 PP + I + \varepsilon I = 0.$$

Remark 8.6. Assumption 8.6 is used to improve the rate of convergence of the estimation error by injecting additional terms (from 2 to m) which depend upon odd powers of the output error.

Observer Convergence Analysis. In order to prove the observer convergence, we analyze the observer error which is defined as $e = x - \hat{x}$. From Eqs. (8.41) and (8.43), the dynamics of the observer is given by

$$\dot{e} = \overline{A}e + F - \sum_{i=2}^{m} K_i (Ce)^{2i-1}$$

where $\overline{A} := A - K_1 C$, and $F := \Psi(x,u) - \Psi(\widehat{x},u)$.

Now, we present a lemma which will be useful in the convergence analysis.

Lemma 8.1 ([61]). *Given the system (8.41) and its observer (8.43), with the error given by* $e := x - \hat{x}$. *If* $P = P^T > 0$ *then:*

$$2\, e^T P \left[\Psi(x,u) - \Psi(\hat{x},u)\right] \leq \varphi^2 e^T PP\, e + e^T e \qquad \square$$

The following proposition proves the observer convergence.

Proposition 8.1. *Let the system (8.41) be algebraically observable and Assumption 8.1 and Assumption 8.2 hold. The nonlinear system (8.43) is an exponential polynomial observer of the system (8.41); that is to say, there exist constants* $\kappa > 0$ *and* $\xi > 0$ *such that*

$$\|e(t)\| \leq \kappa\, exp(-\xi t)$$

where $\kappa = \dfrac{\|e_0\|_P}{\sqrt{\alpha}}$, $\xi = \dfrac{\varepsilon}{2\beta}$, $\alpha = \lambda_{min}(P)$, *and* $\beta = \lambda_{max}(P)$.

Proof. We use the following Lyapunov function candidate $V = e^T Pe$,

$$\Rightarrow \dot{V} = \dot{e}^T Pe + e^T P\dot{e} = e^T \left[\overline{A}^T P + P\overline{A}\right] e + 2e^T PF - 2e^T P \sum_{i=2}^{m} K_i (Ce)^{2i-1}$$

Using Lemma 8.1 we obtain,

$$\dot{V} \leq e^T \left[\overline{A}^T P + P\overline{A} + \varphi^2 PP + I\right] e - 2e^T P \sum_{i=2}^{m} K_i (Ce)^{2i-1}$$

Making some algebraic manipulations on the last term of the above inequality, and taking into account that $Ce \in \mathbb{R}$, we obtain,

$$\dot{V} \leq e^T \left[\overline{A}^T P + P\overline{A} + \varphi^2 PP + I\right] e - 2\sum_{i=2}^{m} (Ce)^{2i-2} e^T PK_iCe$$

For simplicity, we define $M_i = PK_iC$, $i \in \{2,\ldots,m\}$, then we have

$$\dot{V} \leq e^T \left[\overline{A}^T P + P\overline{A} + \varphi^2 PP + I\right] e - \Big\{(Ce)^2 \left[e^T M_2 e + (e^T M_2 e)^T\right] +$$
$$+ (Ce)^4 \left[e^T M_3 e + (e^T M_3 e)^T\right] + \ \ldots + (Ce)^{2m-2} \left[e^T M_m e + (e^T M_m e)^T\right]\Big\}$$

Above expression can be rewritten in a simplified form

$$\dot{V} \leq e^T \left[\overline{A}^T P + P\overline{A} + \varphi^2 PP + I\right] e - \sum_{i=2}^{m} (Ce)^{2i-2} e^T (M_i + M_i^T) e$$

From Assumption 8.2, the second term in the right hand side of the above inequality always will be positive or zero, therefore

$$\dot{V} \leq e^T \left[\overline{A}^T P + P\overline{A} + \varphi^2 PP + I\right] e \tag{8.44}$$

By Assumption 1 (and Remark 8.5), we have

$$\dot{V} \leq -\varepsilon \|e\|^2 \tag{8.45}$$

We write the Lyapunov function as $V = \|e\|_P^2$, then by Rayleigh-Ritz inequality we have that

$$\alpha \|e\|^2 \leq \|e\|_P^2 \leq \beta \|e\|^2 \tag{8.46}$$

where $\alpha := \lambda_{min}(P)$, and $\beta := \lambda_{max}(P) \in \mathbb{R}^+$ (because P is positive definite).
By using (8.46) we obtain the following upper bound of (8.45)

$$\dot{V} \leq -\frac{\varepsilon}{\beta} \|e\|_P^2 \tag{8.47}$$

Taking the time derivative of $V = \|e\|_P^2$ and replacing in inequality (8.47), we obtain

$$\frac{d}{dt} \|e\|_P \leq -\frac{\varepsilon}{2\beta} \|e\|_P$$

Finally, the result follows with

$$\|e(t)\| \leq \kappa \exp(-\xi\, t) \tag{8.48}$$

where $\kappa = \|e_0\|_P / \sqrt{\alpha}$, and $\xi = \varepsilon / 2\beta$. □

8.4.2 Asymptotic Reduced Order Observer

Now, let us consider the nonlinear system described by (8.1). The unknown states of the system can be included in a new variable $\eta(t)$ and the following new augmented system is considered

$$\begin{aligned}\dot{x}(t) &= f(x,u,\eta)\\ \dot{\eta}(t) &= \Delta(x,u)\\ y(t) &= h(x)\end{aligned} \tag{8.49}$$

where $\Delta(x,u)$ is a bounded uncertain function. The problem is to reconstruct the variable $\eta(t)$. This problem is overcome by using a reduced order observer [47]. Before proposing the corresponding observer for reconstructing the variable $\eta(t)$ we introduce some hypotheses:

Assumption 8.7. $\eta(t)$ satisfies the AOC (Definition 8.2).

Assumption 8.8. γ is a C^1 real-valued function.

Assumption 8.9. Δ is bounded, i.e., $|\Delta| \leq M < \infty$.

Assumption 8.10. For t_0, sufficiently large, there exists $K > 0$, such that, $\limsup_{t\to t_0} \frac{M}{K} = 0$.

Next Lemma describes the design of a proportional reduced order observer for system (8.49).

Lemma 8.2 ([47]). *If Assumptions 8.7 to 8.10 are satisfied, then the system*

$$\dot{\hat{\eta}} = K(\eta - \hat{\eta}) \tag{8.50}$$

is an asymptotic reduced order observer of free-model type for system (8.49), where $\hat{\eta}$ denotes the estimate of η and $K \in \mathbb{R}^+$ determines the desired convergence rate of the observer. □

Remark 8.7. To reconstruct $\eta(t)$ by using an auxiliary state $\hat{\eta}(t)$ sometimes we need to use the output time derivatives, but these may be unavailable. To overcome this fact, an auxiliary function completely artificial γ is defined in such away that it cancels out all nonmeasurable terms. This action defines a differential equation for γ. This equation is solved, then, γ is substituted in the differential equation of the estimated state and finally the estimate of η is obtained.

We give the following immediate corollary.

Corollary 8.1 ([47]). *The dynamic system (8.50) along with*

$$\dot{\gamma} = \psi(x,u,\gamma), \quad \textit{with} \quad \gamma_0 = \gamma(0) \quad \textit{and} \quad \gamma \in C^1$$

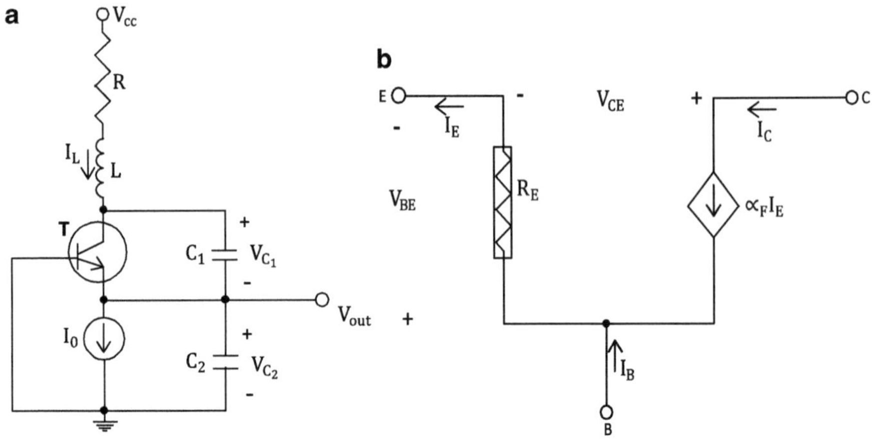

Fig. 8.5 Colpitts oscillator. (**a**) Circuit configuration. (**b**) Model of the Bipolar Junction Transistor (BJT)

constitute a proportional asymptotic reduced order observer for system (8.49), where γ is a change of variable which depends on the estimated state $\hat{\eta}$, and the state variables. □

8.4.3 Experimental Results

These proposals are applied to a Colpitts oscillator [37]. The Colpitts oscillator has been widely considered for the synchronization problem, see for instance [18, 28]. We can also mention one previous work with the Colpitts system using the differential algebraic approach [51], in that work the authors only report a numerical simulation study and not a real-time experiment. The intention of choosing the Colpitts system example is to clarify the proposed methodology and to highlight the simplicity and flexibility of the present approach.

In this work we consider the classical configuration of the Colpitts oscillator [45]. The circuit contains a bipolar junction transistor 2N2222A as the gain element (Fig. 8.5b), and a resonant network consisting of an inductor and two capacitors (Fig. 8.5a).

The Colpitts circuit is described by a system of three nonlinear differential equations, as follows:

$$\begin{aligned} C_1\dot{V}_{C_1} &= -f(V_{C_2}) + I_L \\ C_2\dot{V}_{C_2} &= I_L - I_0, \\ L\dot{I}_L &= -V_{C_1} - V_{C_2} - RI_L + V_{CC} \end{aligned} \tag{8.51}$$

where $f(\cdot)$ is the driving–point characteristic of the nonlinear resistor. This can be expressed in the form $I_E = f(V_{C_2}) = f(-V_{BE})$. In particular, we have $f(V_{C_2}) = I_S \exp(-V_{C_2}/V_T)$.

We introduce the dimensionless state variables (x_1, x_2, x_3), and we choose the operating point of (8.51) to be the origin of the new coordinate system. In particular, we normalize voltages, currents and time with respect to $V_{ref} = V_T$, $I_{ref} = I_0$ and $t_{ref} = 1/w_0$, respectively, where $w_0 = 1/\sqrt{LC_1C_2/(C_1+C_2)}$, is the resonant frequency of the unloaded *L–C* tank circuit. Then, the state equations for the Colpitts oscillator can be rewritten in the next form:

$$\begin{aligned}\dot{x}_1 &= -a\ \exp(-x_2) + ax_3 + a\\ \dot{x}_2 &= bx_3\\ \dot{x}_3 &= -cx_1 - cx_2 - dx_3\end{aligned} \tag{8.52}$$

where, $a = b\frac{C_2}{C_1}$, $b = \frac{I_0}{w_0 C_2 V_T}$, $C = \frac{V_T}{w_0 L I_0}$, $d = \frac{R}{Lw_0}$.

According to Definition 8.2, it is evident that system (8.52) is algebraically observable with respect to the output $y = x_2$, since the unknown states x_1 and x_3, can be rewritten as

$$x_3 = \frac{\dot{x}_2}{b} = \frac{\dot{y}}{b} \tag{8.53}$$

$$x_1 = -\frac{1}{c}\left[\frac{1}{b}\ddot{y} + \frac{d}{b}\dot{y} + cy\right] \tag{8.54}$$

hence, Colpitts oscillator is algebraically observable with respect to the selected output $y = x_2$.

The system parameters (a, b, c, d) are not exactly known, but as it can be easily verified, they can be reconstructible or identifiable in the sense of Definition 8.2, that is to say, they satisfy differential equations in $\mathbb{R}\langle u, y\rangle$. Indeed by considering available the complete state vector[2] $(y_1 = x_1, y_2 = x_2, y_3 = x_3)$, we can obtain the following relationships from (8.52):

$$a = \frac{\dot{y}_1}{-\exp(-y_2) + y_3 + 1}; \qquad b = \frac{\dot{y}_2}{y_3}$$

$$c = \frac{-\dot{y}_3^2 + y_3\ddot{y}_3}{\dot{y}_3(y_1 + y_2) - (\dot{y}_1 + \dot{y}_2)}; \qquad d = -\frac{(\dot{y}_1 + \dot{y}_2)\dot{y}_3 - (y_1 + y_2)\ddot{y}_3}{(\dot{y}_1 + \dot{y}_2)y_3 - (y_1 + y_2)\dot{y}_3} \tag{8.55}$$

with $(-\exp(-y_2) + y_3 + 1) \neq 0$, $y_3 \neq 0$, $[\dot{y}_3(y_1 + y_2) - (\dot{y}_1 + \dot{y}_2)] \neq 0$ and $[(\dot{y}_1 + \dot{y}_2)y_3 - (y_1 + y_2)\dot{y}_3] \neq 0$.

[2]This is a realistic assumption taking into account that in the physical circuit these variables are related to currents and voltages, which can be measurable.

The Colpitts oscillator (8.52) is Liouvillian (Definition 8.3), since the parameter a is expressed in terms of an exponential of the output $y_2 = x_2$, or equivalently, $\exp(-\int \dot{y}_2) = \exp(-y_2)$.

8.4.3.1 Synchronization of the Colpitts Oscillator Employing the Exponential Polynomial Observer

For the implementation of the observer we first rewrite (8.52) in the form (8.41),

$$\dot{x} = \begin{bmatrix} 0 & 0 & a \\ 0 & 0 & b \\ -c & -c & -d \end{bmatrix} x + \begin{bmatrix} -a\exp(-\widehat{x}_2) + a \\ 0 \\ 0 \end{bmatrix}$$

$$y = \begin{bmatrix} 0\,1\,0 \end{bmatrix} x \tag{8.56}$$

Applying Proposition 8.1, we have

$$\dot{\widehat{x}} = \begin{bmatrix} 0 & 0 & a \\ 0 & 0 & b \\ -c & -c & -d \end{bmatrix} \widehat{x} + \begin{bmatrix} -a\exp(-\widehat{x}_2) + a \\ 0 \\ 0 \end{bmatrix} + \sum_{i=1}^{m} \begin{bmatrix} k_{1,i} \\ k_{2,i} \\ k_{3,i} \end{bmatrix} \left(\begin{bmatrix} 0 & 1 & 0 \end{bmatrix} e \right)^{2i-1}$$

Hence, the state observer is rewritten as,

$$\begin{aligned} \dot{\widehat{x}}_1 &= a\widehat{x}_3 - a\exp(-\widehat{x}_2) + a + k_{1,1}e_{1,2} + k_{1,2}(e_{1,2})^3 + \ldots + k_{1,m}(e_{1,2})^{2m-1} \\ \dot{\widehat{x}}_2 &= b\widehat{x}_3 + k_{2,1}e_{1,2} + k_{2,2}(e_{1,2})^3 + \ldots + k_{2,m}(e_{1,2})^{2m-1} \\ \dot{\widehat{x}}_3 &= -c\widehat{x}_1 - c\widehat{x}_2 - d\widehat{x}_3 + k_{3,1}e_{1,2} + k_{3,2}(e_{1,2})^3 + \ldots + k_{3,m}(e_{1,2})^{2m-1} \end{aligned} \tag{8.57}$$

It is clear that in order to have a useful implementation of (8.57) we need a fairly good knowledge of the model parameters, (a,b,c,d). In theory we know this parameters from the modeling and the relations of the circuit components, but for some experimental results using data from the Colpitts oscillator is not enough. System parameters could be estimated by using (8.55), however, we can not guarantee that the denominator in each relation is different to zero. In this context, relations given in (8.55) are not defined for all time.

In order to estimate (a,b,c,d) in any time, we define

$$\begin{aligned} \hat{a} &= \frac{\dot{y}_1}{\chi_1(0.5-\psi_1)+\psi_1}; & \hat{c} &= \frac{-\dot{y}_3^2 + y_3\ddot{y}_3}{\chi_3(0.5-\psi_3)+\psi_3} \\ \hat{b} &= \frac{\dot{y}_2}{\chi_2(0.5-\psi_2)+\psi_2}; & \hat{d} &= -\frac{(\dot{y}_1+\dot{y}_2)\dot{y}_3 - (y_1+y_2)\ddot{y}_3}{\chi_4(0.5-\psi_4)+\psi_4} \end{aligned} \tag{8.58}$$

where: $\psi_1 = -\exp(-y_2) + y_3 + 1$, $\psi_2 = y_3$, $\psi_3 = \dot{y}_3(y_1 + y_2) - (\dot{y}_1 + \dot{y}_2)$, $\psi_4 = (\dot{y}_1 + \dot{y}_2)y_3 - (y_1 + y_2)\dot{y}_3$, and χ_i is the characteristic function defined by

$$\chi_i = \begin{cases} 1 & \text{if } |\psi_i| \leq 0.5 \\ 0 & \text{if } |\psi_i| > 0.5, \quad \text{for } i \in \{1,\ldots,4\}. \end{cases}$$

The system (8.58) requires the knowledge of the time derivatives of the measurement y_j, $1 \leq j \leq 3$. Let us consider the following time derivatives to be estimated

$$\eta_j = \dot{y}_j$$
$$\overline{\eta}_j = \ddot{y}_j, \quad \text{for } 1 \leq j \leq 3$$

Using the reduced order observer proposed in [51],

$$\dot{\hat{\eta}}_j = K_j(\eta_j - \hat{\eta}_j) \tag{8.59}$$

introducing the change of variable γ_j

$$\hat{\eta}_j = \gamma_j + K_j y_j \tag{8.60}$$

and from (8.59) and (8.60) we can get $\dot{\gamma}_j = -K_j\hat{\eta}_j$, then again from (8.60)

$$\dot{\gamma}_j = -K_j\gamma_j - K_j^2 y_j \tag{8.61}$$

then, (8.61) together with (8.60) constitute an asymptotic observer for $\eta_j = \dot{y}_j$.

In the same manner, the corresponding asymptotic observer for $\overline{\eta}_j = \ddot{y}_j$ is given by

$$\dot{\overline{\gamma}}_j = -\overline{K}_j\overline{\gamma}_j - \overline{K}_j^2\hat{\eta}_j$$
$$\hat{\overline{\eta}}_j = \overline{\gamma}_j + \overline{K}_j\hat{\eta}_j$$

We verified the real time performance of the exponential observer by using the WINCON platform. To achieve the synchronization in real time, in WINCON was implemented the scheme (8.57) in the master–slave configuration. Figure 8.6 shows the real implementation of the Colpitts circuit. The circuit parameters are: L = 100 μH; $C_1 = C_2 = 47$ nF, R = 45 Ω, $I_0 = 5$ mA. Using the circuit parameters we obtain $a = b = 6.2723$, $c = 0.0797$, and $d = 0.6898$.

The nonlinear term $\Psi(x)$ in (8.56), satisfies the Lipschitz condition and is considered as follows

$$\Psi(x) = \begin{bmatrix} -a\exp(-x_2) + a \\ 0 \\ 0 \end{bmatrix}$$

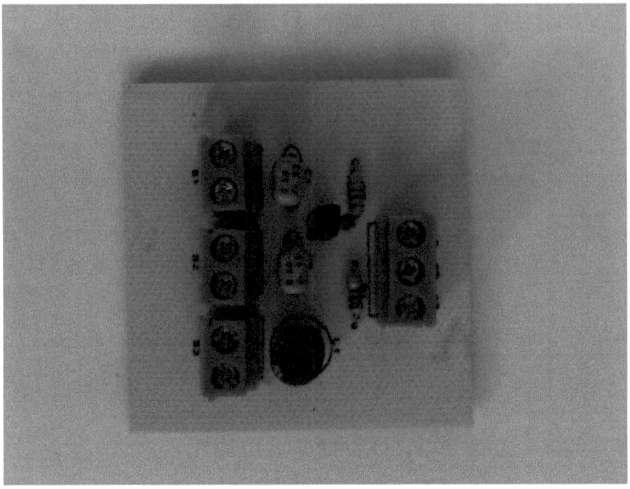

Fig. 8.6 Implementation of the Colpitts circuit (master system)

It is necessary to calculate the Lipschitz constant φ introduced in (8.42) over the bounded set

$$\Omega = \{x \in \mathbb{R}^3 | \ |x_1| < M_1, \ |x_2| < M_2, \ |x_3| < M_3\} \tag{8.62}$$

Considering the Jacobian of $\Psi(x)$ as

$$\left[\frac{\partial \Psi(x)}{\partial x}\right] = \begin{bmatrix} 0 & a\exp(-x_2) & 0 \\ 0 & 0 & 0 \\ 0 & 0 & 0 \end{bmatrix} \tag{8.63}$$

it can be concluded that[3]

$$\left\| \frac{\partial \Psi(x)}{\partial x} \right\|_\infty \leq 3\max\{0, a\exp(-x_2)\}, \quad a \in \mathbb{R}^+. \tag{8.64}$$

From (8.62), it is obvious that the following conditions hold for all the points in the bounded set Ω

$$a\exp(-x_2) < a\exp(M_2) = \max\{a\exp(-x_2)\}, \quad a \in \mathbb{R}^+. \tag{8.65}$$

[3]Let us consider the matrix $\mathscr{A} = [a_{ij}]_{1\leq i,j\leq n}$, then (see Chap. 5 in [60])

$$\|\mathscr{A}\|_\infty := n \max_{1\leq i,j\leq n} |a_{ij}|$$

Hence

$$\left\| \frac{\partial \Psi(x)}{\partial x} \right\|_{\infty} \leq 3\ a\ \exp(M_2) \tag{8.66}$$

With (8.62)–(8.66) a Lipschitz constant that satisfies the Lipschitz condition (8.42) is defined as follows

$$\varphi = 3\ a\ \exp(M_2)$$

In this case $M_1 = 3$, $M_2 = 0.1$, $M_3 = 6$, and $a = 6.2723$, $\Rightarrow \varphi = 20.7959$.

Following the observer 8.57, for $m = 2$, and solving the LMI given by Assumption 8.5, the observer gains K_1 and K_2, and a positive definite matrix P are as follows

$$K_1 = \begin{bmatrix} 10.2130 \\ 16.1211 \\ 10.1500 \end{bmatrix}, \quad K_2 = \begin{bmatrix} 3 \\ 2 \\ 3 \end{bmatrix}, \quad P = \begin{bmatrix} 38.8560 & -36.7794 & 19.4606 \\ -36.7794 & 37.9331 & -20.7898 \\ 19.4606 & -20.7898 & 16.4869 \end{bmatrix} > 0,$$

with eigenvalues $\lambda_1(P) = 1.3151$, $\lambda_2(P) = 5.2514$, and $\lambda_3(P) = 86.7095$.

The performance index (quadratic synchronization error) of the corresponding synchronization process is calculated as [50]

$$J(t) = \frac{1}{t + 0.001} \int_0^t |e(t)|^2_{Q_0}, \quad Q_0 = I$$

Figures 8.7a–c show the obtained results by using the exponential polynomial observer (8.57), it is clear that the synchronization is achieved fairly acceptable even with the noisy measurements. The Colpitts circuit starts in $x(0) = [0\ \ 0\ \ 0]^T$ and the arbitrary initial conditions for the observer are $\hat{x}(0) = \left[1.506\ -0.1\ 2.1\right]^T$. Figure 8.7d shows the performance index of the synchronization, which depicts an exponential behavior.

8.4.3.2 Synchronization of the Colpitts Oscillator by Means of the Asymptotic Reduced Order Observer

Let us consider the normalized system of the Colpitts oscillator. We assume that the output system is $y = x_2$. Therefore, the slave system consists in two estimation structures to achieve synchronization with the master system. Such structures are obtained as follows. Firstly, verify that the master system (Colpitts oscillator) is algebraically observable, and then, by using (8.50), construct the observer for the unknown states. Previously, we have verified that master system (Colpitts oscillator) is algebraically observable – see Eqs. (8.53) and (8.54). Then, both unknown states of the master system are algebraically observable, and, therefore, we can construct the observers based on Lemma 8.2 and Corollary 8.1.

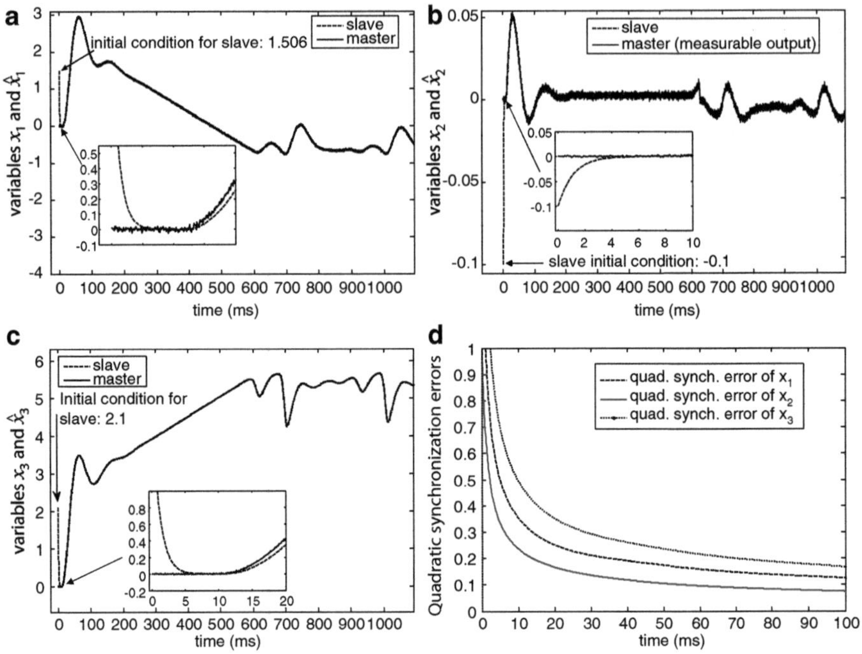

Fig. 8.7 Real-time synchronization of Colpitts oscillator employing observer (8.57): (**a**) synchronization of x_1, (**b**) synchronization of x_2, (**c**) synchronization of x_3, and (**d**) performance index

For x_3 the observer is given by,

$$\dot{\gamma}_3 = -\frac{K_3^2}{b}y - K_3\gamma_3$$

$$\hat{x}_3 = \frac{K_3}{b}y + \gamma_3 \tag{8.67}$$

and for x_1, we have

$$\dot{\gamma}_4 = -K_4[\gamma_4 + K_4 y]$$

$$\dot{\gamma}_5 = [K_1 - d]\frac{K_1}{cb}[\gamma_4 + K_4 y] - K_1 y - K_1\gamma_5$$

$$\hat{x}_1 = -\frac{K_1}{cb}[\gamma_4 + K_4 y] + \gamma_5 \tag{8.68}$$

Therefore, (8.67) and (8.68) constitute the slave system. Now, we present some experimental results for the synchronization of the Colpitts oscillator by using the asymptotic reduced order observer (8.67)–(8.68). Figure 8.8a, b show the obtained

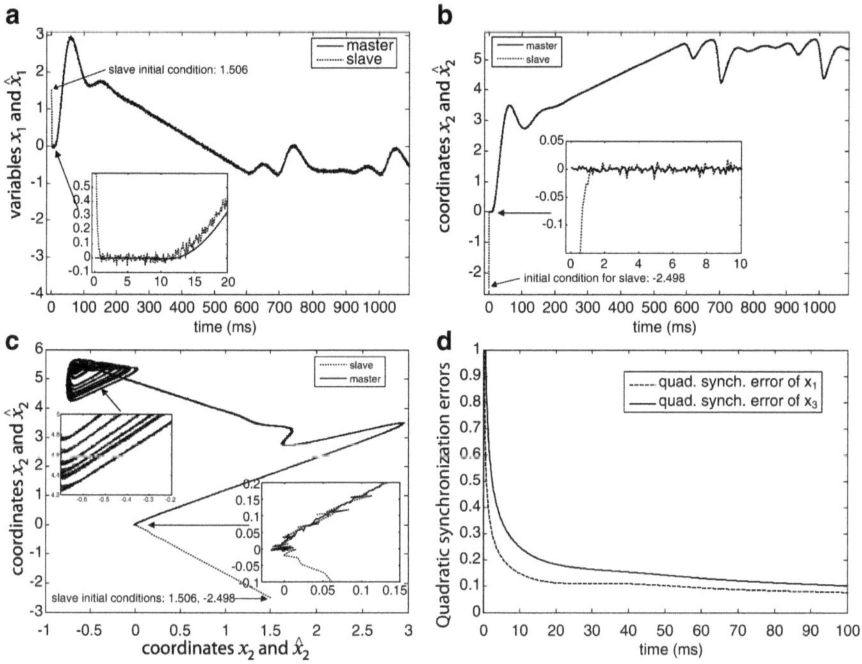

Fig. 8.8 Real-time synchronization of Colpitts oscillator using reduced-order observer (8.68)–(8.67): (**a**) synchronization of x_1, (**b**) synchronization of x_3, (**c**) Phase portrait of the master system (x_1 versus x_3) and the slave system ($\hat{x}_1$ versus $\hat{x}_3$), and (**d**) performance index

results for the initial conditions $\hat{x}_1 = 1.506$ and $\hat{x}_3 = -2.498$ in the schemes (8.67) and (8.68), respectively. As we can note, the synchronization results achieved with the reduced order observer are good. Figure 8.8c presents the phase portrait, where clearly is observed the chaotic behavior of the Colpitts oscillator. Finally, Fig. 8.8d illustrates the performance index, which has a tendency to decrease.

8.5 Application to Secure Communications

Synchronization can be classified into mutual synchronization (or bidirectional coupling) [69] and master-slave synchronization (or unidirectional coupling) [58]. The chaos-based secure communications have updated their fourth generation [68]. The continuous synchronization is adopted in the first three generations while the impulsive synchronization is used in the fourth generation. Less than 94 Hz of bandwidth is needed to transmit the synchronization signal for a third-order chaotic transmitter in the fourth generation while 30 kHz bandwidth is needed to transmit the synchronization signals in the other three generations [72].

Information signal embedded in a chaotic transmitter can be recovered by a receiver if it is a replica of the transmitter. In this work, a new aspect of chaotic communication is introduced. A sliding-mode observer replaces the conventional chaotic system at the receiver side, which does not need information from the transmitter. So the uncertainties in the transmitter and the transmission line do not affect the synchronization, the proposed communication scheme is robust with respect to some disturbances and uncertainties. Duffing equation is provided to illustrate the effectiveness of the chaotic communication.

Linear and nonlinear observers in control theory literatures can be applied to design receivers. The receiver is regarded as a chaotic observer, which has two parts a duplicated chaotic system of the transmitter and an adjustable observer gain [43]. Some modifications were made when it is difficult to obtain a replica of the synchronization. For example, the transmitter and the receiver are set into the same chaotic structures, parameter identification methods can be used to construct the chaotic receiver [32]; when there are uncertainties in synchronization (the transmitter is not known exactly, there is noise in the transmission line, etc.), the transmitter and the receiver could be established in the same fuzzy models, fuzzy model-based design method was applied to reach synchronization [42]; stability analysis of observer-based chaotic communication with respect to uncertainties can be found in [6,47].

Robust control techniques and many traditional schemes have been applied in robust synthesis for chaotic synchronization, e.g. robust observer and H_∞ technique are used in [66, 67]. Since sliding-mode observer contains a sliding-mode term, it provides the robustness against an inaccurate modeling of measurements and output noises. The early works dealing with sliding mode observers which consider measurement noise were proposed in [13], where is discussed the state estimation using sliding mode technique. In [12] is discussed the variable structure control as a high-speed switched feedback control resulting in a sliding mode. In [4] is treated an analysis of systems with sliding mode in the presence of noises. In [64], successfully designed, so named, sliding-mode approach to construct observers which are highly robust with respect to noises in the input of the system. But, it turns out that the corresponding stability analysis cannot be directly applied in the situations with the output noise (or, mixed uncertainty) presence. So, it is still a challenge to suggest a workable technique to analyze the stability of identification error generated by sliding-mode (discontinuous nonlinearity) type observers [46].

In this work, a novel design approach for chaotic communication is proposed, where the receiver is a pure sliding-mode observer. The main difference with the above methods is that the receiver is no longer a chaotic system. The uncertainty of the transmitter will not affect the synchronization. The proposed communication scheme can be more robust than both transmitter and receiver employed in chaotic systems. But the information may be recovered by the observer who does not have knowledge about the transmitter, this is a big challenge to secure communication by means of chaos.

8.5.1 Chaotic Communication Based on Sliding-Mode Observer

In normal chaotic communication, the transmitter and the receiver are chaotic systems. They can be described in the form of the following nonlinear system

$$\begin{aligned} \dot{\xi} &= f(\xi) + g(\xi)u \\ y &= h(\xi) \end{aligned} \tag{8.69}$$

where $\xi \in \mathbb{R}^n$ is the state of the plant, $u \in \mathbb{R}$ is a control input, $y \in \mathbb{R}$ is a measurable output, f, g and h are smooth nonlinear functions. Most chaotic systems have uniform relative degree n, i.e.

$$L_g h(\xi) = \cdots = L_g L_f^{n-2} h(\xi) = 0, \;\; L_g L_f^{n-1} h(\xi) \neq 0$$

So there exists a mapping $\eta = T(\xi)$ which can transform the system (8.69) into the following normal form [33]

$$\begin{aligned} \dot{\eta}_i &= \eta_{i+1}, \quad i = 1, 2, n-1 \\ \dot{\eta}_n &= \Phi(\eta, u) \\ y &= \eta_1 \end{aligned} \tag{8.70}$$

where $\Phi(\cdot)$ is a continuous nonlinear function.

In this paper, *chaos modulation* [6, 43] is used for communication, where the information signal s is embedded into the output of the chaotic transmitter. The transmitter is a slight modification of the normal chaotic systems (8.70) as follows:

$$\begin{aligned} \dot{\eta}_i &= \eta_{i+1}, \quad i = 1, 2, n-1 \\ \dot{\eta}_n &= \Phi(\eta, u) \\ y &= \eta_1 + s \end{aligned} \tag{8.71}$$

where the output $y = \eta_1 + s$ is chaotic masking.

Now, we discuss a new observer-based receiver, we propose the following sliding-mode observer for the receiver

$$\begin{aligned} \dot{\hat{\eta}}_i &= \hat{\eta}_{i+1} + m_i sign(y - \hat{y}), \quad i = 1, 2, n-1 \\ \dot{\hat{\eta}}_n &= m_n sign(y - \hat{y}) \end{aligned} \tag{8.72}$$

where $\hat{\eta}_j$ are the states on the receiver side and $\hat{y} = \hat{\eta}_1$ the estimate of the output y, $m_j = m^j \ \tau^{-j}, \forall\, 1 \leq j \leq n$, m and τ are small positive parameters, $m > 0$, $0 < \tau < 1$ and

$$sign(y - \hat{y}) = \begin{cases} 1 & \text{if } (y - \hat{y}) > 0 \\ -1 & \text{if } (y - \hat{y}) < 0 \\ \text{undefined} & \text{if } (y - \hat{y}) = 0 \end{cases}.$$

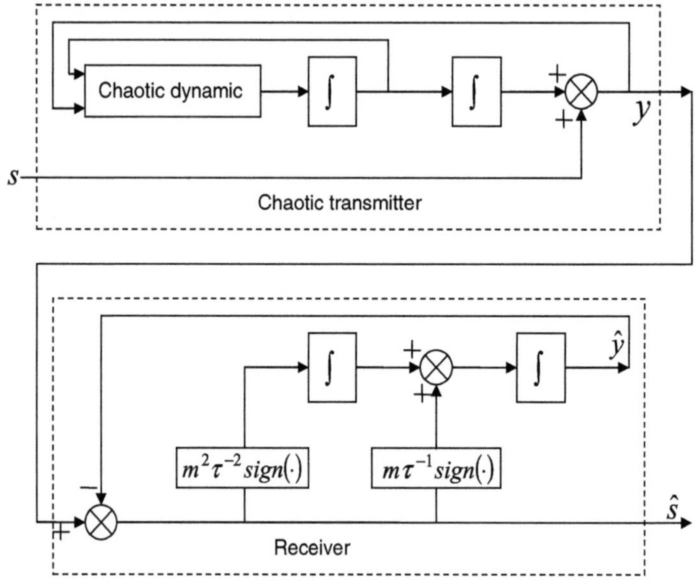

Fig. 8.9 Sliding-mode chaotic communication

The schematic diagram of the chaotic communication based on sliding-mode observer for $n = 2$ is shown in Fig. 8.9.

Let us define the synchronization errors as

$$e_1 = \eta_1 - \hat{\eta}_1,\ e_i = (\eta_i - \hat{\eta}_i)/m,\ i = 2, ..., n, \tag{8.73}$$

The recovered signal at receiver is

$$\hat{s} = y - \hat{y} = e_1 + s$$

By (8.71) and (8.72) the synchronization error $e = \begin{bmatrix} e_1 & ... & e_n \end{bmatrix}^T$ can be formed as

$$\dot{e} = A_{\bar{\mu}}e - Ksign(Ce + s) + \Delta f \tag{8.74}$$

where $\bar{\mu} > 0$ is a regularizing parameter, $C = \begin{bmatrix} 1 & 0 & ... & 0 \end{bmatrix}$, $A_{\bar{\mu}} = \begin{bmatrix} -\bar{\mu} & m & 0 & ... & 0 \\ 0 & -\bar{\mu} & m & & 0 \\ 0 & 0 & -\bar{\mu} & & \vdots \\ & & & \ddots & m \\ 0 & 0 & 0 & ... & -\bar{\mu} \end{bmatrix}$,

$K = \begin{bmatrix} m_1 \\ m_2 \\ ... \\ m_n \end{bmatrix}$, and $\Delta f = \begin{bmatrix} \bar{\mu}e_1 \\ ... \\ \bar{\mu}e_{n-1} \\ \Phi + \bar{\mu}e_n \end{bmatrix}$ is an uncertainty term.

The following assumptions are used for our theoretical result:

Assumption 8.11. There exist nonnegative constants L_{0f}, L_{1f}, such that the following generalized quasi-Lipschitz condition holds

$$\|\Delta f\| \leq L_{0f} + (L_{1f} + \|A_{\bar{\mu}}\|) \|e\| . \tag{8.75}$$

Assumption 8.12. Information signal is assumed to be bounded as $\|s\|_{\Lambda}^2 = s^T \Lambda s \leq (\bar{s})^2 < \infty$ where Λ is a symmetric definite positive matrix.

Assumption 8.13. There exists a positive definite matrix $Q_0 = Q_0^T > 0$, such that the following matrix Riccati equation

$$PA_{\bar{\mu}} + A_{\bar{\mu}}^T P + PRP + Q = 0 \tag{8.76}$$

with

$$R := \Lambda_f^{-1} + 2\|\Lambda_f\| L_{1f} I, \ \Lambda_f = \Lambda_f^T > 0,$$

$$Q := Q_0 + 2(L_{1f} + \|A_{\bar{\mu}}\|^2) I$$

has a positive definite solution $P = P^T > 0$. Since $P > 0$, there exists $k > 0$ such that $K = kP^{-1}C^T$.

To calculate the solution to the Riccati equation (8.76), the following parameters have been selected

$$\Lambda_f = \lambda_f I, \ L_{1f} = \bar{\mu}, \ R = (\lambda_f^{-1} + 2\lambda_f \mu) I, \ Q_0 = q_0 I, \ Q = (q_0 + 8\bar{\mu}^2) I. \tag{8.77}$$

Theorem 8.2. *The sliding-mode observer-based receiver (8.72) can recover the information signal s which is embedded in the chaotic transmitter (8.71), the signal recovery error $\tilde{s} = s - \hat{s}$ converges to the following residual set*

$$D_{\varepsilon} = \{\tilde{s} \| \|\tilde{s}\|_P \leq \tilde{\mu}(k)\}$$

where P is a solution of the Riccati equation (8.76)

$$\tilde{\mu}(k) = \left(\frac{\rho(k)}{\sqrt{(k\alpha_P)^2 + \rho(k)\alpha_Q} + k\alpha_P} \right)^2$$

with

$$\rho(k) = 2\|\Lambda_f\|L_{0f}^2 + 4k\left(\sqrt{(}n\Lambda_f^{-1})\right)\bar{s}$$

$$k\alpha_P = k\left(\lambda_{min}(P^{-1/2}C^TCP^{-1/2})\right)$$

$$\alpha_Q = \lambda_{min}(P^{-1/2}C^TCP^{-1/2}) > 0$$

and n is the dimension of the chaotic system.

Proof. Taking the Lyapunov function candidate V (e) as $V(e) \triangleq e^TPe = \|e\|_P^2$, $0 < P = P^T \in R^{r\times r}$, and using the matrix inequality

$$X^TY + Y^TX \leq X^T\Lambda_f X + Y^T\Lambda_f^{-1}Y$$

which is valid for any $X, Y \in R^{n\times m}$, $0 < \Lambda_f = \Lambda_f^T \in R^{r\times r}$ [60], then it follows that

$$\dot{V}(e) \leq e^T\left(PA_{\bar{\mu}} + A_{\bar{\mu}}^TP + PRP + Q\right)e - e^TQe + 2\|\Lambda_f\|L_{0f}^2 - 2k(Ce)^T\operatorname{sign}(Ce+s)$$

by using

$$x^T\operatorname{sign}[x+z] \geq \sum_{i=1}^{n}|x_i| - 2\sqrt{n}\|z_i\|$$

Then

$$\dot{V}(e) \leq -e^TQe + 2\|\Lambda_f\|L_{0f}^2 - 2k\left(\sum_{i=1}^{n}|(Ce)_i| - 2\sqrt{n}\|s\|\right)$$

$$\leq -e^TQe - 2k\sum_{i=1}^{n}|(Ce)_i| + \rho(k)$$

where

$$\rho(k) = 2\|\Lambda_f\|L_{0f}^2 + 4k\left(\sqrt{n\Lambda_f^{-1}}\right)\bar{s}$$

Thus

$$\dot{V}(e) \leq -\|e\|_Q - 2k\alpha_P\|e\|_P + \rho(k)$$

where

$$\left(\sum_{i=1}^{n}|(Ce)_i|\right)^2 \geq \sum_{i=1}^{n}(|(Ce)_i|)^2 = \|Ce\|^2 = \|CP^{-1/2}P^{-1/2}e\|^2 \geq \alpha_P e^TQe$$

with

$$\alpha_P = \lambda_{min}(P^{-1/2}C^T CP^{-1/2}) \geq 0$$

So that, from (8.77) we obtain

$$\dot{V}(e) = -\alpha_Q V(e) - \vartheta\sqrt{(}V(e)) + \beta$$

where

$$\alpha_Q = \lambda_{min}(P^{-1/2}QQ^T P^{-1/2}) > 0, \quad \vartheta = 2k\alpha_P, \quad \beta = \rho(k)$$

By the theorem proposed in [46] and Assumptions 8.11–8.13, we can formulate the following

$$\left[1 - \frac{\tilde{\mu}}{V}\right]_+ \to 0$$

where the function $[\cdot]_+$ is defined as in Theorem 8.1. The other part of the proof is similar as [46]. □

8.5.2 Numerical Simulation

We use the Duffing chaotic system as transmitter, the information signal s is embedded in the output of the transmitter. The Duffing equation is [9]

$$\begin{aligned}
\dot{\eta}_1 &= \eta_2 + \frac{1}{\tau}s \\
\dot{\eta}_2 &= -1.1y - y^3 - 0.4\eta_2 + 2.1\cos(1.8t) + \frac{1}{\tau^2}s \\
y &= \eta_1 + s, \qquad \eta(0) = [0,0]^T
\end{aligned} \tag{8.78}$$

Now we design the sliding-mode receiver as (8.72). We choose $m = 0.1$, $\tau = 0.01$. The sliding-mode observer-based receiver is

$$\begin{aligned}
\dot{\hat{\eta}}_1 &= \hat{\eta}_2 + 10\,\text{sign}(y - \hat{y}) \\
\dot{\hat{\eta}}_2 &= 10^2\,\text{sign}(y - \hat{y}) \\
\hat{y} &= \hat{\eta}_1, \qquad \hat{\eta}(0) = [1,1]^T
\end{aligned} \tag{8.79}$$

The information signal s is chosen as sinusoidal signal with frequency of 100 Hz as in [6, 43], i.e.

$$s = 0.05\sin(200\pi t)$$

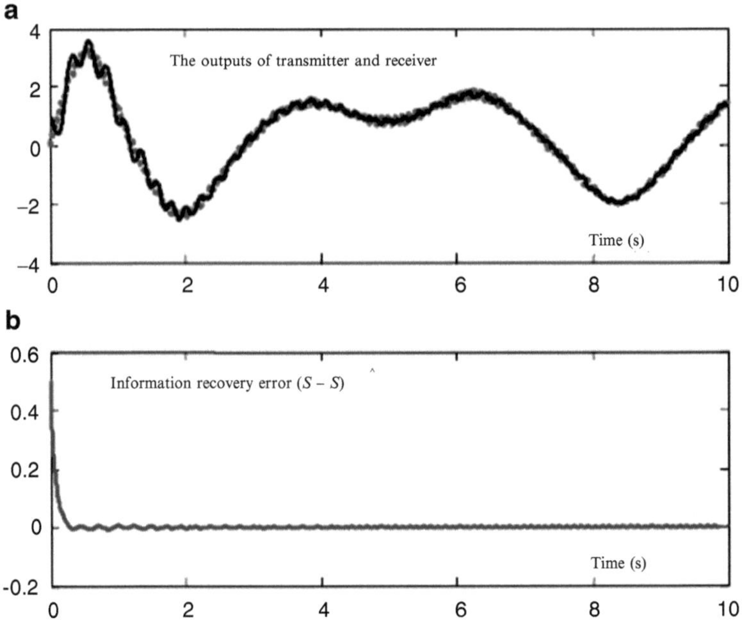

Fig. 8.10 Duffing equation for chaotic communication

Figure 8.10 shows the communication process with Duffing chaotic transmitter and the receiver, here the waveform of the transmitted signal y is shown in subplot (a), the convergence behavior of $s-\hat{s}$ is shown in subplot (b).

After transient process ($t > 0.1$ s), the maximum relative error is defined as

$$e_{\max} = \frac{\max(|s-\hat{s}|)}{\max(|s|)}$$

In this case, $e_{\max} \approx 1.5\%$ which is acceptable for signal communication.

Acknowledgements Juan L. Mata-Machuca and Andrés Rodríguez-Bollain are grateful with CONACyT (Mexico) for the corresponding postgraduate scholarships.

References

1. Aguilar-Ibañez, C., Martínez-Guerra, R., Aguilar-López, R., Mata-Machuca, J.L.: Synchronization and parameter estimations of an uncertain Rikitake system. Phys. Lett. A **374**, 3625–3628 (2010)
2. Aguilar-López, R., Femat, R., Martínez-Guerra, R.: Importance of chaos synchronization on technology and science. In: Banerjee, S. (ed.) Chaos Synchronization and Cryptography for Secure Communications: Applications for Encryption, pp. 210–246, IGI Global (2010)

3. Aguilar-López, R., Martínez-Guerra, R.: Synchronization of a class of chaotic signals via robust observer design. Chaos, Solitons Fractals **37**, 581–587 (2008)
4. Anulova, S.V.: Random disturbances of the operation of control systems in the sliding mode. Autom. Rem. Contr. **47**, 474–479 (1986)
5. Ayati, M., Khaloozadeh, H.: Stable chaos synchronisation scheme for nonlinear uncertain systems. IET Contr. Theor. Appl. **4**, 437–447 (2010)
6. Boutayeb, M., Darouach, M., Rafaralahy, H.: Generalized state observers for chaotic synchronization and secure communication. IEEE Trans. Circ. Syst. I **49**, 345–349 (2002)
7. Bowong, S.: Stability analysis for the synchronization of chaotic systems with different order: application to secure communications. Phys. Lett. A **326**, 102–113 (2004)
8. Chelouah, A.: Extensions of differential flat fields and Liouvillian systems. Proceedings of the 36th IEEE Conference Decision Control, San Diego, CA, pp. 4268–4273 (1997)
9. Chen, G., Dong, X.: On feedback control of chaotic continuous-time systems. IEEE Trans. Circ. Syst. **40**, 591–601 (1993)
10. Cherrier, E., Boutayeb, M., Ragot, J.: Observers-based synchronization and input recovery for a class of nonlinear chaotic models. IEEE Trans. Circ. Syst. I **53**, 1977–1988 (2006)
11. Cuomo, K.M., Oppenheim, A.V., Strogatz, S.H.: Synchronization of Lorenz-based chaotic circuits with applications to communications. IEEE Trans. Circ. Syst. I **40**, 626–633 (1993)
12. De Carlo, R., Zak, S., Drakunov, S.: Variable structure and sliding mode control. Control Handbook, Electrical Engineering Handbook Series (1996)
13. Drakunov, S.V., Utkin, V.: Sliding mode observers: Tutorial. In: Proceedings of the 34th IEEE Conference Decision Control (CDC), pp. 3376–3378 (1995)
14. Elabbasy, E., Agiza, H., El-Dessoky, M.: Global chaos synchronization for four scroll attractor by nonlinear control. Sci. Res. Essay **1**, 65–71 (2006)
15. Emadzadeh, A., Haeri, M.: Global Synchronization of two different chaotic systems via nonlinear control. In: Proceedings of the ICCAS, Gyeonggi-Do, Korea (2005)
16. Feki, M.: Observer-based exact synchronization of ideal and mismatched chaotic systems. Phys. Lett. A **309**, 53–60 (2003)
17. Feki, M.: An adaptive chaos synchronization scheme applied to secure communication. Chaos, Solitons Fractals **18**, 141–148 (2003)
18. Fotsin, H.B., Daafouz, J.: Adaptive synchronization of uncertain chaotic Colpitts oscillator based on parameter identification. Phys. Lett. A **339**, 304–315 (2005)
19. Fradkov, A.L.: Cybernetical physics: from control of chaos to quantum control. Springer, Berlin (2007)
20. Fradkov, A.L., Andrievsky, B., Evans, R.J.: Adaptive observer-based synchronization of chaotic systems with first-order coder in the presence of information constraints. IEEE Trans. Circ. Syst. I **55**, 1685–1694 (2008)
21. Fradkov, A.L., Nijmeijer, H., Markov, A.: Adaptive observer-based synchronisation for communications. Int. J. Bifurc. Chaos **10**, 2807–2814 (2000)
22. Garfinkel, S., Spafford, G.: Practical unix and internet security. O' Reilly & Associates Inc., Sebastopol, CA (1996)
23. Gauthier, J., Hammouri, H., Othman, S.: A simple observer for nonlinear systems: aplications to bioreactors. IEEE Trans. Autom. Contr. **37**, 875–880 (1992)
24. Gavalas, G.R.: Non-linear Differential Equations of Chemically Reacting Systems. Springer, New York (1968)
25. Ghosh, D., Banerjee, S., Chowdhury, A.: Synchronization between variable time-delayed systems and cryptography. Europhys. Lett. **80**(30006), 1–6 (2007)
26. Ghosh, D., Chowdhury, A., Saha, P.: On the various kinds of synchronization in delayed Duffing-Van der Pol system. Commun. Nonlinear Sci. Numer. Simulat. **13**, 790–803 (2008)
27. Gray, P., Scott, S.K.: Chemical Oscillations and Instabilities. Clarendon Press, Oxford (1990)
28. Guo-Hui, L.: Synchronization and anti-synchronization of Colpitts oscillators using active control. Chaos, Solitons Fractals **26**, 87–93 (2005)
29. Harb, A., Ahmad, W.: Chaotic systems synchronization in secure communication systems. Proc. World Congress Computer Science Computer Engineering, and Applied Computing. Las Vegas (2006)

30. He, Z., Li, K., Yuang, L., Sui, Y.: A robust digital structure communications scheme based on sporadic chaos synchronization. IEEE Trans. Circ. Syst. I **47**, 397–403 (2000)
31. Hua, C., Guan, X.: Synchronization of chaotic systems based on PI observer design. Phys. Lett. A **334**, 382–389 (2005)
32. Huijberts, H., Nijmeijer, H., Willems, R.: System identification in communication with chaotic systems. IEEE Trans. Circ. Syst. I **47**, 800–808 (2000)
33. Isidori, A.: Nonlinear Control Theory. Springer, New York (1995)
34. Jorgensen, D.V., Aris, R.: On the dynamics of a stirred tank with consecutive reactions. Chem. Eng. Sci. **38**, 45–53 (1983)
35. Kalman, R.E.: A new approach to linear filtering and prediction problems. Trans. ASME J. Basic Eng. Series D **82**, 35–45 (1960)
36. Keller, H.: Non-linear observer design by transformation into a generalized observer canonical form. Int. J. Contr. **46**, 1915–1930 (1987)
37. Kennedy, M.P.: Chaos in Colpitts oscillator. IEEE Trans. Circ. Syst. I **41**, 771–774 (1994)
38. Khalil, H.: Nonlinear Systems, 3rd edn. Englewood Cliffs, NJ: Prentice–Hall (2002)
39. Krener, A.J., Isidori, A.: Linearization by output injection and nonlinear observers. Syst. Contr. Lett. **3**, 47–54 (1983)
40. Levant, A.: Universal SISO sliding mode controllers with finite tine convergence. IEEE Trans. Autom. Contr. **46**, 1447–1451 (2001)
41. Li, D., Lu, J., Wu, X.: Linearly coupled synchronization of the unified chaotic systems and the Lorenz systems. Chaos, Solitons Fractals **23**, 79–85 (2005)
42. Lian, K.Y., Chiang, T.S., Chiu, C.S., Liu, P.: Synthesis of fuzzy model-based designs to synchronization and secure communication for chaotic systems. IEEE Trans. Syst. Man Cybern. B **31**, 66–83 (2001)
43. Liao, T.L., Huang, N.S.: An observer-based approach for chaotic synchronization with applications to secure communication. IEEE Trans. Circ. Syst. II **46**, 1144–1150 (1999)
44. Luenberger, D.: An introduction to observers. IEEE Trans. Autom. Contr. **16**, 592–602 (1971)
45. Maggio, G.M., De Feo, O., Kennedy, M.P.: Nonlinear analysis of the Colpitts oscillator and applications to design. IEEE Trans. Circ. Syst. I **46**, 1118–1130 (1999)
46. Martínez-Guerra, R., Aguilar, R., Poznyak, A.: A new robust sliding-mode observer design for monitoring in chemical reactors. Trans. ASME J. Dyn. Syst. Meas. Contr. **126**, 473–478 (2004)
47. Martínez-Guerra, R., Cruz, J., Gonzalez, R., Aguilar, R.: A new reduced-order observer design for the synchronization of Lorenz systems. Chaos, Solitons Fractals **28**, 511–517 (2006)
48. Martínez-Guerra, R., Diop, S.: Diagnosis of nonlinear systems: An algebraic and differential approach. IEE Proc. Contr. Theor. Appl. **151**, 130–135 (2004)
49. Martínez–Guerra, R., Mendoza–Camargo, J.: Observers for a class of Liouvillian and, non-differentially flat systems. IMA J. Math. Contr. Inf. **21**, 493–509 (2004)
50. Martínez-Guerra, R., Poznyak, A., Díaz, V.: Robustness of high-gain observers for closed-loop nonlinear systems: theoretical study and robotics control application. Int. J. Syst. Sci. **31**, 1519–1529 (2000)
51. Martínez-Guerra, R., Rincón Pasaye, J.J.: Synchronization and anti-synchronization of chaotic systems: A differential and algebraic approach. Chaos, Solitons Fractals **28**, 511–517 (2009)
52. Martínez-Guerra, R., Yu, W.: Chaotic communication and secure communication via sliding-mode observer. Int. J. Bifur. Chaos **18**, 235–243 (2008)
53. Martínez-Guerra, R., Yu, W., Cisneros-Saldaña, E.: A new model-free sliding observer to synchronization problem. Chaos, Solitons Fractals **36**, 1141–1156 (2008)
54. Min, L., Jing, J.: A new theorem to synchronization of unified chaotic systems via adaptive control. Chaos, Solitons Fractals **24**, 1363–1371 (2004)
55. Morgül, O., Feki, M.: A chaotic masking scheme by using synchronized chaotic systems. Phys. Lett. A **251**, 169–176 (1999)
56. Morgül, O., Solak, E.: Observed based synchronization of chaotic systems. Phys. Rev. E **54**, 4803–4811 (1996)
57. Nijmeijer, H., Mareels, I.M.Y.: An observer looks at synchronization. IEEE Trans. Circ. Syst. I **44**, 882–890 (1997)

58. Pecora, L.M., Carrol, T.L.: Synchronization in chaotic systems. Phys. Rev. A **64**, 821–824 (1990)
59. Pérez, G., Cerdeira, H.A.: Extracting messages masked by chaos. Phys. Rev. Lett. **74**, 1970–1973 (1995)
60. Poznyak, A.S.: Advanced Mathematical Tools for Automatic Control Engineers: Deterministic Techniques, vol. 1, pp. 77–212. Elsevier (2008)
61. Raghavan, S., Hedrick, J.: Observer design for a class of nonlinear systems. Int. J. Contr. **59**, 515–528 (1994)
62. Röbenack, K., Lynch, A.F.: High-gain nonlinear observer design using the observer canonical form. IET Contr. Theor. Appl. **1**, 1574–1579 (2007)
63. Schuler, H., Schmidt, C.: Calorimetric-state estimator for chemical reactors diagnosis and control: Review of methods and applications. Chem. Eng. Sci. **47**, 899–908 (1992)
64. Slotine, J., Hedricks, J., Misawa, E.: On sliding observers for nonlinear systems. J. Dyn. Meas. Contr. **109**, 245–252 (1987)
65. Soroush, M., Tyner, D., Grady, M.: Adaptive temperature control of multiproduct jacketed reactors. Ind. Eng. Chem. Res. **38**, 4337–4344 (1999)
66. Suykens, J.A.K., Curran, P.F., Chua, L.O.: Robust synthesis for master-slave synchronization of Lures systems. IEEE Trans. Circ. Syst. I **46**, 841–850 (1999)
67. Suykens, J.A.K., Curran, P.F., Vandewalle, J., Chua, L.O.: Robust nonlinear H_∞ synchronization of chaotic Lure's systems. IEEE Trans. Circ. Syst. I **44**, 891–940 (1999)
68. Tao, Y.: Chaotic secure communication systems history and new results. Telecommun. Rev. **9**, 597–634 (1999)
69. Ushio, Y.: Synthesis of synchronized chaotic systems based on observers. Int. J. Bifurc. Chaos **9**, 541–546 (1999)
70. Wang, C., Ge, S.: Adaptive backstepping control of uncertain Lorenz system. Int. J. Bifurc. Chaos **11**, 1115–1119 (2001)
71. Wang, F., Liu, C.: A new criterion for chaos and hyperchaos synchronization using linear feedback control. Phys. Lett. A **360**, 274–278 (2006)
72. Yang, T., Chua, L.O.: Impulsive stabilization for control and synchronization of chaotic systems theory and application to secure communication. IEEE Trans. Circ. Syst. I **44**, 976–988 (1997)
73. Young, J., Farrel, J.: Observer based backstepping control using online approximation. Proceedings of the IEEE American Control Conference, Chicago, IL, pp. 3646–3650 (2000)
74. Zhu, F.: Observer-based synchronization of uncertain chaotic systems and its application to secure communications. Chaos, Solitons Fractals **40**, 2384–2391 (2009)

Chapter 9
Nonlinear Dynamics for Information Processing

Y.V. Andreyev, A.S. Dmitriev, A.N. Miliou, and A.N. Anagnostopoulos

9.1 Relation Between Dynamics and Information

The role that dynamic chaos plays in processing information by human and animal brains is extensively investigated in the last decades. The very existence of chaotic modes in the brain is considered doubtless, and the efforts of the researchers are now concentrated on the study of those special brain functions, for which chaos is either necessary, or has some advantages compared to simple dynamics.

The last few decades have been witnessing an excessive interest towards processing, memorizing and storing information in live systems. Unlike the addressed memory now used in computers, the memory of humans and animals is associative, i.e., both storing and retrieval of information are based not on the index of a memory cell but on the content [37].

In the literature, there are quite a number of different approaches realizing the association principle. One of the most popular among them is that of the neural network models [17, 30, 33]. Such models are described as dynamical systems and the objects being memorized or recognized are related to basic attractors, viz. stable modes. The basin of attraction defines the limits of recognition of each image.

Y.V. Andreyev
Kotelnikov Institute of Radio Engineering and Electronics of RAS, Moscow, Russia
e-mail: chaos@mail.cplire.ru

A.S. Dmitriev
Kotelnikov Institute of Radio Engineering and Electronics of RAS, Moscow, Russia
e-mail: dmitr@chaos.msk.su

A.N. Miliou (✉)
Department of Informatics, Aristotle University of Thessaloniki, Greece
e-mail: amiliou@csd.auth.gr

A.N. Anagnostopoulos
Department of Physics, Aristotle University of Thessaloniki, Greece
e-mail: anagnost@physics.auth.gr

S. Banerjee et al. (eds.), *Applications of Chaos and Nonlinear Dynamics in Engineering – Vol. 1*, Understanding Complex Systems, DOI 10.1007/978-3-642-21922-1_9,

The neural network models are attractive and computationally efficient, but still they possess a number of features which are not characteristics of live neural networks. In particular, available experimental data reveal, in a straightforward manner, the importance of complex dynamics and chaos in the information analysis of live neural systems [53]. To this end, investigation of the principles of information handling based on the nonlinear dynamics appears to be very interesting.

The functional role of cortical chaos appearing due to thalamo-cortical interaction is discussed in [44], and [47]. Chaos is considered as a possible mechanism of self-referential logic, and as a machine for a short-term memory based on this logic.

W.J. Freeman observed chaotic activity in the learning process of the rabbit's olfactory system [25, 27, 28, 53, 64]. He has found, that the rabbit remembers a known smell by coding it in a spatially coherent and temporally periodic activity of the olfactory potential. In the case that the animal feels a new smell the coding mechanism ceases to work and the activity of the olfactory bulb becomes a low-dimensional chaos, as if it were a "novelty" filter, forming the state "I don't know".

Based on the analysis of human electro-encephalograms a hypothesis was suggested by [12, 13, 19], that the functional role of chaos is determined by the property of chaotic dynamics to increase the resonance capacity of the brain, giving a chance for extremely rich responses to an external stimulus.

Further hypotheses about the functional role of chaos comprise a nonlinear pattern classifier [28, 64], a catalyst of learning [53], a stimulus interpreter [58], a memory searcher [61], etc. A more thorough list of possible roles of cortical chaos in information processing can be found in [59], along with a rich reference base of research between 1970s and 1990s.

Studies of the systems with dynamic chaos demonstrate the close connection between the theory of dynamical systems and information processing. A number of basic results in dynamic theory have been formulated with regard to objects coupled with information. For example, Sharkovskii's Theorem [51] proves the existence of a countable set of cycles of certain structure, in 1-D dynamical systems.

Countable sets of periodic motions appear also in continuous-time systems [1,52, 56]. To describe the behavior of such systems, methods of symbolic dynamics are used [35,40], which are based on notions of complexity and information [16,38,39].

Procaccia expressed a few ideas [50], indicating a connection between chaos, unstable periodic orbits and information properties of dynamical systems. Chaotic orbits can be organized around the skeleton of unstable periodic orbits. Each periodic orbit (or a point) can be universally encoded, using symbolic dynamics which employ a "grammar" defining permitted "words". Such a grammar can be universal which means that different dynamical systems, belonging to the same universal class, have the same distribution of periodic orbits in the corresponding space. Periodic orbits and their eigenvalues can be derived directly from experimental data as shown by the algorithm given in [11]. Details of the above ideas are also discussed in [18, 34].

The problem of information streams in 1-D maps is analyzed in [63]. The authors refer to the studies [26, 54, 55], in which information is argued to be a fundamental concept in the theory of dynamical systems and chaos. In particular, sensitivity to initial conditions rigorously refers to information production. Further, the authors consider a dynamical system described by a mapping f of an interval into itself and show that its iterates produce a special process which they call information stream.

In [43] the rates of information streams are estimated, and restrictions imposed by computers are discussed. The rate of information stream is interpreted by the volume of new data per unit time. For coupled 1-D maps derived from experimental data in the Belousov-Zhabotinsky reaction, the authors showed that the data rate is equivalent to Kolmogorov entropy K_S.

Another reason to consider dynamic chaos, from the information viewpoint, is the existence of natural objects with deterministic chaotic dynamics [10, 62] or with mixed dynamics, containing both deterministic chaos and a random process. As a rule, there is a 1-D signal, which could be processed in order to obtain more or less detailed information about the object's dynamics. Such a process reveals a method of getting information about the object using the chaotic process that takes place in it.

In [29] Jeff Hawkins analyzes contemporary knowledge on human brain from the viewpoint of technical implementation of the main principles of brain functioning. He claims that intelligence must be treated as memory, rather than behavior, and points that the brain is purely a dynamical system with nothing static in it. According to his opinion, the brain operates as predictive hierarchical associative memory, in which patterns are stored as dynamic structures (e.g., cycles), and the input signal to the brain is time-varying. Neo-cortex stores pattern sequences; calls patterns back associatively; stores them in an invariant form, and hierarchically. Associative memory is activated (accessed) by means of applying time-varying patterns at the input (or parts of patterns or distorted patterns). Activation goes up the cortex and retrieves a sequence of the stored patterns. At this point, we would like to stress the idea that the carriers of information in such a model of the brain are dynamic objects rather than fixed points of any kind.

Thus, experimental investigations of electric activity of the brain and its certain neural sub-systems, simulation of various neural networks and qualitative analysis of information processes in the brain allowed us to suggest and to prove, to some extent, several hypotheses about the role of chaos in the brain activity.

The use of different approaches, models and methods in the study of the functional role of chaos leads us to the idea of the existence of general principles of information processing in chaotic systems, independently of their concrete nature and realization. This allows us to hope that we can investigate the main relations of information processing using simple models. However, we face the problem of the proper choice of a convenient dynamical system, i.e. a system which is simple enough, allows thorough description, and at the same time exhibits complex and chaotic behavior.

The approach that we follow implies that information processing in a dynamical system is associated with the notion of an attractor in the system phase space carrying information. Information processing, e.g., recognition, is associated with structural transformations of the attractors (bifurcations) and essential change in the system's behavior.

The first step in information processing, storing, is coupled with the synthesis of a nonlinear dynamical system with a specially structured phase space, i.e. with attractors corresponding to stored information. This approach is used in neural networks where for a given set of images a neural network is synthesized (trained) in such a way that these images correspond to equilibrium states of this dynamical system. Therefore, the simplest type of attractor, a stable point in the system's phase space, is used as the carrier of information.

There are also reports in the literature (e.g., [14, 59, 60]) using more complicated attractors, such as cycles (periodic orbits) and strange attractors, to carry information in neural networks. But the enormous complexity of the cooperative motion of neurons in conventional neural networks makes the direct synthesis (calculation) of these networks very hard or even practically impossible. To design such a dynamical system, one has to use time-consuming procedures of training, which obscure investigations of the general principles of information processing.

Our effort is to derive some simple mathematical models allowing easy and thorough description and exhibiting complex or even chaotic behavior. Issuing from the above concept of the existence of general principles in information processing independent of the concrete dynamical system, we used a class of discrete time one-dimensional systems, namely, piecewise linear maps of an interval into itself $x_{n+1} = f(x_n)$. The efforts were concentrated on the synthesis of dynamical systems with prescribed cycles in the system phase space. As a result, a method of storing information using stable limit cycles of 1-D maps as information carriers was proposed [20, 22].

(The term "stable limit cycle", means a discrete finite-period cycle which is the limit for any phase trajectory starting from any point within a certain vicinity of a cycle. Similar considerations can be applied to the term "unstable limit cycle").

The use of more complicated attractors, i.e., cycles rather than equilibrium points, offers new capabilities in information processing, for example, associative memory [20, 22]. Further investigations have shown that this method of storing information can be applied in practice to storing pictures, texts, signals, etc., [4–6, 21, 24]. The method was also extended to store information in 2-D and multi-dimensional maps and to store multi-dimensional information sequences (cycles of vectors) [7].

In this chapter, we describe the original method along with its developments and generalizations, including the use of chaotic systems and unstable limit cycles for storing information, and discuss opportunities of information processing in such systems.

In Sect. 9.2, we briefly describe the original method [20, 22] of storing information in 1-D maps, as stable and unstable cycles, the realization of associative memory, and the estimation of the storage capacity.

In Sect. 9.3, we investigate dynamic properties of the map with stored information.

In Sect. 9.4, we discuss the use of unstable limit cycles for memory scanning.

In Sect. 9.5, we demonstrate that information can be stored as chaotic attractors.

In Sect. 9.6, we investigate 1-D maps as recognition machines. Information is retrieved from 1-D maps using various methods of map function modification. "Long-term" and "short-term" memories are shown in the adaptive memory model.

In Sect. 9.7, we demonstrate how the original method is generalized to store information in multi-dimensional maps.

In Sect. 9.8, we give some examples of information processing applications based on the described theory.

Finally, we summarize the information processing functions realized in the proposed dynamical systems, and draw some conclusions on the role of chaos in the discussed models.

9.2 Storing Information as Dynamic Attractors of Nonlinear Systems

Processing information with dynamical systems includes the:

- Choice of attractor types, suitable for processing;
- Choice of dynamic phenomena, necessary to implement basic operations of information processing;
- Development of principles of the unambiguous relation of information with the trajectories of the dynamical system;
- Development of concrete mathematical models allowing to process information as map trajectories and to control dynamic phenomena, in order to implement basic processing operations;
- Development of software to simulate dynamic processors;
- Investigation of dynamic processor models;
- Solution of complex problems with the dynamic processor, which are to solve with traditional approaches.

In chaotic maps, instability leads to uncertainty of the location of the phase trajectory. Information on the system's initial state (always known with finite accuracy) is lost in transitions from the previous state to the next one. Therefore, one can talk about information production in such systems. If the map has a regular behavior (e.g., periodic motion), then its future is completely determined, and information production is zero. Starting from arbitrary initial conditions, the trajectory falls into the basin of attraction of the regular attractor (fixed point/limit cycle) and consequently information on initial conditions is lost, since the trajectories from different initial points become indistinguishable. Thus, information "vanishes", and regular attractors can be seen as information "sinks".

Hence, nonlinear maps can either produce or annihilate information. Each trajectory of a dynamical system can be treated as an information signal, so, the set of the map trajectories is a certain information "depository". This "depository" has a number of interesting features, depending on the type of attractors of the dynamical systems.

9.2.1 Synthesis of 1-D Maps with Prescribed Attractors

In [20, 22] we proposed to store information in nonlinear dynamical systems as dynamic attractors (cycles, periodic orbits or even chaotic attractors). A method of storing information as stable cycles of the maps of a segment into itself was proposed. This is actually a method for synthesis of a dynamical system (1-D maps), in phase space at which there exist cycles of prescribed structure. This method is based, in part, on the ideas of symbolic dynamics [35, 40]. We partition the phase space of the dynamical system into adjacent regions and give symbols to them. Then, when the phase trajectory visits some region, we treat this as "appearance" or "production" of the corresponding symbol by the dynamical system. In the theory of symbolic dynamics this partitioning (generatrix) must be precise, in order to provide an unambiguous relation between the system variables and the produced symbols. We design the dynamical system as follows: first, we partition the system phase in a way that we consider convenient to us, then we construct the cycles in the system phase space that run through the necessary space regions, which would mean production of the required symbols in the prescribed order, and finally, design a dynamical system which has these cycles in its phase space.

Let us introduce the main notions and terms on an example of storing two information blocks, finite 1-D strings *babe* and *add*. For simplicity, we use a subset of the Latin characters $A = \{a,b,c,d,e\}$ as the alphabet. The length of the alphabet is $N_A = 5$. Our aim is to design the function of an 1-D map $x_{n+1} = f(x_n)$, such that, stable cycles exist in the phase space of this dynamical system. Each information block of length n stored in the map is unambiguously related to an n-period cycle γ_n. The symbols of the strings are coded by the amplitude of the mapping variable x_n. We will store the words in an 1-D map using second-level storing, which means that each point of the cycles is determined by a pair of successive symbols.

We divide the phase space of the dynamical system (the unit interval $I = [0,1]$) into N_A subintervals of the first level (each with length $1/N_A = 0.2$) and relate them to the elements of the alphabet. Then, we repeat the procedure and divide each of the subintervals of the first level into subintervals of the second level (with length $1/N_A^2 = 0.04$) and also relate them to the alphabet elements, as shown in Fig. 9.1.

Now, we design two cycles $\gamma_n = \{x_1, x_2, \ldots x_n\}$ unambiguously related to the stored information blocks. Three cycle points for the word *add* are related with the block fragments (pairs) *ad*, *dd*, and *da* (the information block is mentally closed in a loop). The cycle point corresponding to the fragment *ad* is the center of the second-level subinterval corresponding to the symbol d located within the first-level subinterval corresponding to the symbol a. Other cycle points are created similarly. The cycle points corresponding to block babe are determined by the pairs *ba*, *ab*, *be*, *ea*.

Having created the cycles γ_n in 1-D phase space, we construct a dynamical system possessing the phase space with such a structure. In order to obtain a map with a cycle of period n passing through the points $x_1, \ldots, x_n$, it is necessary to plot the points $(x_1,x_2), (x_2,x_3), \ldots, (x_n,x_1)$ on the plane (X_m, X_{m+1}) and draw the curve

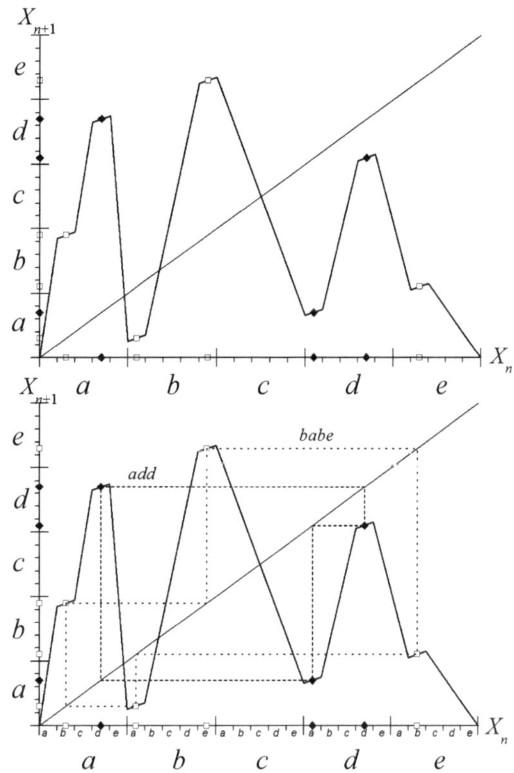

Fig. 9.1 Storing two information blocks *babe* and *add* in an 1-D map. Cycle points are designated with squares and diamonds, respectively. Storage level $q = 2$, $s = 0.5$. (**a**) The map function. (**b**) Information carrying cycles

$y = f(x)$ passing through these points. This permits us to design a map containing an arbitrary finite number of stable cycles with prescribed period lengths and structures.

In the plane (X_m, X_{m+1}) we plot the pairs of successive points (x_i, x_{i+1}) for all the cycles. These points form the "skeleton" of the map function $f(x)$. Through these points, we then draw short straight-line segments (called information regions), all with the same fixed slope s. We will control the stability of the cycles by changing the slope of these segments (turning them around the central point which coincides with the cycle point lying on this segment).

It is known that the stability of a cycle is determined by its eigenvalue λ. In the case of an 1-D map $x_n = f(x_{n+1})$, the eigenvalue for the cycle $\gamma_n = \{x_1, x_2, \dots x_n\}$ is equal to

$$\lambda = f^{(n)'}(x_1) = f'(x_1) \cdot f'(x_2) \cdot \ldots \cdot f'(x_n) \tag{9.1}$$

Here, $\lambda = s^n$. If $|\lambda| < 1$ (i.e., $|s| < 1$) the cycle is stable, otherwise it is unstable. To complete the synthesis of the piecewise linear map function $f(x)$, we connect the information regions and the unit interval endpoints in series with straight-line segments, which we will call non-information segments. The plot of the map with the information cycles is shown in Fig. 9.1.

Iterates of the designed map produce the output information stream: an occurrence of the system variable x_i in a first-level subinterval is treated as "generation" of the corresponding alphabet element. Mathematically, it is $m_i = int(N_A \cdot x_i)$, where m_i is the order number of this element in the alphabet, and $int(\cdot)$ denotes the integer part of a number. Thus, the motion of the phase trajectory over a cycle in the system phase space is accompanied by continuous reproduction of the corresponding information block.

Storing information as stable limit cycles allows easy associative access to the stored information. If an equilibrium point is used as an information carrier, all information or most of its part, is necessary to access the point and to retrieve the information. If an image is stored as a stable cycle, as in our case, each point of the cycle is related to only a part of the image, and only a piece of the original information is necessary to get a point near the cycle and to retrieve the whole image by iterating the dynamical system. Thus, associative access to the stored information becomes possible, yet in the expense of iteration time. Indeed, if we take an excerpt $a_i\, a_{i+1}\, \ldots\, a_j$ of an information block of length greater than or equal to the storage level q, then we can apply the same procedure as before and get a point lying at the corresponding cycle: we take the first-level subinterval related to a_i, then the second-level subinterval within it corresponding to a_{i+1}, and so on q times. At last, due to the map design procedure, we get a point within the q-th level subinterval, in an information region related to this information block. Nothing else is necessary now, but to iterate the map until it returns to the same point, and then the entire stored image is recovered. Actually, we can begin from any point within this information region and the output information stream will be the same.

Note that this is a direct access to information, because the offered excerpt is not compared to all the images. Instead, an initial point lying at the required cycle is directly calculated, so the access to information is very fast.

9.2.2 Storing Texts, Signals, Images in 1-D Maps

Not only simple letter sequences can be stored in 1-D maps but any kind of information that permits representation in the form of a sequence of symbols of certain alphabet. Texts, digitized signals, vocal sheet music, they can all be represented by certain symbols from a finite set, i.e., they can be stored as cycles of an 1-D map.

Moreover, the method can be applied to store more complex data, e.g., 2-D graphic images, since they can also be transformed into 1-D symbol sequences.

To store a picture on the map, we first need to divide a two-dimensional picture into a set of elements. Then, we choose an alphabet of these elements, and introduce the order of scanning, to generate a string of symbols. Each of these steps can be done in different ways. We will proceed as follows.

The first step consists of a spatial digitization, in order to transform the initial picture into a pattern with $m \times n$ cells. The color of each of these cells is digitized to the nearest color from a palette of a finite number of colors. This color palette makes the basis of the alphabet. Let there be N_c colors. The pattern corresponding to the picture consists of $m \times n$ elements from this alphabet. We can express this pattern as a two-dimensional matrix:

$$A = [a_{ij}], \quad i = 1, \ldots, m \quad \text{and} \quad j = 1, \ldots, n \tag{9.2}$$

where each a_{ij} is an element of the alphabet.

The next step is the construction of the information block. As was noted before, this could be done in different ways (depending on the particular problem), but the transformation of the $m \times n$ pattern to an information block and the inverse transformation must both be one-to-one.

We will transform the two-dimensional pattern (matrix A) into an information block by reading it line by line from top to bottom. As a result we obtain an one-dimensional sequence of $m \times n$ symbols:

$$a_{11} \;\ldots\; a_{1n} \;\ldots\; a_{21} \;\ldots\; a_{2n} \;\ldots\; a_{m1} \;\ldots\; a_{mn} \tag{9.3}$$

As the cycle of period p repeats itself after p iterations (here, $p = mn$), all cyclic permutations are equivalent and generate the same trajectory. We can say that the system loses its initial phase when converging towards the cycle.

This situation is undesirable when working with pictures, since a picture has a definite beginning and a definite end, and when a cyclic permutation of such an information block is transformed back into a two-dimensional array, the result is difficult to recognize. Therefore, we mark the beginning of the information block by assigning a label (a special symbol of the alphabet). Correspondingly, when restoring the pattern from the values of the mapping variable on the limit cycle, we consider the first element after the label as the beginning of the picture (the label itself is not displayed). Thus, the alphabet is augmented with the special symbol for the label. The label element is used only once in each information block.

As an example, let us make an information block from a picture of the letter E. The procedure is shown in Fig. 9.2. We express this picture as a sequence of binary elements (black and white) and add an element corresponding to the label at the beginning of the sequence. The picture is represented here by an 8×8 array. Assume that the alphabet consists of three elements 0, 1 and 2, where 0 corresponds to black, 1 corresponds to white, and 2 corresponds to the label. Then the information block is written as
2111111100110001001101000011100001101000011000101111111000000000.

Now we determine the minimum number of levels necessary to store the image unambiguously. Observe that the above string contains two identical substrings (001101000011) of length 12. Hence, the number of storage levels cannot be less than 13.

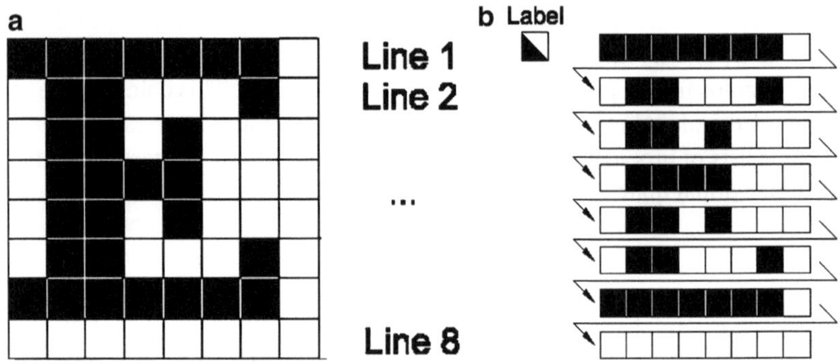

Fig. 9.2 Representation of the letter E. Label symbol is added to the string of black and white elements

9.2.3 *Information Capacity of Memory on 1-D Maps*

The question of information capacity of the method of storing on 1-D maps comes naturally, when it is compared with other methods of storing and retrieval of information.

The discussed method has some restrictions which could be illustrated with the following example: Let the alphabet be an ordered set of digits $\{0, 1, \ldots, 9\}$. Obviously, information block 174 can be stored at the first level. On the same map another information block 268 can also be stored. However, it is impossible to store three blocks 174, 268 and 173 on one map, because it becomes ambiguous. At the same time, storing these blocks at the third or higher levels is possible. Thus, the requirement for single-valued map imposes certain restrictions on the amount and structure of the stored information.

From the method of designing 1-D maps we can deduce that the absence of information blocks of identical fragments with length equal or exceeding the storage level determines the possibility of storing images. If this provision is not fulfilled, the map becomes multi-valued. This is the condition of "orthogonality" of the stored images.

Information capacity of the memory is the maximum number of images (information blocks), that can be simultaneously stored in a dynamical system. It follows from the above analysis that the question of information capacity is reduced to the count of the number of "q-orthogonal" images for the alphabet of length N.

For certainty, we consider storage of equal-length blocks (without labels). First, we find the number of various images of length l which obviously is N^l. Note, that the cycle for an information block coincides with the cycles of all cyclic permutations of this block (there are l of them). That is, blocks 174, 741, 417 are indistinguishable for this method. Since there are images that coincide with some of their cyclic permutations (e.g., an image composed of l equal symbols), the total number of different images of length l is greater than or equal to N^l/l.

We then consider the first case of $q \geq l$, i.e., the storage level is greater than or equal to the image length. Let p be the number of information blocks coinciding with some of its cyclic permutations. These cannot be stored in 1-D maps as length-l images. In principle, all information blocks can be stored, because two different images of length l can have identical fragments of length less than l and consequently, less than q. So, these images comply with the "orthogonality" conditions. The total number of information blocks that can be stored in this case is $(N^l - p)/l \leq N^l/l$.

In the second case of $q < l$, the storage level is less than the image length. Let an information block $C = c_1c_2 \dots c_n$ be stored, where c_i are alphabet elements. In order to check, whether C complies with "q-orthogonality", it is necessary to consider all the fragments of information blocks of length q, $C_i = (c_ic_{i+1} \dots c_{i+q-1})$, $i = 1, \dots, l$. If all the fragments are different, then the information block can be stored. However, these fragments cannot be used in any other stored image.

The total number of all possible fragment of length q in the discussed case is N^q. To store one information block, l of them are used, whereas all of them are "orthogonal". Other blocks containing these fragments cannot be stored. The stored information block of length l contains l different length-q fragments. So, of all possible information blocks of length l (there are no less than N^q of them) at most N^q/l can be stored.

Thus, information capacity E of the method of storing images as limit cycles of 1-D maps is:

$$E \leq \begin{cases} N^l/l, & \text{if } l \leq q \\ N^q/l, & \text{if } l \geq q \end{cases} \tag{9.4}$$

As an example let us estimate how many six-letter words could be stored using three levels. The alphabet consists of 26 elements. The total number of information blocks (words), with cyclic permutations is considered the same, and it is larger than $26^6/6$. If we store the word *cipher*, then we would not be able to store information blocks containing *cip*, *iph*, *phe*, *her*, *ere* and *rci* as substrings. Having stored the word *coffee* along with the word *cipher*, we exclude six more fragments (*cof*, *off*, *ffe*, *fee*, *eec* and *eco*) from the list of available fragments. With a computer we processed all six-letter words ($N = 26, l = 6, q = 3$) and found that the capacity of the method at the third level is at most 2929.

It is evident from Eq. (9.4), that $E \leq 1$ when $l = N^q$, and if $l > N^q$ then no images can be stored. Therefore, for a given N and n, the minimum number q of storage levels is given by:

$$q = int(log_N l) = int\left(\frac{ln l}{ln N}\right) \tag{9.5}$$

The *relative capacity* E_{rel} of the method is defined as the ratio of the maximum number of images that can be stored unambiguously to the total number of images and is given by:

$$E_{rel} \leq \begin{cases} 1, & \text{if } l \leq q \\ 1/N^{l-q}, & \text{if } l \geq q \end{cases} \tag{9.6}$$

The relative capacity depends only on the difference of q and l, and it drops exponentially with a decrease of the number of storage levels with respect to the length of the images.

9.2.4 Coding of Information Blocks

It is evident from the above estimates that the information capacity of 1-D maps as data depositories is not very large. Besides, it is difficult to store similar information blocks with the original method. The number of the storage levels must be greater than the length of identical fragments, but an increase in the number of storage levels leads to exponential decrease of the size of information region. Finite accuracy of computer calculations puts a limit on the potential number of storage levels.

To overcome this difficulty, a development of the original method is proposed in [5, 6], where it is shown that the amount of information that can be stored and retrieved could be essentially increased by a special encoding of the initial information blocks based on compression (elimination of redundancy).

Analysis of information sequences (pictures, texts, etc.) shows that the main difficulties in storing are associated with long uniform pieces of data, although the common sense tells us that little information is hidden in them. This leads us to an idea of compressing information before storing it, but the compression method must match the requirements of storing (e.g., at level q). We call this process of compressing data (elimination of redundancy) "orthogonalization", in order to stress that the aim here is to provide the possibility to store the images with our method [6, 7, 9].

One of the possible algorithms is encoding the information using an alphabet of repeating fragments. The idea is as follows: if the information blocks have more than one identical fragments of length q, then the alphabet is extended by one new symbol, representing this fragment, and in the information blocks these fragments are substituted by this new symbol. Therefore, new information blocks of less length are obtained, and the alphabet length becomes $(N+1)$. The procedure is repeated until the blocks become "orthogonal", i.e., containing no identical fragments of length q.

As a result, shorter orthogonal encoded information blocks are obtained, together with an extended alphabet, which consists of the original alphabet of length N and the added alphabet, i.e. the symbols used to substitute the fragments of length q.

Thus, the encoding procedure means elimination of information redundancy by means of coding the repeating fragments with short symbols. Unlike the above approach, in which storing is provided by means of increasing its level, here the storage level is fixed and the alphabet is extended. This gives us in principle the possibility to store any set of information blocks at any level, beginning from the second.

The coding method is reversible, i.e., lossless. To decode information, the symbols of the added alphabet are substituted by length-q fragments. Possibly, this procedure must be repeated several times, until the fragments of the elements of the added alphabet are expressed through the original alphabet elements.

Obviously, the coding procedure is not unique. The results of encoding the symbols of the encoded sequences and the extended alphabet are determined by the order of alphabet extension with identical fragments, i.e., by the procedure of their search.

9.2.4.1 Information Compression

As was noted above, the method of coding (orthogonalization) information sequences is developed with the aim to store an arbitrary set of information blocks in 1-D maps. Its "side effect" is lossless information compression, i.e., a decrease of the total amount of data, necessary for complete and exact description of information objects, e.g., texts or images.

At present, there are several known methods for information compression. These can be divided into two large groups: reversible (lossless) and lossy methods. Lossy compression methods allow greater compression at the expense of more losses. Typical compression ratio of reversible methods ranges between 2 and 5, whereas that of lossy methods ranges between 10 and 30, depending on the information structure. Lossy methods are applied only to pictures and audio signals, because in other information areas (processing texts, data storage and transmission) losses are unacceptable.

From the viewpoint of information compression, the described orthogonalization method is very close to known lossless compression methods, such as arithmetic, Lempel-Ziv (LZ), etc. As a result, the compression ratios and other parameters of the described method are very close to the ones of these methods.

9.2.4.2 Associative Information Retrieval of Coded Data

To implement associative memory, i.e., recovery of a stored image by its arbitrary fragment, one must set the initial conditions on the related attractor. For this, it is required a fragment $(a_j\, a_{j+1}\, \dots\, a_{j+q-1})$ of length q of information block, which is used to calculate the initial point x_0:

$$x_0 = \sum_{j=1}^{q} \frac{a_{j+k-1}}{N^k} \tag{9.7}$$

Associative memory is organized as follows.

In order to be stored at level q, information blocks are orthogonalized, and then the encoded blocks are stored as limit cycles of 1-D maps. If a fragment of the encoded information block of length q is known, then according to expression (9.7) the initial point x_0 can be calculated, the encoded information block can be recovered, and the original information block can be decoded back. However, recovery of the original block by a fragment of the original block (not the coded block) is much more interesting.

To realize associative access in this case, we should be able to transform a given request in initial alphabet into the corresponding encoded fragment. The first step is encoding this request using the present added alphabet, i.e., the table of fragments obtained by orthogonalization of information blocks.

The problem is that the fragment given as request can begin with an arbitrary symbol of the original information block, including those that don't exist in the encoded block, because they were incorporated into new symbols. For example, a block $(abcdefghijk)$ after encoding it at the third level becomes $(xdyhz)$, where $x = (abc)$, $y = (efg)$, $z = (ijk)$, and the associative memory system is presented as a fragment of the initial block $(cdefghi)$. When this fragment is encoded from the initial to the new alphabet, a new sequence $(cdyhi)$ is obtained, that contains a "correct" piece of encoded block (dyh) in the center and "garbage" elements in the beginning and the end.

To get rid of them, we use the following procedure. For the first q elements of the encoded requested fragment we calculate the initial point using expression (9.7). Then, we determine if this point fits in any information interval. We can do this, because we know the map function. If x_0 doesn't fit in an information interval, then at least the first symbol of the encoded fragment is "wrong". In this case, we proceed to the next element and so on, until we find that the initial point x_0 fits in an information interval. This means that an existing fragment $(b_1\ b_2\ \ldots\ b_q)$, from which the initial point x_0 was obtained, is a candidate for the beginning of a "correct" part of the encoded fragment. In order to check this, we iterate the map several times. The initial point x_0 gives rise to a sequence of iterates $x_0, x_1 = f(x_0), x_2 = f(x_1) = f(f(x_0)), \ldots$, and to the corresponding information streams $c_0 = int(N_{x_0}), c_1, c_2, \ldots$. If $(b_1\ b_2\ \ldots\ b_q)$ is really a correct piece, and the "garbage" elements, in the beginning, are already removed, then $c_0 = b_1, c_1 = b_2$, etc. If this is not the case, then the fitting of the information interval was random, and the element b_1 of the sequence is also wrong. In this way if there is a correct piece of length q (at least) in the encoded requested fragment, it will be found. If its length is not less than q, it could be used to get an initial point on the attractor. When an initial point x_0 is obtained, the entire information sequence can be easily restored and, consequently, the entire image.

If the whole encoded fragment was examined and the initial point on the attractor could not be found, this would mean that the length of the requested fragment was insufficient and after encoding there was no correct fragment of length q in it.

As a result, the system of associative memory based on the described principles, in response to the offered information block or part of it, practically forms immediately one of the following two answers: (a) it either returns the initial point x_0 on the corresponding attractor (limit cycle) from which the entire block can be restored; or (b) it responds that the offered information is insufficient to unambiguously locate the initial point on the attractor, or the requested block is not stored in the map. When a block fragment is offered to the system of associative memory, the system either, without error, restores the information block, or informs the user about its inability to perform the task. An error is impossible due to unambiguity of the requested encoding procedure, the "orthogonality" of the stored information blocks, the single-valuedness of the map itself, and due to the above-described procedure that compares the encoded fragment with the information stream generated by it, during the search for an initial point on the attractor.

Note that the formation of an initial point on the attractor and, consequently, the recovery of the initial image by its fragment takes place without a comparison of the requested fragment with all stored images. After the encoding of the request and calculation of the initial point x_0 just a few iterates are necessary in order to determine its presence on the map attractor. The time for each iteration is proportional to the logarithm of the volume of stored information.

Thus, the described system of associative memory has the property of a very fast correlator.

In the case of software realization, the amount of the stored information is limited only by the volume of computer RAM, which means that Gigabytes of information (texts, signals, pictures) can be stored in ordinary personal computers while retaining the property of very fast associative access [21, 24].

9.3 Bifurcations of the Maps with Stored Information

To effectively process information in a nonlinear dynamical system, we must have knowledge of the behavior of the trajectories and the existence and stability of dynamic structures in its phase space [8]. In this section we discuss the phenomena that occur when the stability of limit cycles with stored information is lost and when the slope s is increased above unity.

Bifurcation diagrams of 1-D maps with stored information are analyzed using the results obtained in [41, 42]. As it is shown in these papers, in piecewise linear maps beside the ordinary cycles of points $\gamma_m = \{x_1, x_2, \dots, x_m\}$ there can be cycles of intervals[1] $\Gamma_m = \{I_1, I_2, \dots, I_m\}$, i.e., chaotic attracting sets composed of a finite number of intervals I_k. When moving on this attractor, the trajectory goes through the points $x_1, x_2, , x_m$, which are located sequentially on the intervals $I_1, I_2, \dots, I_m$. In [41, 42] the theory of piecewise-linear maps with a single extremum was developed and birth mechanisms of cycles of chaotic intervals and dynamical properties of these attractors were investigated.

[1]Cycles of intervals, also known as "noisy intervals", can also take place in smooth maps. For example, the attractor of logistic maps $x_{n+1} = \lambda x_n(1 - x_n)$ at $\lambda = 4$ with continuous chaotic set on segment $[0, 1]$ can be treated as a cycle of chaotic intervals of period 1.

Prior to commenting on bifurcation diagrams of the maps with stored information, let us consider schematically the phenomena in the vicinity of information regions during the bifurcation of stability loss of limit cycles at $s = 1$ (cycle eigenvalue $\mu = +1$).

The stability of period-m motion of the map $f : I \to I$ is convenient in order to be analyzed by the map f^m (which is m times the applied function f). Limit cycle $\gamma_m = \{x_1, x_2, \ldots, x_m\}$ of period m of map corresponds to the fixed point x_c of the map f^m, where x_c is one of the points of the cycle γ_m. The stability of the limit cycle γ_m is unambiguously related with the stability of the fixed point x_c of the map f^m; cycle γ_m is stable, if x_c is stable, i.e., $|f^{m'}(x_c)| < 1$, and vice versa, whereas $|f^{m'}(x_c)| = |f'(x_1) \cdot \ldots \cdot f'(x_m)| = |\mu|$, i.e., the cycle's eigenvalue.

In the critical point $s = 1$ all the points of the information region at $f^m : I \to I$ are fixed points and they lie on the bisector $x_{n+m} = x_n$. When s goes through $+1$, at the border of information and non-information segments it appears a "corner", similar to that depicted in Fig. 9.3a, where the right side of the information region is drawn (x_c is in the middle of the information segment, i.e., the cycle point O_R is the vortex and x_R coincides at first with the right border of the information region).The slope of the segment $x_c O_R$, which is the image of the right side of information segment of the map f, is equal to s^m, and the slope of segment $O_R y_R$ is equal to $s^{m-1} p$, where $p \approx$ const, is the slope of the non-information segment of the map f, adjacent to the information segment of the previous cycle point ($|p| > 1$). Since the slopes of both segments at the corner are greater than unity, no stable limit cycles can exist here. As it is shown in [31], a chaotic attractor of the type of interval cycle $\Gamma = \{[x_a x_R]\}$ of period 1 appears, where $x_a = f^m(x_R)$.

This dynamic object consists of the interval $[x_a x_R]$ and is attracting since its birth while its motion is chaotic with positive Lyapunov exponent. This chaotic attractor of period 1 of the map f^m is related to the interval cycle Γ_m of period m of the map f, which is obtained by iterations of the interval $[x_a x_R]$ by the map f. This attracting cycle goes through the border areas of the corresponding information regions which is now the unstable limit cycle γ_m. The evolution of the interval cycle Γ_R of the map f^m depends on the slope of the segments $[x_c O_R]$ and $[O_R y_R]$, constituting the corner (Fig. 9.3a–c).

In the moment of birth, the interval cycle coincides with the corner vortex O_R. As s increases, the slopes of the corner segments also increase, and the attractor $\Gamma = \{[x_a x_R]\}$ extends to both sides of O_R. At some critical value of $s = s_1$, when the condition $f^m(x_R) = x_c$ holds, the interval cycle occupies all the right half of the information interval of the original map and part of the adjacent non-information one (Fig. 9.3b). With further increase of s a "hole" $[x_1 x_2]$ appears in the attractor, where $f^m(x_1) = x_R$ and $f^m(x_2) = x_R$, through which the phase trajectory leaves the attractor (Fig. 9.3c) and goes to the left half of the information segment.

This can be treated as a crisis of the attractor, i.e., transformation of the chaotic attractor into transient (metastable) chaos due to its collision with the fixed point x_c.

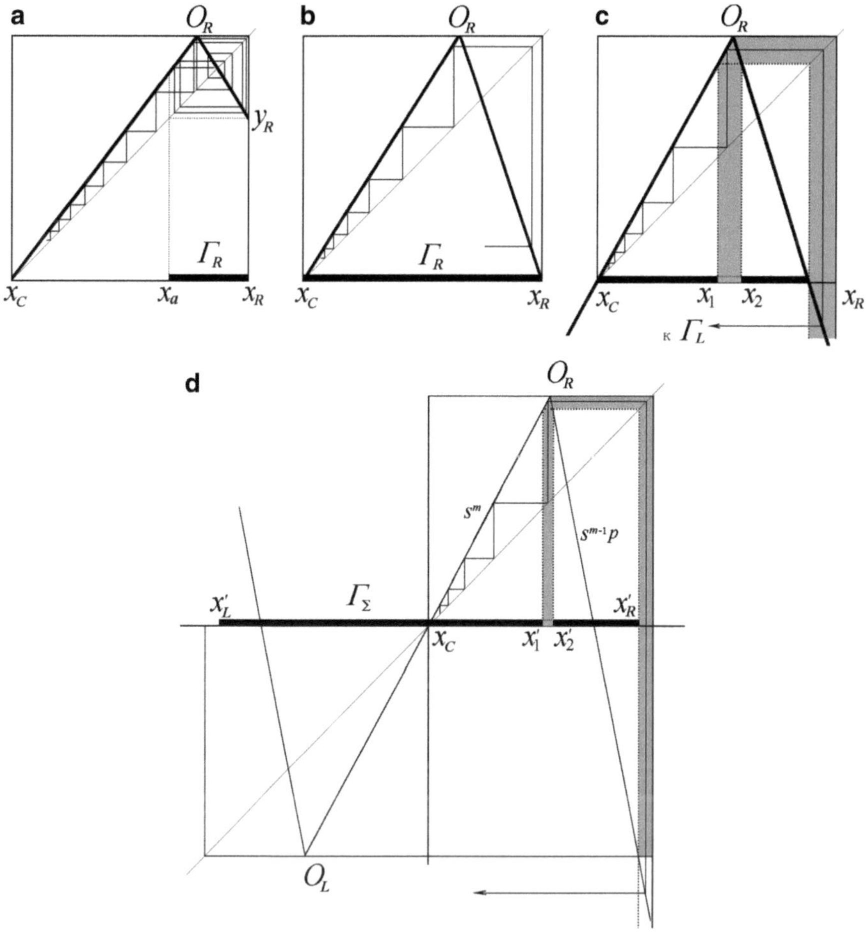

Fig. 9.3 Bifurcation phenomena in maps with stored information (**a**)–(**d**)

Following the approach of [31], the average time of the phase trajectory's presence on the metastable attractor Γ_R can be estimated as:

$$T \sim P^{-1} = \left(\frac{|x_2 - x_1|}{|x_R - x_c|}\right)^{-1} = \left(\frac{s^m}{1 + s/|p|} - 1\right)^{-1} \tag{9.8}$$

where P is the probability of the trajectory to leave the interval cycle, whereas $P = 0$ at $s = s_1$. For s close to s_1

$$T \sim P^{-1} = (m(s - s_1))^{-1} \tag{9.9}$$

Note that at $s = s_1$ the phase trajectory leaves the attractor and gets into the center of information segment (point x_c), and at higher s it gets closer to its left

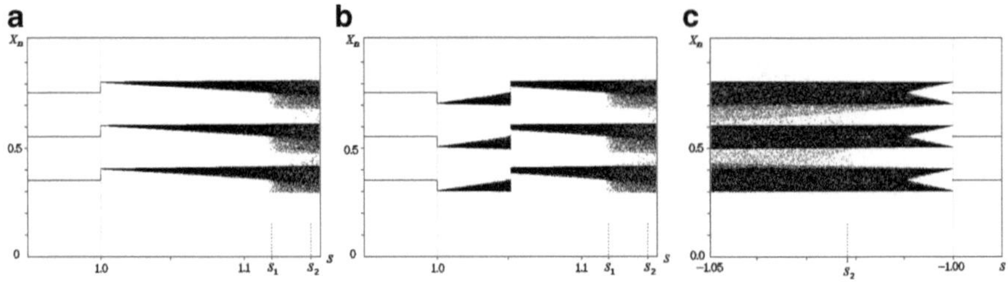

Fig. 9.4 Bifurcation diagrams of the map with a single cycle (**a**)–(**c**)

border. If at the left side of the information segment similar phenomena take place with the attractor Γ_L, then we have the combined attractor Γ_Σ, that unites Γ_R and Γ_L on both halves of the information segment. This attractor is stable, because the channels through which the trajectory can leave Γ_R or Γ_L are directed to each other. As s further increases, it also loses stability: at a certain $s = s_2$ a hole $[x'_1 x'_2]$ appears, where $f^m(x'_1) = x'_R$ and $f^m(x'_2) = x'_R$, and the trajectory leaves the vicinity of the information segment and goes wandering over the phase space (Fig. 9.3d). The dependence of the average time T of the presence of phase trajectory on the unstable set Γ_Σ is similar to (9.8).

Let us illustrate these qualitative speculations with the analysis of the bifurcation diagrams of a map with single information block 375 stored at the first level (Fig. 9.4). The alphabet here is the set of digits {0, 1, 2, 3, 4, 5, 6, 7, 8, 9}, i.e., $N = 10$. If the phase trajectory at the moment of bifurcation birth of the interval cycles at $s = 1$ is at the right side of the information segments with respect to the limit cycle points (Fig. 9.4a), it is attracted to the interval cycle that goes through the right half of the information segments ("right" attractor Γ_R). The Lyapunov exponent for this attractor is $\lambda \approx 0.3$. By increasing s, the measure of this interval cycle on the interval [0, 1] increases, and at $s_1 \approx 1.12$ the attractor reaches the center of the information segment x_c. This is the moment of the stability loss of the "right" interval cycle. At $s > 1.12$ the trajectory, through the appeared channel, leaves the vicinity of the attractor.

However, as it follows from Fig. 9.3c, the nature of the stability loss is such that the phase trajectory leaves the attractor on the right half of the information segment and gets to its left half. If the attractor on the left border of the information segment ("left" attractor Γ_L) was stable at that moment, then the trajectory would remain in it. But it had lost stability even earlier (at $s \approx 1.05$), then, after some wandering the trajectory goes back to the "right" attractor, etc.

Thus, at $s = s_1 \approx 1.12$ a united chaotic attractor Γ_Σ of period 3 appears, that embraces all information regions. This attractor is composed of two unstable interval cycles Γ_R or Γ_L, where the only channel of leaving the attractor is directed towards the other, thus any trajectory that enters Γ_Σ remains in it.

As s increases, the running-away channel at the end of the information segment becomes "deeper" (Fig. 9.3d) and at $s = s_2 \approx 1.14$ the united attractor Γ_Σ loses

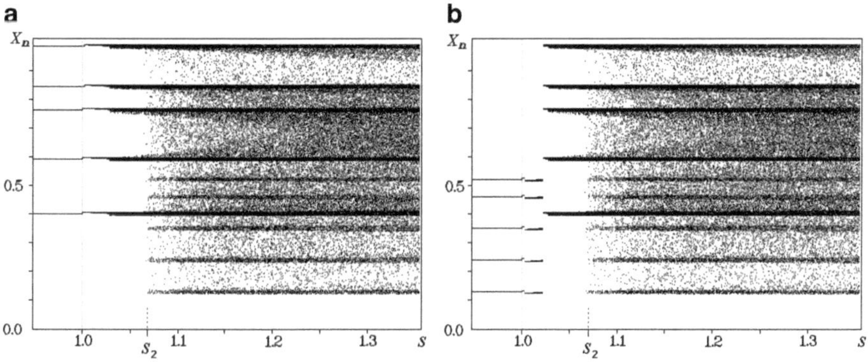

Fig. 9.5 Bifurcation diagram of the maps with two information blocks (**a**)–(**b**)

stability and the trajectory leaves it. By s_2 we denote the moment of stability loss of all interval cycles of the map. At higher s, global chaos is observed in the system.

Figure 9.4b depicts the bifurcation diagram when the phase trajectory was at the left side of the information cycle with respect to limit cycle point at the moment of bifurcation, $s = 1$. In this case it is attracted to the interval cycle Γ_L with Lyapunov exponent $\lambda \approx 0.2$. However, its lifetime is essentially short: it loses stability at $s \approx 1.05$. At higher values of the slope s of the trajectory goes from Γ_L to stable Γ_R, while further events coincide with those described in Fig. 9.4a.

The nature of bifurcation phenomena at the cycle's eigenvalue $\mu = -1$ (at $s = -1$ and odd length of the information cycle) is essentially different of the discussed cases: the loss of stability leads to the birth of a doubled interval cycle (Fig. 9.4c), since for the map $f^m : I \to I$ the slope of the segment with the fixed cycle point x_c, becomes equal to $s^m = -1$, and by each iteration of f^m the borders of the information segments change places with each other. The actual period of the interval cycle becomes $2m$ and not m. Instead of bifurcation with the merge of interval cycles from both borders of the information segment with increasing $|s|$ (at $s_1 \approx 1.009$), the entire information segment is filled due to the extension of the interval cycle, which can be interpreted as inverse bifurcation with period doubling of the interval cycle.

If several information blocks are stored in the system, when $|\mu|$ crosses unity, in the vicinity of each of the corresponding limit cycles, either two attracting interval cycles (at $\mu = 1$), or one doubled cycle (at $\mu = -1$) are born independently. Their evolution with increasing s coincides, in general, with that described above. The slopes of the non-information segments, adjacent to different information segments, are different, so, the "lifetimes" and stability conditions of the parameter s are different for each of these interval cycles. This is shown in Fig. 9.5, in which the bifurcation diagrams for a map with two information blocks 12345 and 97583, stored at the second level, are depicted. The size of the information regions is $N = 10$ times less than in the previous case, and it is equal to 0.01. In Fig. 9.5a, a diagram is depicted for the case, when by the stability loss bifurcation of the limit cycle, carrying the information block 97583, the phase trajectory is attracted to the right interval cycle. At $s \approx 1.03$ the interval cycle at the right border loses stability and the trajectory leaves through a "hole" similar to $[x_1\ \ x_2]$ in Fig. 9.3c. However, it goes towards the attractor on the left border. If it was stable at that moment, the trajectory would remain there. But due to similar instability reasons, the trajectory remains there for some time, then returns to the first attractor, and so on. Here, the birth of a united interval cycle, that embraces the entire information segments of the cycle 97583, is observed.

Figure 9.5b illustrates the case when during the bifurcation of the limit cycle the stability is lost, corresponding to the information block 12345 and the trajectory is attracted to the newly-born stable interval cycle that goes through the right border of the information segments. But it loses stability practically immediately, and the trajectory goes to the interval cycle at the other border of the information segments. At $s \approx 1.015$ the united interval cycle $\Gamma_{5,\Sigma}$ is born and when it also loses stability (at $s \approx 1.025$), there are still unstable interval cycles connected with the other information block 97583, and the phase trajectory is attracted to it.

At $s = s2 \approx 1.07$ the united interval cycle for block 97583 loses stability, i.e., a hole appears in the attractor, through which the trajectory can leave it, as is shown in Fig. 9.3d. Thus, no stable structures remain in the phase space of the dynamical system. In this case, the phase trajectory starts wandering over the system phase space. When the holes first appear on the interval cycles are small, and the trajectory spends most of its time in the vicinity of the interval cycles, but with increasing slopes the holes extend, and the map variable distribution becomes more uniform.

The analysis of the bifurcation diagrams of the maps with stored information shows that at $|s| < 1$ the only stable attractors in this oscillating system are the information limit cycles while at $1 < |s| < s_2$ – corresponds to chaotic interval cycles. At $|s| > s_2$ transition to global chaos, through intermittency, takes place in nonlinear dynamical systems.

9.4 Random Access Memory in 1-D Maps

In Sect. 9.2 we described the method to store information on 1-D maps using the stable limit cycles. We will show in this section the possibility to realize random access in one-dimensional associative memory, in the sense that the trajectory visits all regions which correspond to the information stored in the map. It is clear that we cannot use the stable limit cycles for this purpose, because the trajectory of any point in the basin of attraction of a stable cycle will converge to this cycle and remain on it.

We want the trajectory to visit every region (or just the regions where the information is stored) of the phase space from time to time. For this purpose, we can use a strange attractor with a nonzero distribution of the system variable in every point on the interval [0, 1]. However, in general, a strange attractor visits each region only briefly.

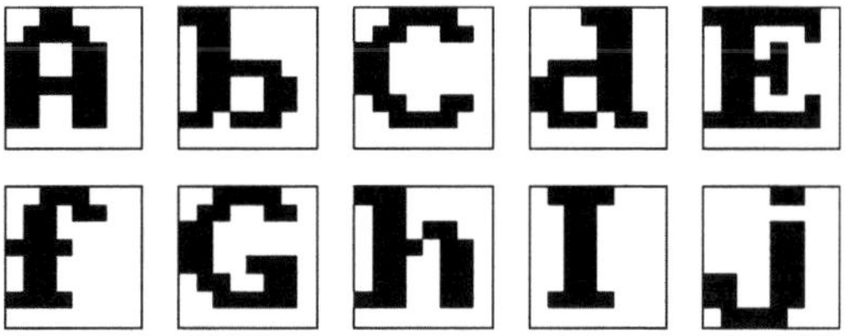

Fig. 9.6 Set of letters stored on an 1-D map

For the random access memory, we exploit the phenomenon of intermittency. The idea of memory scanning using intermittency was discussed qualitatively by Nicolis [45, 46]. In the intermittency mode, the phase trajectory "lingers" in some definite regions of the phase space. So, we want the trajectory to be in the vicinities of the limit cycles corresponding to the information blocks.

"Random access" in associative memory can therefore be realized using the unstable cycles to store the information blocks. Each cycle can be easily be made unstable by letting the absolute value of the cycle's eigenvalue (1) (i.e., the product of the slopes in the cycle points) to be slightly greater than 1. A trajectory starting from the vicinity of a cycle will leave the cycle after some time and begin to wander in the phase space until it gets into the neighborhood of the same limit cycle or another limit cycle with the same stability property. Then, the trajectory will linger near that limit cycle for some time before leaving it again. If, independently of the initial condition, the trajectory visits the neighborhoods of all points, intermittency will then be observed over short but random time intervals with respect to all cycles corresponding to the stored information. It must be noted that intermittency takes place in a rather narrow region of the space parameter only.

Let us demonstrate random-access memory with an example of two-dimensional data, i.e., graphic images.

Ten information blocks, corresponding to ten 8×8 patterns (depicting letters A, b, C, d, E, f, G, h, I, j), are stored in 1-D map at the second level (Fig. 9.6). The information blocks were preliminary compressed. The map is designed in the same way as it was described earlier, with the only exception that the slopes s of the information segments are slightly greater than 1 ($s = 1.02$). Hence, the cycles are unstable. The sequence of snapshots in Fig. 9.7 demonstrates the intermittent appearance of all ten images.

As can be seen in Fig. 9.7, the trajectory visits all districts with the stored information and lingers long enough in the vicinity of the unstable information cycles. By means of switching the slopes of the information segments and making them less than 1, the unstable cycle can be made stable.

Intermittency in chaotic neural-like systems with stored images was also exhibited in [3], and the possibility of using intermittency to implement memory scanning in neural-like systems with chaotic dynamics using an external oscillator similar to the pacemaker, was shown in [23].

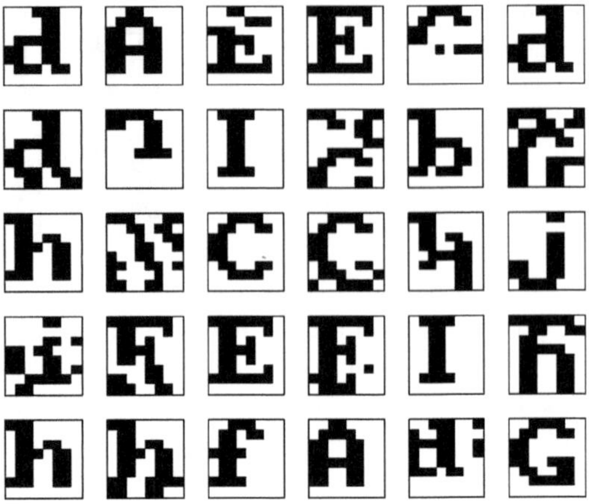

Fig. 9.7 Intermittency between cycles representing letters. Snapshots are taken at arbitrary moments

9.5 Storing Information as Chaotic Attractors

The motion over the interval cycle is chaotic; however, the cycle itself is stable and confined.

The interval cycle consists of a finite number of continuous intervals, that embrace the information intervals of the corresponding information cycle (or parts of them) and small parts of the adjacent non-information intervals. The order of going round the intervals of this chaotic attractor coincides with the round order of the information limit cycle points, located on these intervals. Therefore, if we consider the information stream $a_i\ a_{i+1}\ a_{i+2}\ \ldots$ (where $a_i = int(N_{xi})$, $int(x)$ is the integer part of x) produced by the map $x_{n+1} = f(x_n)$ during its motion on this chaotic attractor, it appears to be a reproduction of the stored information block.

According to the algorithm of storing information in 1-D maps, in order to reproduce an element of the information block, it is sufficient that the trajectory x_i hits the corresponding 1-level region of the map. It is clear from the phase trajectory distribution (e.g., Fig. 9.8a), that this condition may not be fulfilled when storing at the first level occurs, because the interval cycle incorporates parts of the non-information intervals. However, this difficulty can be simply eliminated by means of diminishing the length of the information interval.

Thus, the information, stored as stable limit cycles of 1-D maps at $|s| < 1$, is not lost at the moment of bifurcation birth of the corresponding chaotic attractor. It rather attains a qualitative new carrier and can be retrieved in the same way as that of the stable limit cycles of points. The class of dynamic objects and the information carriers for the method of information storing in 1-D dynamical systems is proposed in [20, 22], and can be extended due to the chaotic attractors. In this case, the region of the parameter s used to store information as stable attractors is also extended.

Figure 9.8 presents some examples of the use of chaotic attractors (interval cycles) for information storing. Phase portraits, phase trajectory distributions and

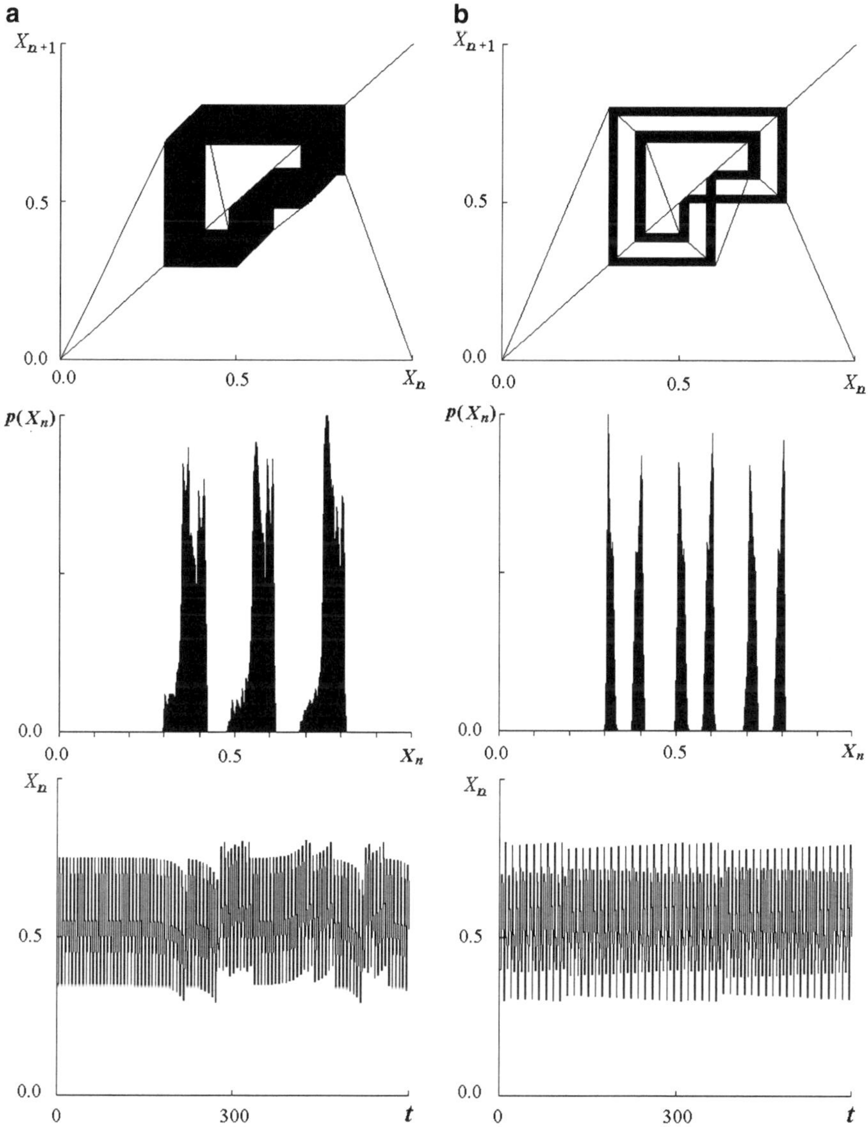

Fig. 9.8 Interval cycles of the map with information block 375 stored at the first level

time series are also given for the case of information block 375 stored at the first level. In Fig. 9.8a,b, solutions for $s = 1.125$ and $s = -1.005$ are depicted respectively.

At $s < 0$ and odd periods of the information cycle, the interval cycle appears doubled (Fig. 9.8b). The corresponding information stream is a repetition of the stored information block, because the sequential values of the iteration variable fit the first-level regions corresponding to digits 3, 5 and 7.

9.6 Unstable Cycles and Recognition

Storing information as periodic motions in 1-D maps is an easy and efficient method of associative memory organization. However, an analysis of the papers devoted to investigate the information processing in natural brains indicates an essentially complicated behavior in natural neural networks, (e.g., see a review in [59]. In particular, the periodic motion often indicates some "degenerate" states in the brain while the "usual" state is chaotic [13, 53, 64]. Besides, the existence of the stable limit cycles in the 1-D map, phase space leads to competition of the cycles. This means that iterates from an arbitrary initial point may converge to any cycle of the system regardless of the images presented for recognition, because the attraction basins of the cycles have fractal structure [20, 21]. This is a definite drawback of the model with the stable cycles discussed above, and we propose to overcome it by storing information as unstable cycles. The systems that we propose below present new capabilities for information processing.

In the above described procedure of information storing, the cycles may be easily become unstable. All that is necessary is to change the slope of the information regions in the piecewise linear map function: if $|s| > 1$, then the eigenvalues for all the cycles are $\lambda > 1$, and the cycles become unstable (see relation (9.1)).

Then, there are no stable information cycles in the map. If there is a "parasitic" stable cycle, this means that it passes through a non-information segment with a slope less than 1 in magnitude, which makes its eigenvalue also less than one. Such cycles can be excluded with a local correction of the map function, while the structure of the unstable cycles with the stored information remain undisturbed. Because of the absence of stable limit cycles in the map, the phase trajectory wanders chaotically with mapping iterates over the phase space visiting most of its part.

9.6.1 Direct Map Function Control

There are a number of methods that can be used to retrieve information from a map. In this section we demonstrate the method of direct control of the map function (slope switching) in order to make the desired cycle stable while retaining others unstable, and we apply this method to image recognition [5].

According to the map design procedure, the phase space of the map contains a skeleton of unstable periodic orbits coupled with the stored information blocks and passing through the information regions of the map function. It also contains unstable periodic orbits coupled only with the non-information segments of the map which will be ignored in further discussion.

We want to derive a regular procedure of the map function f deformation, such that if an image coincides with one of the stored images (or is close enough),

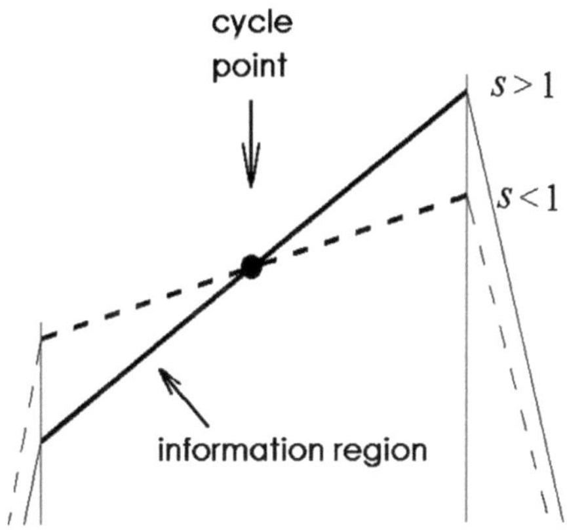

Fig. 9.9 Switching the slope of an information region

the corresponding cycle becomes stable and attracting. Then the phase trajectory converges to this stable periodic orbit and the stored image is reproduced by the system, which can be treated as recognition. Otherwise, no stable cycles appear, and the motion in the dynamical system remains chaotic. Thus, the character of the motion in the modified dynamical system, regular or chaotic, indicates the result of recognition.

Let M information blocks be stored in a map at the q-th level, and an image is presented for recognition in the form of a string with L symbols. The question that the system has to answer is, does this image correspond to any of the stored information blocks or not?

The procedure is as follows. We look through L fragments of this string, each q symbols long, and change the slopes of those information regions of the map function that correspond to the fragments, so that they become less than one in magnitude, as in Fig. 9.9. If the presented image coincides with an image stored in the system, all information regions of the map coupled with this image become switched, and the cycle's eigenvalue becomes less than one. Hence, the cycle becomes stable. If we now iterate the modified map beginning from an arbitrary initial point, we find that after some time the phase trajectory converges to this single stable limit cycle, and the system's behavior becomes regular.

If the presented image has nothing in common with the stored images, the map function f is not distorted, and the motion in the dynamical system remains chaotic.

This procedure is not as complicated as it may seem. To represent the piecewise linear map function $f(x)$, we need to remember only the coordinates of the centers of the information regions from which the cycles pass through (x_i, y_i), $i = 1, \dots N$, where N is the number of the information regions in the map, each with its own value of the slope s. We also have to recall the storage level q. The cycle points (x_i, y_i) are sorted by abscissa and kept in ascending order. When making a decision on whether a fragment $a_i\ a_{i+1}\ \dots\ a_{i+q-1}$ corresponds to an information region, we do not compare it to all the q-long substrings of all the information blocks, but take the central point x of the related subinterval of q-th level, and search for it in the set of the cycle points (using dichotomy). Thus, for each fragment of the presented image $\approx log_2(N)$ comparisons should be made, where N is the number of the cycle points in the map function.

It is important, that only one attracting cycle appears in the system's phase space. Therefore, if the distribution of the phase trajectory values on the global chaotic attractor intersects the information regions of the map, then in the modified map the trajectory will inevitably "fall" in after time onto an information region of the stable cycle and converge to this cycle. Pictorially, this can be described as an appearance of a "hole" in the chaotic set, through which the trajectory "leaks" out from the chaotic to the regular mode. This is the case of the loss of stability of a chaotic set and the transition from stable to metastable (transient) chaos. The appearance of this transition (crisis of chaotic attractor) allows us to realize the recognition of information.

Indeed, the above considerations imply that the changes in the map function are small; so that the system's variable distribution remains roughly the same. The studies have shown that this condition is satisfied for storage at levels higher than the first, because the size of information regions becomes small compared to the unit interval, and switching the slope of an information segment of the map function has a weak effect on the slopes of the neighbor segments (see Fig. 9.9).

The stable limit cycle that appears because of the crisis is unique, so the recognition process is practically independent of the initial conditions of the phase trajectory. The choice of an initial point determines only the duration of the transient process from the metastable chaotic set to the stable limit cycle, i.e., the rate of recognition.

The dynamic properties of the system with information stored as unstable cycles are illustrated by an example with three information blocks 97583, 14568 and 123 stored at the second level (Fig. 9.10). The alphabet here consists of 10 symbols, the digits 0 to 9. The slope of the information regions of the map is $s = 1.5$. Note, that the phase trajectory of the initial map with unstable cycles visits all information regions of the map (Fig. 9.10b). This property is assumed to remain, if the disturbances to the map caused by the change of the slopes are small.

If we now present an image stored in the system, the corresponding stable cycle will appear in the system's phase space, and the trajectory will converge to it after some time. Moreover, the described method can also be applied to the cases when the information on the image is incomplete, or partially incorrect or distorted, or only some parts of an image are available. Indeed, the cycle can become stable even if some of the information regions which it passes through, remain unchanged (with the slope greater than one), because its eigenvalue (9.1) is the product of the corresponding slopes.

To illustrate this opportunity, the strings 1231, 135681 and 946839 were used, i.e., a stored information block without errors in the first case, with one error in the second case, and two errors in the third (all blocks closed in a loop). In all three cases a stable limit cycle appeared, related to the corresponding information block, and after a transient process, caused by the wandering of the phase trajectory over the metastable set, the trajectory fell onto the corresponding stable limit cycle

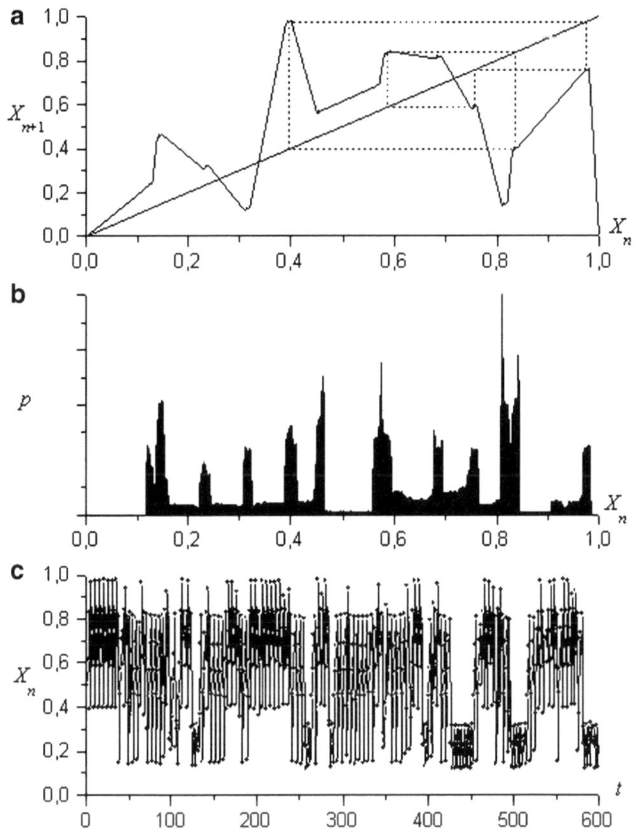

Fig. 9.10 Chaotic dynamics of a map with information stored as unstable cycles. Three blocks 123, 14568 and 97583 are stored at the second level, $s = 1.5$. (**a**) The map function. The cycle for 97583 is shown. (**b**) Distribution of the phase trajectory over the phase space. (**c**) Time series

(Fig. 9.11). The slopes here were switched from 1.5 to 0.5, and in all three cases the iterates have begun from the same initial point $x_0 = 0.6$. The difference in the time of convergence, approx. 180 iterates for the string 1231 and 2 iterates for 946839, is casual and is determined by the initial point.

We can easily estimate the percentage of erroneous symbols in the presented strings for which the recognition is still successful using expression (9.1). Assume, that an information block with length n is stored in an 1-D map as an unstable cycle with the slopes of its information regions $s = s_u > 1$. Then assume that an image is presented, and the slopes of the k information regions of the corresponding cycle are switched to $s = s_s < 1$, and $(n-k)$ remained unchanged. The condition for the cycle stability is:

$$|\lambda| = |s_s|^k \cdot |s_u|^{n-k} < 1 \tag{9.10}$$

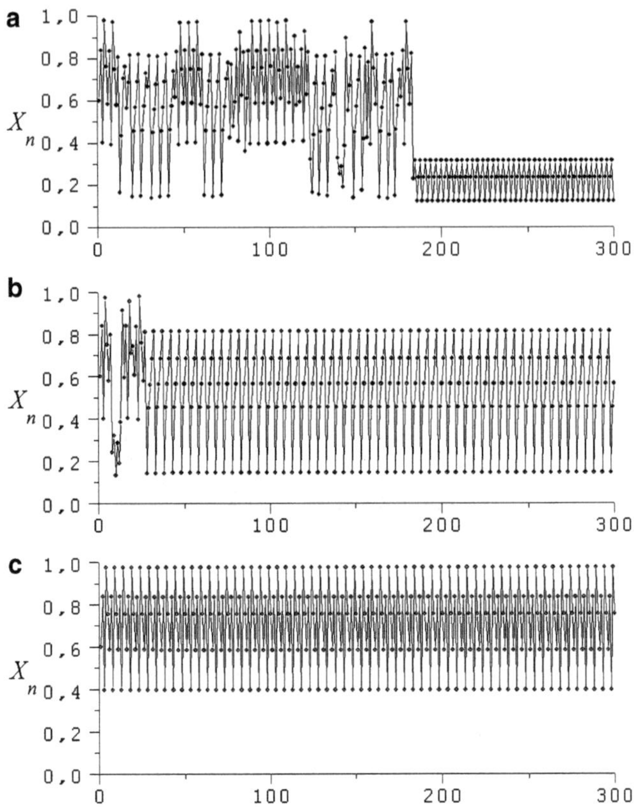

Fig. 9.11 Image recognition in the map from Fig. 9.10. Convergence of the phase trajectory to corresponding cycle after the presentation of an image. The related information region slopes are switched from $s = 1.5$ to $s = 0.5$. (**a**) String 123 is presented. Cycle 123 is stabilized. (**b**) String 13568 is presented. Cycle 14568 is stabilized. (**c**) String 94683 is presented. Cycle 97583 is stabilized

After simple transformations we find the condition for the necessary number of switched regions that provide stable cycle is:

$$k = n\frac{\log s_u}{\log s_u - \log s_s} = \frac{n}{1 - \log s_s / \log s_u} \tag{9.11}$$

Relation (9.10) allows us to estimate the admissible number of errors in a block, or helps to calculate the slope s_s for a prescribed level of errors, by which the related cycle will become stable. For instance, the eigenvalues of the cycles from Figs. 9.8a–c are $0.5^3 = 0.125$, $0.5^3 \cdot 1.5^2 = 0.28125$ and $0.5^2 \cdot 1.5^3 = 0.84375$, respectively.

A corollary from (9.10): for $s_s = 0$, cycle is absolutely stable, $\lambda = 0$.

The method of direct map function control that was designed to retrieve information from the map is based on the knowledge of the concrete map construction, but some general methods, such as the cycle stabilization after the OGY procedure [48], or the chaotic synchronization [2,49], also seem applicable for this purpose.

9.6.2 *Adaptive Model and Recognition*

The above discussed possibility of retrieving the information stored as unstable cycles of an 1-D map is an intermediate step in creating a model of an "ON-LINE" system processing a continuous external information stream and capable of selecting ("recalling") the information images stored in the system. If there are excerpts of the "known" information objects in the input information stream, the system "recalls" them and reproduces them thoroughly (because of the associative property) in the output information stream; if there are no "known" objects in the input stream, the system must return to the initial state (restore chaotic behavior). These different states of the system can be related to "short-term" memory ("inspiration") and "long-term" memory ("storage") inherent in the brain systems.

The realization of these properties is possible with an adaptive model, that we introduce in this section which is a generalization of the above model. It represents an 1-D map with information stored as unstable cycles, and the form of the map function is controlled by an external signal (Fig. 9.12). The external signal here is an endless sequence of symbols fed to the system's input, e.g., a random sequence, or a sequence containing successive repetitions of some of the stored information blocks. All the elements of the external signal belong to the same alphabet as the stored information blocks. They are fed to the system's input synchronously, one per iteration. The input signal doesn't influence the system's variable x_n, but it directly controls the system's function f by changing the slopes of the information regions. The slopes in this model are not switched, but oscillate between two boundaries. Now, when an image at the input is not known immediately as a whole, the map function is modified permanently at each step according to the piece of information available at the moment. If the input contains pieces of information related to some information regions, their slopes are rapidly decreased with iterates, or slowly return to the initial value.

The system's operation cycle consists now of two main stages. In the first, we modify the system's function (each segment) according to the symbol that is present currently at the input. At the second stage, we iterate the modified map.

If the storage was made at q-th level, the input symbols are accumulated to form a fragment of length q and, consequently, a point X in the unit interval, the central point of the q-th level subinterval related to this fragment. If for i-th time moment the signal at the input corresponds to an information region j, i.e., q last symbols of the input stream form a fragment related to the information region j, the slope of this region s_i^j is decreased according to some rule and may become less than one in magnitude.

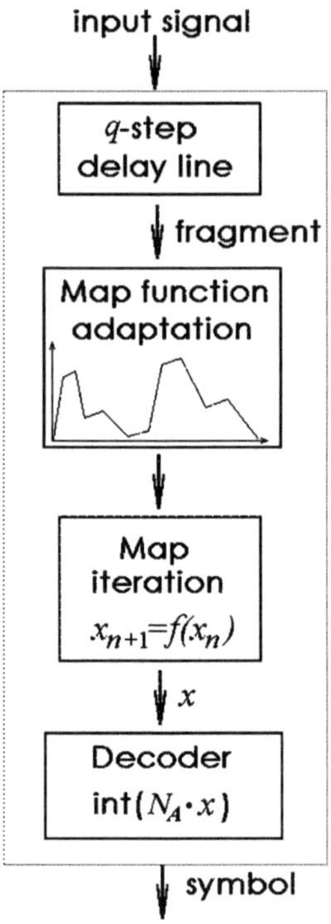

Fig. 9.12 Block diagram of the adaptive recognition model

Besides, backward relaxation is introduced: if at a moment i the input signal doesn't correspond to the information region j, then the slope s_i^j returns to the initial state, i.e., to the value of the slope in the absence of the external signal.

Thus, at each time moment only one information region in the map function can be turned downwards (decreased), all the others are turned upwards. If the input sequence consists of successive repetitions of a stored image, the related information regions will eventually become modified (their slopes set less than one in magnitude), and a stable cycle will appear in the system. Thus, the system permanently adapts itself to the input signal.

Let us denote the upper boundary for the information regions by $S_u(>1)$, and the lower boundary by $S_s(<1)$. Then the relaxation to S_u for the j-th information region at i-th time step (no signal for this region) is described by the equation:

$$s_{i+1}^j = \mu s_i^j + S_u(1-\mu) \tag{9.12}$$

where μ determines the rate of relaxation. The convergence of the information region slope to the state S_s is described with a similar equation:

$$s_{i+1}^j = \alpha s_i^j + (1-\alpha)S_s \tag{9.13}$$

where the coefficient α determines the rate of convergence. In numerical experiments we used the values $\alpha = 0.1$ and $\mu = 0.9$, which provided fast convergence to S_s and slow return to S_u. Then a general equation describing the dynamics of s_i^j is given by:

$$s_{i+1}^j = [\alpha s_i^j + (1-\alpha)S_s]\delta_{ij} + [\mu s_i^j + S_u(1-\mu)](1-\delta_{ij}) \tag{9.14}$$

where $\delta_{ij} = 1$ if $i = j$, otherwise $\delta_{ij} = 0$. It is taken into account in (14), that at each time moment each information region is turned only in one direction.

As follows from (9.14), the convergence to S_s in the presence of the corresponding external signal may take place only if the condition

$$0 < |n\alpha| << |\mu| < 1 \tag{9.15}$$

is satisfied, where n is the length of the corresponding cycle. The simplest case, $\alpha = 0$ corresponds to one-step convergence. The larger the magnitudes of α and μ, the slower the processes of convergence to S_s and return to S_u would be.

9.6.3 *Models of "Long-Term Memory" and "Short-Term Memory"*

In this section we show how the notions of "long-term memory" and "short-term memory", widely used in the study of the principles of memory functioning in living systems (e.g., [36]), are applicable to the behavior of the adaptive systems.

Let us begin with the long-term memory. After the information blocks are stored, they are present in the system at all times, and the carriers of information are the unstable cycles. Therefore, such a system may be interpreted as a long-term memory. The information is present in the system, but an external stimulus is necessary to retrieve it.

The external stimulus is a signal containing information blocks stored in the system, precise or with some errors. In general, the system doesn't respond to other information, remaining in a chaotic state. If the external signal with the stored information is fed to the system's input, a stable limit cycle appears replacing one of the unstable cycles. When the external stimulation is ceased, this cycle remains stable until the slopes of the corresponding information regions return to their initial values, i.e., while the condition for the cycle's stability $\lambda < 1$ is fulfilled.

The dynamics of the system is determined by the initial conditions. Consider two typical cases.

First case: We begin iterations just after the cease of the external stimulation by the signal containing a stored image. A point at the related cycle is taken for the initial conditions. In this case the trajectory occurs directly at the stable limit cycle, and the system "recalls" (recognizes) the stored information. Then the external signal is absent, the slopes of the information regions gradually return to the initial state S_u.

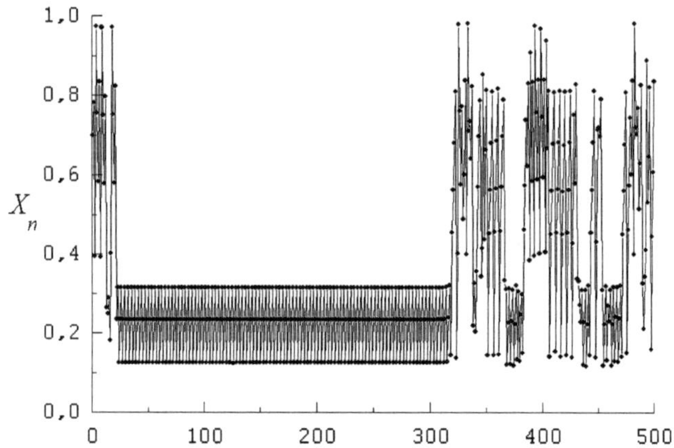

Fig. 9.13 Time series of an adaptive system with the blocks 123, 14568 and 97583 stored at the second level for the external signal representing the block 123

The phase trajectory remains in the vicinity of the cycle for some time (this time is composed of the time of the cycle's stability loss and the time of "running away" from the unstable cycle), and then leaves it and begins wandering over the phase space, leading to the chaotic behavior of the system. Thus, for some time the system "remembers" the information actively, and then "forgets" it. The stored information again becomes passive, i.e., transferred to the long-term state. The active phase, a period of time when the system's phase trajectory is in the vicinity of the limit cycle, may be considered as "short-term memory".

Second case: The initial point is arbitrary. The iterations are beginning simultaneously with the external signal. A stable limit cycle appears in the system for some time, but this doesn't automatically lead the system's trajectory to fall onto the cycle. The trajectory may still wander over the phase space. If the invariant measure of the autonomous system (without the external stimulation) has such a property, that the trajectory hits from time to time upon a map's information region, belonging to the vicinity of this stable cycle, then a "hole" appears in the system's global chaotic attractor, and it becomes metastable. If the limit cycle remains stable for sufficiently long time, the trajectory will eventually fall onto it.

The above discussion is illustrated by the example of three information blocks, i.e., 123, 14568, 97583, stored at the second level, as in Sect. 9.3. An external signal, consisting of successive repetitions of the information block 123 (100 repetitions) is given at the input of the system (Fig. 9.13). Iterations begin with initial condition $x_0 = 0.7$. The input signal is present for 300 iterates, then it is stopped. Approximately 22 iterations after the start, the phase trajectory falls onto the appeared stable cycle. It remains in the vicinity of the cycle for 318 iterates, while the external signal is active plus the time for the limit cycle to loose stability after the cease of the external signal and the time for the trajectory to "run away" from the unstable cycle. The information region slopes oscillate here between $S_u = 1.5$ and $S_s = 0.5$.

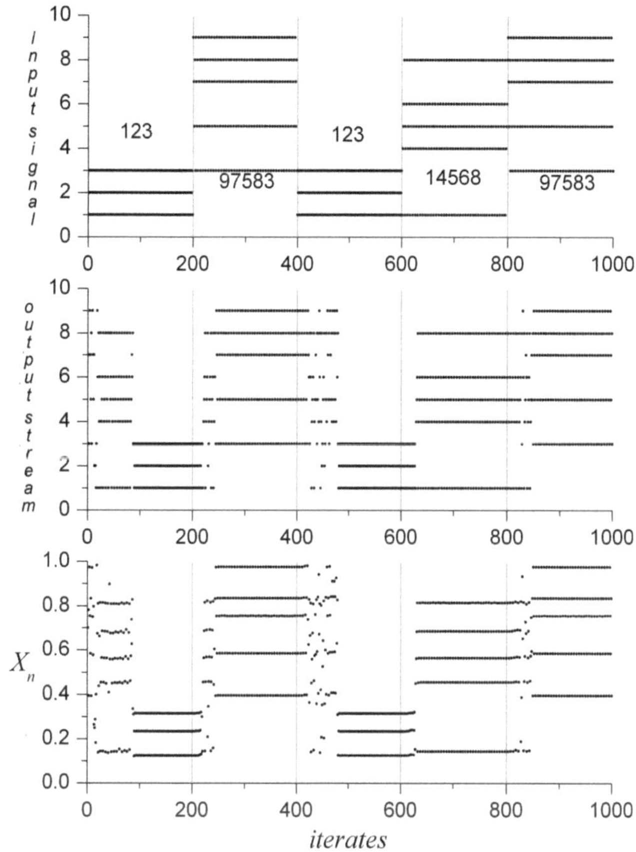

Fig. 9.14 "ON-LINE" recognition of the input stream representing repetitions of the stored information blocks

It should be noted, that sometimes in numerical experiments the system's trajectory converges so close to the stable cycle, that precisely coincides with it at last; when the cycle becomes unstable the trajectory remains at the cycle and never leaves it. This is a computer effect associated with the finite accuracy of calculations. An external random noise was added to the right-hand side of the system's equation in order to avoid this effect: The noise amplitude was 0.01 of the information region length, i.e., 10^{-4}.

An example of an "ON-LINE" system is given in Fig. 9.14, where the dynamical system with the 1-D map from Fig. 9.13 demonstrates switching between three stored information blocks. The input signal, the output information stream and the map trajectory are plotted. The input signal is pieces of the oscillations representing the stored images, 200 points each. It is evident from the figure that the phase trajectory x_n follows (though with a time delay) the input images, so the dynamical system successfully recognizes the images in the input signal. A certain delay in switching between the cycles in the output stream is associated with the discussed effect of the short-time memory.

Let us discuss now the role of the dynamical chaos in the above described chaotic models. We can distinguish the following important properties that emerge with chaos.

A competition of the stable information cycles caused by the strong dependence on the initial conditions disappears, because there is a single attracting cycle in the recognition process.

Chaos can be treated here as a reservoir containing "useful" trajectories (along with many other ones). The main role of global chaos in these systems seems to be the global mixing, providing guarantee though random access to all the stored images: independently of the initial point the system's phase trajectory will sooner or later occur in the vicinity of any required cycle (in the properly designed map).

9.6.4 Use of Synchronization to Retrieve Information

The idea of using synchronization to retrieve the information stored as unstable cycles is based on the possibility to synchronize an arbitrary trajectory of any map by means of acting on the map with a signal that is represented by all or certain samples of that trajectory [57]. The effect of the external signal on the map consists in mixing this signal with the values of the map's variable. The weight coefficient of the external mixing signal is called coupling coefficient. It can be shown that for a given trajectory there is a minimum value of coupling, at which synchronization is still possible. This minimum (threshold) value is unambiguously determined by the first (largest) Lyapunov exponent of this trajectory. It is important that for a stable trajectory the threshold coupling value is zero, i.e., synchronization is possible at any value of coupling, but for an unstable trajectory the threshold has positive value, which increases the instability of the trajectory.

Thus, by means of acting on the map carrying the information stored as unstable cycles with a signal that is represented by all or certain samples of one of those cycles, we can synchronize this cycle, which corresponds to retrieval of information stored in it.

9.7 Storing Information in 2-D and Multi-Dimensional Maps

To store information at the qth level with the original method, we used nested subdivisions of the system's phase space. The size of the information region corresponding to a length-q fragment of information block was N^q. Here we discuss an alternative method of data storage, using an increase of data space dimension to implement the increase of the storage level.

9.7.1 Synthesis of 2-D Maps

The principles of designing 2-D maps with prescribed stable or unstable limit cycles were proposed in [7]. Let, as in the case of storing information on limit cycles of 1-D maps, an information block $(a_1,\ldots,a_n)$ ready to be stored. In order to store this information block at the second level of an 1-D map it has to be transformed into a sequence of pairs of symbols:

$$(a_1,a_2),(a_2,a_3),\ldots,(a_n,a_1) \tag{9.16}$$

The pairs of symbols are assumed to be unique (no repetitions), otherwise, preliminary sequence coding is necessary, as was shown above.

Consider now the storage of this block on a 2-D map of a square $[0,1]\times[0,1]$ into itself. Each element j of the alphabet of length N is related to an interval of axes X and Y:

$$I_j^x = \left[\frac{j}{N},\frac{j+1}{N}\right),\quad I_j^y = \left[\frac{j}{N},\frac{j+1}{N}\right),\quad j=1,2,\ldots,N \tag{9.17}$$

and the central points $(j+0,5)/N$. In this case, a pair of elements (a_m,a_{m+1}) from the sequence (16) is related to a square:

$$I_{a_m}\times I_{a_{m+1}} = \left[\frac{a_m}{N},\frac{a_m+1}{N}\right)\times\left[\frac{a_{m+1}}{N},\frac{a_{m+1}+1}{N}\right) \tag{9.18}$$

and to a point in the middle of this square (x_m,y_m). The information block can be related to a cycle that goes through the points $(x_1,y_1),(x_2,y_2),\ldots,(x_n,y_n)$. Note that $x_{m+1}=y_m$. Consequently, the necessary 2-D maps must have the following form:

$$\begin{aligned} x_{m+1} &= y_m \\ y_{m+1} &= f(x_m,y_m) \end{aligned} \tag{9.19}$$

Let us choose such a function $f(x,y)$, so that the map (9.19) has a single limit cycle corresponding to the stored information block.

Let the symbol a_{m-1} of the original sequence be the i-th alphabet element, a_m is j-th, and a_{m+1} is k-th element of the alphabet. Let the point (x,y) belongs to the square $I_i\times I_j$ (such squares will be further called informative). We define the function $f(x,y)$ in this square as a plane S^{ij} that goes through the points $\left(\frac{i}{N},\frac{j+1}{N},\frac{k+0.5-\Delta}{N}\right)$, $\left(\frac{i}{N},\frac{j}{N},\frac{k+0.5-\Delta}{N}\right)$, $\left(\frac{i+1}{N},\frac{j+1}{N},\frac{k+0.5-\Delta}{N}\right)$ of space XYZ. In this square $I_i\times I_j$, the function is:

$$f(x,y) = 2\Delta\left(x-\frac{i+0.5}{N}\right)+\frac{k+0.5}{N} \tag{9.20}$$

The value 2Δ defines the slope of the plane S^{ij} to plane XY.

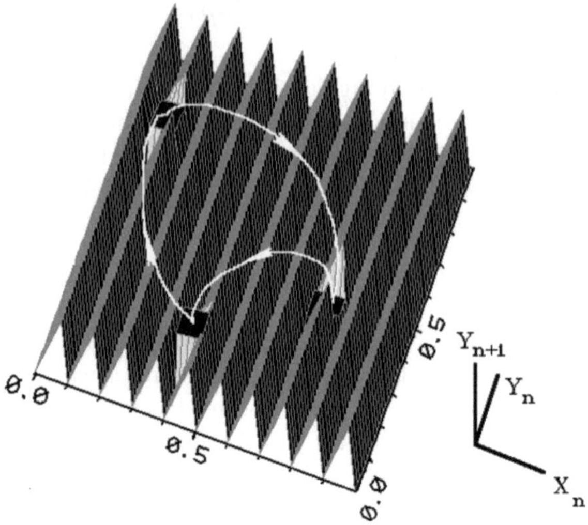

Fig. 9.15 Function of 2-D map with single information cycle

Assume now that the point (x,y) lies within the non-information square $I_i \times I_k$, i.e., in sequence (9.16) there is no such pair of i-th and k-th alphabet elements. Then the function $f(x,y)$ within this square is defined by the plane S^{ij}, that goes through the points $\left(\frac{i}{N}, \frac{k+1}{N}, \frac{\varepsilon}{N}\right)$, $\left(\frac{i}{N}, \frac{k}{N}, \frac{\varepsilon}{N}\right)$, $\left(\frac{i+1}{N}, \frac{k+1}{N}, \frac{\varepsilon}{N}\right)$. In this case:

$$f(x,y) = \frac{x - i/N}{1/N} \cdot \left(1 - \frac{1}{N}\right) + \frac{\varepsilon}{N} \tag{9.21}$$

where the small value ε ($\approx 10^{-2}$) is introduced in order to prevent an appearance of false fixed points of the map (9.19) in the point $(0,0)$.

Thus, the function $f(x,y)$ of map (9.19) in the information squares is defined by the function (9.20) and in the non-information squares by the function (9.21). It is evident that the stability of the cycles stored in this map requires that the slopes of the function $f(x,y)$ in the information squares with respect to plane XY be less than unity, i.e., $2\Delta < 1$.

More than one block can be written in similar way.

Example: Let the alphabet be composed of 10 digits 0, 1, 2, …, 9, and one information block 174 is stored in a 2-D map. The constructed map is stored in Fig. 9.15.

9.7.2 *Evaluation of the Information Capacity for the Method of Storing Information in 2-D Maps*

Previously we discussed the question of the information capacity for the storing method. It was shown that the maximum amount of information that can be stored

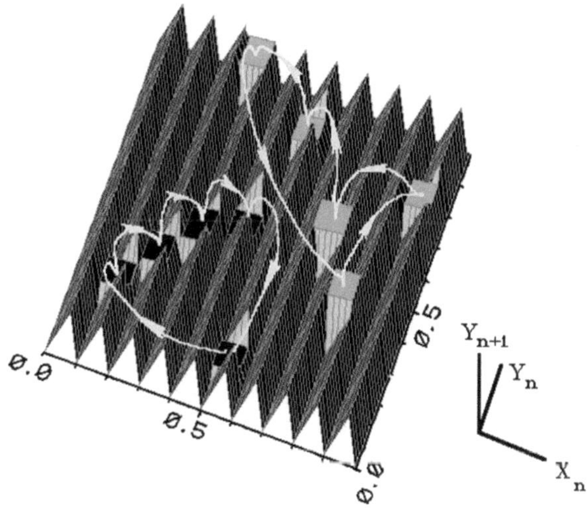

Fig. 9.16 A map with two information cycles

on the map is N^q symbols of log_2N bits each. The length of the information segment of an 1-D map is equal to $(1/N)^q$, in the case of ordinary-precision calculations it must be no less than 10^{-8}, and the storage capacity ranges between 10^5 and 10^6 bytes. In the case of 2-D maps the length of the side of the information square is $1/N$, so, the total capacity increases to 10^{11}–10^{12} bytes.

9.7.3 *Fractal Attraction Basins of Cycles of 2-D Maps*

It is known that in the phase space of a nonlinear dynamical system several attractors can coexist. Even in the case of the simplest 1-D and 2-D maps the set of all initial conditions that converge to a concrete attractor can have complex structure and it can't be a region in a rigorous mathematical definition. So, in order to designate the attraction points, the term "basin of attraction" is introduced. The attraction basins of the attractors of nonlinear dynamical systems can be fractal, even when the attractors are stable equilibrium points [32]. An important place among fractals takes the self-affine fractal, which can be decomposed into parts that are obtained from the whole fractal using the affine transformations (including rotation, translation, compression or extension). Each such part has also the self-affine property[2].

Let us illustrate this by an example of the attraction basins of two cycles of a 2-D map with information blocks 97583 and 12345 stored. The mapping function $f(x,y)$ is shown in Fig. 9.16, and the attraction basins are depicted in Fig. 9.17. Black color denotes the unit square points from which the trajectory is attracted to the

[2] The set S is self-affine with respect to a sequence of N affine transforms α_n, if condition $S = \cup\alpha_n S$ holds, whereas $\alpha_n S \cap \alpha_m S = \emptyset$ at $n \neq m$. A set S is decomposed to N non-overlapping parts, each of them is obtained from the whole using one of the affine transforms.

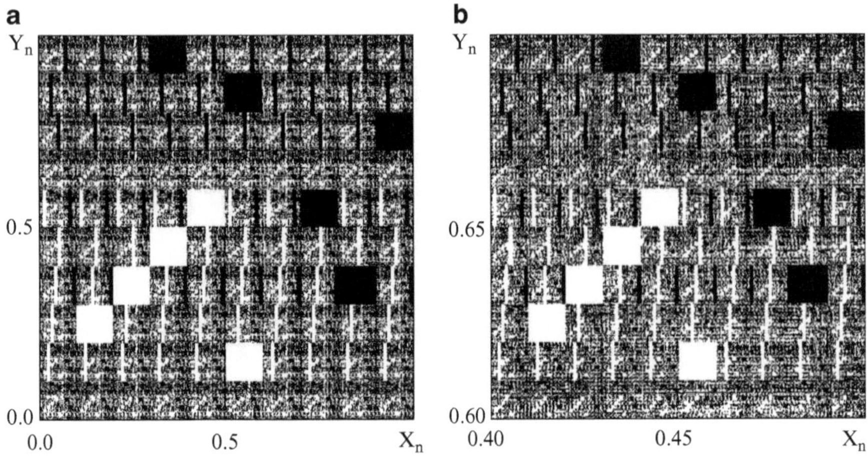

Fig. 9.17 The fractal structure of attraction basins

cycle corresponding to block 97583, and white points – to block 12345. The fractal structure of the attractions basins (Fig. 9.17) clearly exhibits the scaling similarity properties. For example, the basin structure in the square $[0.4,0.5] \times [0.6,0.7]$ (Fig. 9.17b) is similar by a factor $1/10$ to the basin structure of the entire unit square $[0,1] \times [0,1]$ (Fig. 9.17a). Besides, the unit square with an affine transform consisting of flip (mirror reflection) with respect to diagonal $(0,0)$–$(1,1)$ and compression along axis X, can be transformed to rectangle $[0,0.1] \times [0,1]$ by a factor of 10.

The fractal structure of the attraction basins (Fig. 9.17) can be described with a set of affine transforms which perform shift, rotation and compression of various regions of the unit square.

Let us divide the image in squares: $S_{ij} = [(i-1)/N, i/N) \times [(j-1)/N, j/N)$, with $i, j = 1, \ldots, N$.

Define a set of unit square regions, including:

- Rectangles $R_i = [(i-1)/N, i/N) \times [0,1), i = 1, \ldots, N$;
- Any black information square, e.g., $R_{N+1} = [(i-1)/N, i/N) \times [(j-1)/N, j/N)$;
- Any white information square, e.g., $R_{N+2} = [(i-1)/N, i/N) \times [(j-1)/N, j/N)$.

If S_{ij} is the information square, i.e., is absolutely black or white, then it can be obtained by means of shifting R_{N+1} (or R_{N+2}) to the corresponding point of the unit square.

Let the square S_{ij} be a non-information one. Then, it is the image of region R_j by the map:

$$x_S = y_R/N$$
$$y_S = x_R \tag{9.22}$$

where (x_{ij}^S, y_{ij}^S) is a point in S_{ij}, and (x_j^R, y_j^R) is a point in R_j.

Thus, we analytically described the fractal basins of attraction of the cycles of 2-D maps with stored information. Now, we can define some initial distribution of black and white points within the unit square, and apply the corresponding mapping to each square point, and after several iterations we obtain the basin structure.

Note that the affine transforms don't depend on Δ (at $\Delta < 1$). The procedure of describing the attraction basins with the affine transforms is similar to the method of the iterative function systems, used by M. Barnsley [15]. This work shows interconnections of the various methods of nonlinear dynamics and fractal geometry used in information processing systems, and the possibility of getting new properties by mutual complementation of these methods.

9.7.4 Synthesis of Multi-Dimensional Maps

Consider a generalization of the presented method to q-dimensional maps. First, from the original information sequence the following sequence is formed:

$$(a_1, a_2, \ldots, a_q), (a_2, a_3, \ldots, a_{q+1}), \ldots, (a_n, a_1, \ldots, a_{q-1}) \tag{9.23}$$

Again, we assume that the sequence (9.23) has no identical elements. If it has, then as in the 2-D case we eliminate them using the alphabet extension. The vector $(a_j, a_{j+1}, \ldots, a_{j+q-1})$ is related with a q-dimensional cube:

$$\begin{aligned} &I_{m_i} \times I_{m_{i+1}} \times \ldots \times I_{m_{i+q-1}} \\ &= \left[\frac{m_i - 1}{N}, \frac{m_i}{N}\right) \times \left[\frac{m_{i+1} - 1}{N}, \frac{m_{i+1}}{N}\right) \times \ldots \times \left[\frac{m_{i+q-1} - 1}{N}, \frac{m_{i+q-1}}{N}\right) \end{aligned} \tag{9.24}$$

The cycle of period n is composed of points:

$$\begin{aligned} &((m_1 - 0.5)/N, (m_2 - 0.5)/N, \ldots, (m_q - 0.5)/N); \\ &((m_2 - 0.5)/N, (m_3 - 0.5)/N, \ldots, (m_{q+1} - 0.5)/N); \\ &\ldots \\ &((m_n - 0.5)/N, (m_1 - 0.5)/N, \ldots, (m_{q-1} - 0.5)/N). \end{aligned} \tag{9.25}$$

The map having this stable cycle is defined as follows:

$$\begin{aligned} x_{j+1}^{(1)} &= x_j^{(1)}; \\ x_{j+1}^{(2)} &= x_j^{(2)}; \\ &\ldots \\ x_{j+1}^{(q)} &= f(x_j^{(1)}, x_j^{(2)}, \ldots, x_j^{(q)}) \end{aligned} \tag{9.26}$$

The function $f(X_i)$, where $X_i = (x_j^{(1)}, x_j^{(2)}, \ldots, x_j^{(q)})$, if the point X_i belongs to the information cube (9.24), is defined by the expression:

$$f(X_i) = \Delta\left(x_j^{(1)} - \frac{m_i}{N}\right) + \frac{m_{i+q} - 0.5}{N} \tag{9.27}$$

and if the point X_i is in the non-information region of q-dimensional unit square, then:

$$f(X_i) = \frac{x_i^{(1)} - m_i/N}{1/N} \tag{9.28}$$

Similarly to 2-D case, the parameter Δ determines the cycle's stability, because its eigenvalues are equal to: $\lambda_i = (-1)\Delta^{n/q}, i = 1, \ldots, n$.

In the 2-D case, the function $f(x,y)$ defines a 2-D surface in a 3D space (x_i, y_i, y_{i+1}) (Fig. 9.16), and $f(X_i)$, described by (9.27) and (9.28), defines a q-dimensional hyper-surface in a $(q+1)$-dimensional space $(X_i, x_{j+1}^{(q)})$.

9.7.5 *Storing and Retrieval of Multi-Dimensional Signals (Vectors) on Maps*

Previously the method of storing the sequence (9.23) on maps was described. In a similar way the following sequence can be stored:

$$(a_1, b_1, \ldots, r_1), (a_2, b_2, \ldots, r_2), \ldots, (a_n, b_n, \ldots, r_n) \tag{9.29}$$

where a_i is an element of the alphabet A^N; b_i is an element of, in general, another alphabet B^L; …; and r_i is an element of the alphabet R^M.

The method of storing multi-dimensional signals on maps, having capabilities of quick associative search, can find practical application in geophysical studies, tomography, construction of hierarchical systems, etc.

9.8 Application Examples

The capabilities of the described method of information processing by dynamical systems can be illustrated with the following examples. These examples are computer programs that simulate text and picture processing with 1-D and 2-D maps.

A Picture Processor software is developed as associative memory for graphic images [9]. It is capable of storing b/w and color images as limit cycles of 1-D or 2-D maps, and it allows the user to access the images by means of applying arbitrary

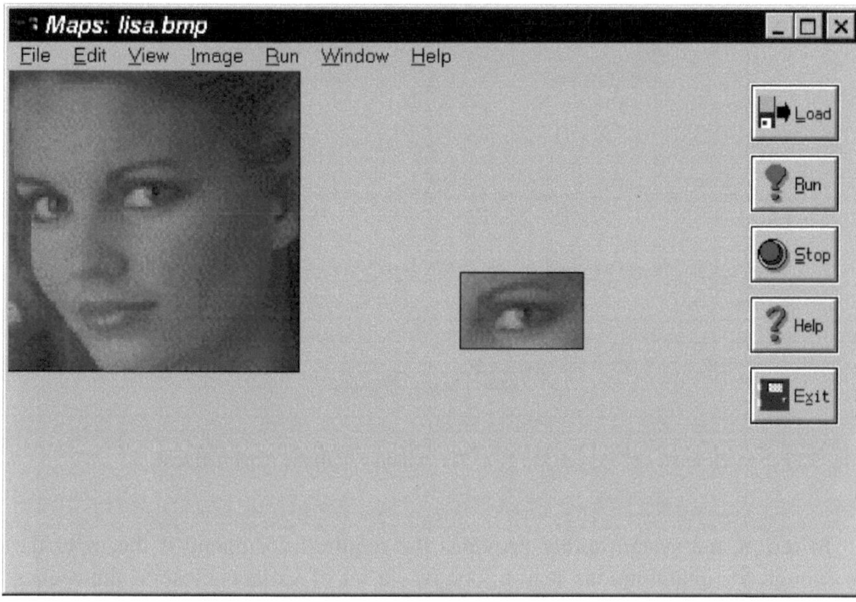

Fig. 9.18 Picture Processor software. A request and the answer

image fragments (Fig. 9.18). The algorithm is immune to errors and provides image recovery in the presence of noise. The number of simultaneously stored pictures in this program could be as high as one hundred.

Picture processing revealed a serious problem: it was rather difficult to form a request picture: we had to use either a fragment of the original stored picture, or a distorted version of it.

So, we turned our attention to processing texts in 1-D or 2-D maps.

The next example is the web-application *Forget-Me-Not*, which is a personal system of managing unstructured facsimile electronic copies of paper documents (e.g., scanned books, faxes, incoming and outgoing business correspondence, etc.) and text documents, with possibility of associative access. The original documents are OCR-ed and the corresponding texts are stored on the map as limit cycles.

The product is used to search the documents using natural-language requests. Unlike many other searching systems, the main search mode here is not the search by keywords but the search using rather large document fragments or even the whole document. Correspondingly, the result of the search is not the documents containing keywords, but the documents most close in content to the given request.

The array of the text documents is processed by the program and transformed into a dynamical archive (cycles of a map). By the analysis of the array, an internal artificial language related to the stored document contents is created. This internal language is based on the extended alphabet, obtained by means of the text compression (see Sect. 9.2.3).

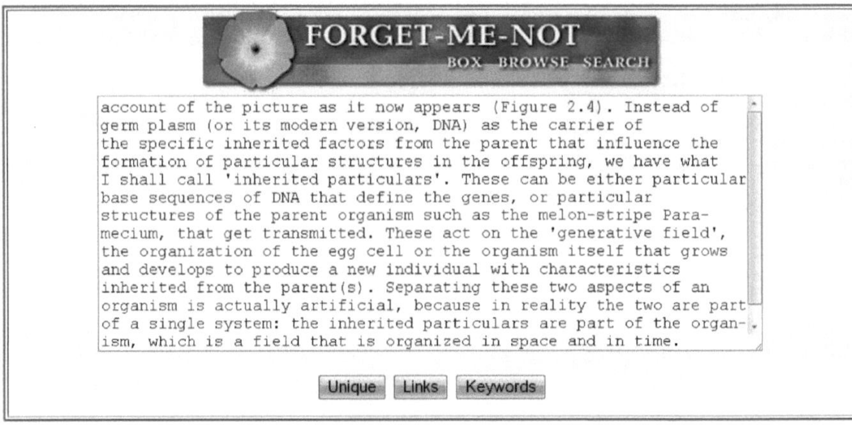

Fig. 9.19 Search modes of Forget-Me-Not. The window with the requested text

In return, the system either provides the required document, if the input data is enough for unambiguous search, or gives a set of variants close to the request. Facsimile copy of the found document is also given. Errors in the request or errors made in the process of the original document OCR-ed into textual form have practically no effect on the search quality (which can be explained by the redundancy of textual data).

Here, the use of the program is illustrated by an example of a book collection. To compile the collection, the books are scanned page by page and OCR-processed. Then the program processes the array of text pages and builds a 2-D map with the pages running on separate orbits in the common phase space.

In this program an electronic book can be read in a way similar to reading a paper-printed one. The "book" pages can be turned over, forward or backward, accessed (browsed) through the table of contents, etc. The user can see facsimile pages, or prefer their text content.

However, the most valuable feature of the *Forget-Me-Not* system is the content-sensitive, associative search for information.

Three different search modes – unique, search for links, and search by keywords – are realized (Fig. 9.19).

In the 'unique' search mode, the system seeks for a document containing the required text fragment. If the required fragment is present in the stored document archive (and is unique), it will be found, even in the case of a certain mismatch between the request and its counterpart in the archive (e.g., some words changed by synonyms, missing or extra words, spelling errors, etc.). Typical minimum of the request length in this search mode is approximately of 1-2 lines of the text.

In the 'links' mode (content-sensitive or associative search), the request is converted using the artificial system language corresponding to the information stored in the archive. Then the request is parsed by the "words" of that internal system language, and the parsing results are displayed. Each element ("word") of

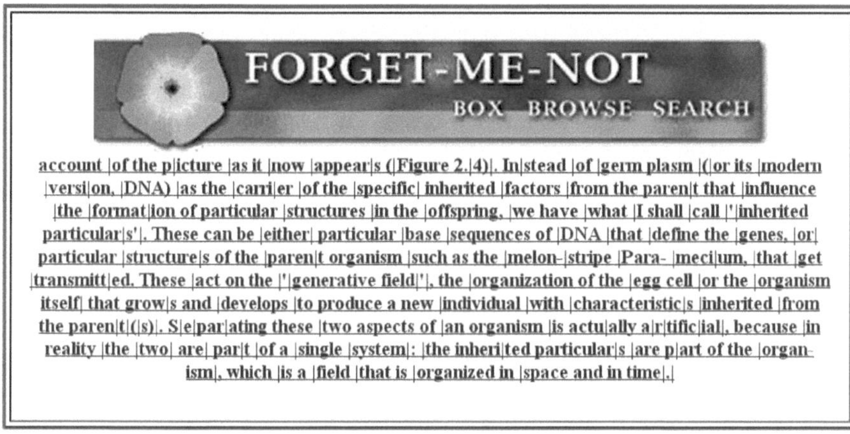

Fig. 9.20 Parsed request

the parsed request is a link to at least one document of the archive (Fig. 9.20). That is, the request is transformed into a sort of hypertext with multiple links. Visual examination of the parsed request allows the user to choose the most informative (from his viewpoint) links. Besides, parsing the request gives useful hints and helps understand which words and combinations can be keys to the documents from this particular archive (Fig. 9.21).

The search by keywords is a standard search mode. The documents are searched for in a usual way for a separate word or a combination of keywords with logical relations.

It is important to note that when the compressed data is stored in the dynamical system, the original data becomes unnecessary, and there is no need to keep it along with the map, because it can be restored by means of map iteration. Actually, information is searched in the compressed data storage (in archive), and it is decompressed only at the stage of displaying the search result. In these conditions, the volume of information stored in the system can be very large. It is mainly restricted by the volume of the computer's RAM.

9.9 Conclusions

The models of nonlinear dynamical systems with information stored as dynamic attractors described in this chapter proved to be a useful tool for studying information processing in chaotic systems. They implement a wide range of information processing functions, such as storing and retrieval, associative memory, memory scanning based on intermittency, image recognition based on storing with unstable cycles and direct map function control, "novelty filter", "long-term" and "short-term" memories, etc.

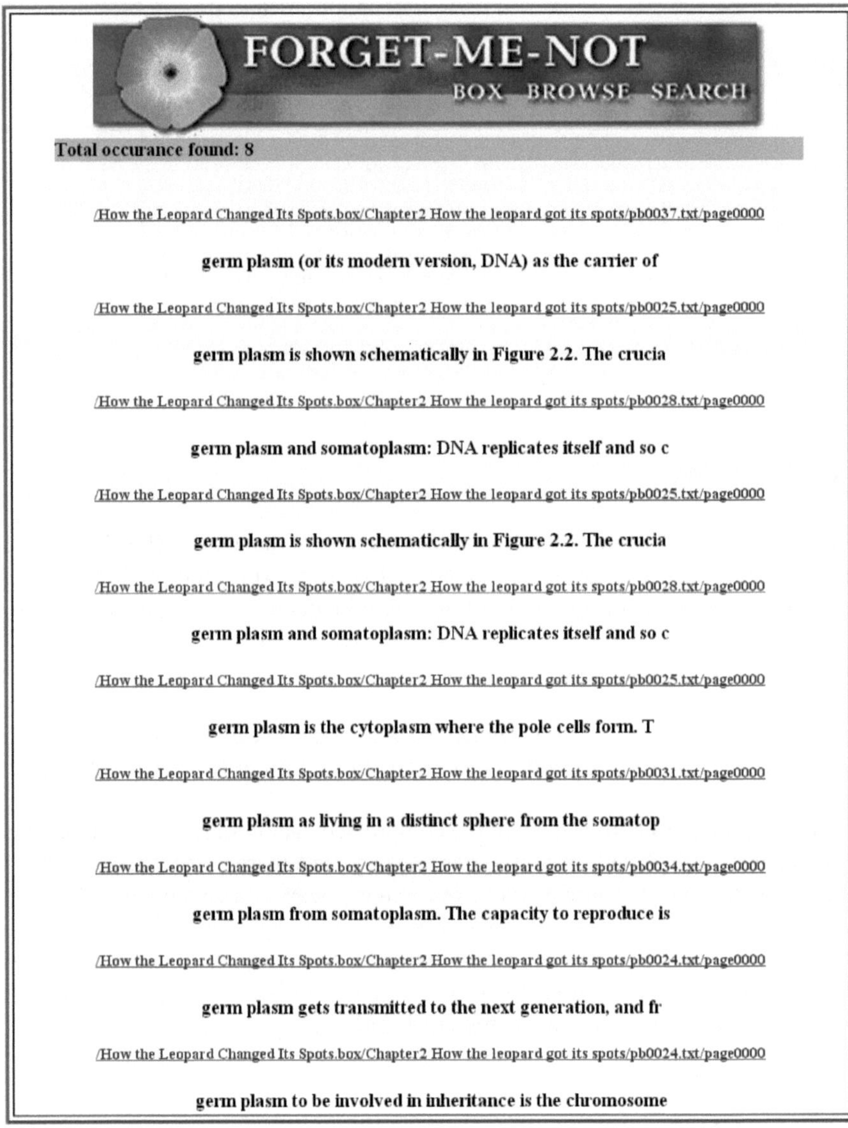

Fig. 9.21 Multiple links branching from "germ plasm"

The associative memory is realized due to storing information as dynamic attractor (stable cycle), which allows access to a stored image by its fragment, i.e., by a small part of the stored image the system "recalls" the place where the image is stored, finds a point on the corresponding cycle and retrieves the whole image by means of iteration (going along the cycle).

Memory scanning or random-access memory is meant here as the mode, in which the phase trajectory wanders chaotically over the phase space and comes "at random" in the vicinity of this or that information cycle, lingers there for some time, sufficient to recover the stored image, and then leaves this vicinity and goes wandering further. This is a "slightly chaotic" mode, with information cycles made unstable but with the cycle eigenvalues (9.1) very close to unity.

The designed systems can operate as "novelty" filter, i.e., when a piece of data is presented to the system with information blocks stored as stable cycles, the system immediately answers the question, whether this data piece is known to the system (i.e., it is stored) or it is "new". For the presented data piece the system calculates a point in the phase space (or a set of points) and checks, whether they fit any cycle point. Since the system is actually described by the ordered sorted set of its cycle points, the answer is very quick. The image recognition by means of direct map function control, "long-term" and "short-term" memory described in Sect. 9.6 becomes possible, when information is stored as unstable cycles and the cycle stability varies depending on the presented request.

When designing the above information processing systems, we had to reconsider certain hypotheses concerning the mechanisms of information processing in nonlinear dynamical systems. For example, the cycles were found to be better information depositories than the strange attractors. Another important point is that when the information is stored on the attractors of a dynamical system, and an external signal is presented for recognition, there is no need to correlate this signal with all the signals stored in the system. The principles of storing and retrieval allow collation of the external signal with its stored copy (if there is any) without exhaustive search. The expense is a certain decrease of robustness, and the gain is a fast correlator, capable of working with large data arrays. For instance, a typical search time of such programs as Picture Processor or Forget-Me-Not is fractions of a second.

Note also surprisingly the high performance of the chaotic processing models, which results in the capability of solving rather complex and large problems by ordinary computer simulation. It seems that this is due to not only good models but also due to the properties of the chaotic systems themselves, in particular, the flexibility and quick reaction to external influence.

In this chapter we provided examples of application of the information processing in nonlinear dynamical systems, which give certain understanding of where and with what purpose such systems can be used. However, the application areas are not limited in the noted directions and are determined by the following features, including:

- Capability of operating with any kind of data (texts, graphs, video, audio, etc);
- No need of preliminary structuring of information;
- Possibility of adding new and deleting unnecessary information;
- Fast associative access to large volumes of information, possibility of searching by arbitrary fragments;
- Immunity to storing errors and operation in the presence of noise.

The discussed models are very simple and allow for complete description, yet they possess considerable information capacity, and may be of practical interest from the viewpoint of information processing technologies using chaos.

References

1. Afraimovich, V.S., Shilnikov, L.P.: On strange attractors and quasiattractors. In: Barenblatt, G.I. (ed.) Nonlinear Dynamics and Turbulence, pp. 1–34. Boston, London, Melburn, Pitmen (1983)
2. Afraimovich, V.S., Verichev, N.I., Rabinovich M.I.: Chaotic synchronization of oscillations in dissipative systems. Izv. VUZov. Radiofizika, **29**, 1050 (1986)
3. Aihara, K.: Chaotic neural networks. In: Kawakami, H. (ed.) Bifurcation Phenomena in Nonlinear Systems and Theory of Dynamical Systems, pp. 143–161. World Scientific (1990)
4. Andreyev, Yu.V.: Attractors and bifurcation phenomena in 1-D dynamical systems with stored information. Izv. VUZov. Prikladnaya nelineinaya dinamika **3**, 3–15 (in Russian) (1995)
5. Andreyev, Yu.V., Belsky, Yu.L., Dmitriev, A.S.: Information processing in nonlinear systems with dynamic chaos. Proceedings of International Seminar Nonlinear Circuits and Systems, Moscow, **1**, 51–60 (1992)
6. Andreyev, Yu.V., Dmitriev, A.S., Chua, L.O., Wu C.W.: Associative and random access memory using one-dimensional maps. Int. J. Bifurc. Chaos, **2**, 483–504 (1992)
7. Andreyev, Yu.V., Belsky, Yu.L., Dmitriev, A.S.: Storing and recognition of information using stable cycles of 2-D and multi-dimensional maps. Radiotekhnika i elektronika **39**, 114–123 (in Russian) (1994)
8. Andreyev, Yu.V, Belsky, Yu.L., Dmitriev, A.S., Kuminov, D.A.: Information processing using dynamical chaos. IEEE Trans. Neural Networks **7**, 290–291 (1996a)
9. Andreyev, Yu.V., Dmitriev, A.S., Kuminov, D.A., Chua, L.O., Wu, C.W.: 1-D maps, chaos and neural networks for information processing. Int. J. Bifurc. Chaos, **6**, 627–646 (1996b)
10. Atmanspacher, H., Scheingraber, H., Voges, W.: Global scaling properties of the chaotic attractor reconstructed from experimental data. Phys. Rev. A **37**, 1314–1322 (1988)
11. Auerbach, D., Cvitanovic, P., Eckmann, J.-P.: Exploring chaotic motion through periodic orbits. Phys. Rev. Lett. **58**, 2387 (1987)
12. Babloyantz, A.: Evidence of chaotic dynamics of brain activity during the sleep cycle. In: Mayer-Kress, G. (ed.) Dimension and Entropies in Chaotic Systems, pp. 252–259. Springer, Berlin (1986)
13. Babloyantz, A., Destexhe, A.: Low-dimensional chaos in an instance of epilepsy. Proc. Natl. Acad. Sci. USA, **83**, 3513–3517 (1986)
14. Baird, B., Eeckman, F.: A normal form projection algorithm for associative memory. In: Hassoun, M.H. (ed.) Associative Neural Memories: Theory and Implementation. Oxford University Press, New York (1992)
15. Barnsley, M.: Fractals Everywhere. Academic (1988)
16. Bowen, R.: Entropy for maps of the interval. Topology **16**, 465–467 (1977)
17. Carpenter, G.A.: Neural network models for pattern recognition and associative memory. Neural Networks **2**, 243 (1989)
18. Cvitanovic, P.: Invariant measurement of strange sets in terms of cycles. Phys. Rev. Lett. **61**, 2729–2732 (1988)
19. Destexhe, A., Sepulchre, J.A., Babloyantz, A.: A comparative study of the experimental quantification of deterministic chaos. Phys. Lett. A **132**, 101–106 (1988)
20. Dmitriev, A.S.: Storing and recognition information in one-dimensional dynamical systems. Radiotekhnika i Elektronika **36**, 101–108 (in Russian) (1991)

21. Dmitriev, A.S.: Chaos and information processing in dynamical systems. Radiotekhnika i elektronika **38**, 1–24 (in Russian) (1993)
22. Dmitriev, A.S., Panas, A.I., Starkov, S.O.: Storing and recognition information based on stable cycles of one-dimensional maps. Phys. Lett. A **155**, 494–499 (1991)
23. Dmitriev, A.S., Jdanova, L., Kuminov, A.: The simplest neural-like systems with chaos, in Proc. Int. Conf. Fluctuations and Noise, Kyoto, Japan (1991)
24. Dmitriev, A.S., Kuminov, D.A., Pavlov, V.V., Panas, A.I.: Storing and processing texts in 1-D dynamical systems. (585), Institute of Radioengineering and Electronics RAS, Moscow (in Russian) (1993)
25. Eisenberg, J., Freeman, W.J., Burke, B.: Hardware architecture of a neural network model simulating pattern recognition by the olfactory bulb. Neural Networks **2**, 315–325 (1989)
26. Farmer, J.D.: Information dimension and the probabilistic structure of chaos. Z. Naturforsch, **37A**, 1304–1325 (1982)
27. Freeman, W.J.: Simulation of chaotic EEG patterns with a dynamic model of the olfactory system. Biol. Cybern. **56**, 139–150 (1987)
28. Freeman, W.J., Yao, Y., Burke, B.: Central pattern generating and recognizing in olfactory bulb. Neural Networks **1**, 277–278 (1988)
29. Hawkins, J., Blakeslee, S.: On intelligence. Times Books, New York (2004)
30. Hopfield, J.J.: Neural networks and physical systems with emergent collective computational abilities. Proc. Natl. Acad. Sci. USA. Apr. **79**, 2554–2558 (1982)
31. Grebogi, C., Ott, E., Yorke, J.A.: Critical exponent of chaotic transients in nonlinear dynamic systems. Phys. Rev. Lett. **57**, 1284 (1986)
32. Grebogi, C., Ott, E., Yorke, J.: Fractal basin boundaries, long-lived chaotic transients, and unstable-unstable pair bifurcation. Phys. Rev. Lett. **50**, 935–938 (1983)
33. Grossberg, S.: Nonlinear neural networks: Principles, mechanisms, and architectures. Neural Networks **1**, 17–61 (1988)
34. Gunaratne, G.N., Procaccia, I.: Organization of chaos. Phys. Rev. Lett. **59**, 1377–1380 (1987)
35. Kitchens, B.: Symbolic Dynamics: One-sided, Two-sided and Countable State Markov Chains. Springer (1998)
36. Klatzky, R.J.: Human Memory. Structures and Processes. Freeman, San Francisco (1975)
37. Kohonen, T.: Content-addressable memories. Springer, Berlin (1980)
38. Kolmogorov, A.N.: Three approaches to the quantitative definition of information. Probl. Inform. Transm. **1**, 1–7 (in Russian) (1965)
39. Li, M., Vitanyi, P.M.B.: An Introduction to Kolmogorov Complexity and its Applications, 2nd edn. Springer, New York (1997)
40. Lind, D., Marcus, B.: An Introduction to Symbolic Dynamics and Coding. Cambridge University Press, Cambridge (1995)
41. Maistrenko, Yu.L., Maistrenko, V.L., Sushko, I.M.: Order for the appearance of attractors in piecewise linear systems. In: Chaos and Nonlinear Mechanics, Series B, **7**, World Scientific (1994a)
42. Maistrenko, Yu.L., Maistrenko, V.L., Sushko, I.M.: Bifurcation phenomena in generators with delay lines. Radiotekhnika i elektronika, Moscow, **8-9**, 1367–1380 (1994a)
43. Matsumoto, K., Tsuda, I.: Calculation of information flow rate from mutual information. J. Phys. A. Math. Gen. **21**, 1405–1414 (1988)
44. Nicolis, J.S.: Should a reliable information processor be chaotic? Kybernets. **11**, 269–274 (1982)
45. Nicolis, J.S.: Dynamics of Hierarchical Systems: An Evolutionary Approach. Springer (1986)
46. Nicolis, J.S.: Chaos and Information Processing: A Heuristic Outline. World Scientific, (1991)
47. Nicolis, J.S., Tsuda, I.: Chaotic dynamics of information processing – the "magic number seven plus-minus two" revisited. Bull. Math. Biol. **47**, 343–365 (1985)
48. Ott, E., Grebogi, C., Yorke, J.A.: Controlling chaos. Phys. Rev. Lett. **57**, 1196–1199 (1990)
49. Pecora, L.M., Carrol, T.L.: Synchronization in chaotic systems. Phys. Rev. Lett. **64**, 821–824 (1990)

50. Procaccia, I.: The organization of chaos by periodic orbits: Topological universality of complex systems. In: Jullien, R. (ed.) Universalities in Condensed Matter. Springer (1988)
51. Sharkovski, A.N.: Coexistence of cycles of a continuous map of the line into itself, Ukrain. Mat. Zh. **16**, 61–71 (1964); English translation: Int. J. Bifur. Chaos Appl. Sci. Engrg. **5**, 1263–1273 (1995)
52. Shilnikov, L.P.: A case of the existence of a denumerable set of periodic motions. Sov. Math. Dokl. **6**, 163–166 (1965)
53. Skarda, C.A., Freeman, W.J.: How brains make chaos in order to make sense of the world. Behav. Brain. Sci. **10**, 161–165 (1987)
54. Shaw, R.: Strange attractors, chaotic behavior and information flow. Z. Naturfursch. **36a**, 80–112 (1981)
55. Schuster, H.G.: Deterministic Chaos. An Introduction. Wiley, New York (2004)
56. Smale, S.: Differentiable dynamical systems, Bull. Amer. Soc. **73**, 747–817 (1967)
57. Starkov, S.O, Andreyev, Yu.V., Dmitriev, A.S., Matveev, M.A., Shirokov, M.Ye.: Synchronization in the map with stored information. Proceedings of 4th Workshop NDES, Seville, Spain, pp. 299–303 (1996)
58. Tsuda, I.: A hermeneutic process of the brain. Progr. Theor. Phys., Supplement **79**, 241–259 (1984)
59. Tsuda, I.: Dynamic link of memory – chaotic memory map in nonequilibrium neural net-works. Neural Networks **5**, 313–326 (1992)
60. Tsuda, I.: Can stochastic renewal of maps be a model for cerebral cortex? Physica D **75**, 165–178 (1994)
61. Tsuda, I., Koerner, E., Shimizu H.: Memory dynamics in asynchronous neural networks. Progr. Theor. Phys. **78**, 51–71 (1987)
62. Voges, W., Atmanspacher, H., Scheingraber, H.: Deterministic chaos in accrecting systems: analysis of the x-ray variability of Hercules X. Astrophys. J. **320**, 794–802 (1987)
63. Wiegrinch, W., Tennekes, H.: On the information flow for one-dimensional maps. Phys. Lett. A **144**, 145–152 (1990)
64. Yao, Y., Freeman, W.J.: Model of biological pattern recognition with spatially chaotic dynamics. Neural Networks **3**, 153–170 (1990)

Chapter 10
A Chaos Based Secure Communication Scheme for Hybrid Message Logging and Asynchronous Checkpointing for Mobile Computing

Santo Banerjee and S. Mukhopadhyay

10.1 Introduction: Mobile Computing Environment and System Model

The rapid development in cellular technologies and wireless telecommunications has made mobile communications an important mode of communication. Mobile communication has created a lot of buzz during the last few years. Apart from communication of voice signals, mobile technology is rampantly being used to send and receive data across these cellular networks. This is the underlying principle of mobile computing.

Wireless connection ties the mobile device to centrally located information or application software through the use of battery powered, portable and wireless computing and communication devices and users have access to information service through a shared infrastructure regardless of their physical location or movement behavior. Such a new environment introduces technical challenges [19] in area of information access such as the following:

- Resource constraints (limited bandwidth, battery power etc.)
- Dynamically adaptation of MHs to resources during migration
- Faults
- Handoff
- Security of data communicated over wireless link
- Intermittent disconnections

S. Banerjee (✉)
Politecnico di Torino, Corso Duca degli Abruzzi 24, 10129 Torino, Italy

Micro and Nanotechnology Unit, Techfab s.r.l., Chivasso, Italy
e-mail: santo.banerjee@polito.it

S. Mukhopadhyay
Army Institute of Management, Judge's Court Road, Alipore, Kolkata-700 027, India
e-mail: sumona.mukhopadhyay@gmail.com

S. Banerjee et al. (eds.), *Applications of Chaos and Nonlinear Dynamics in Engineering – Vol. 1*, Understanding Complex Systems, DOI 10.1007/978-3-642-21922-1_10,

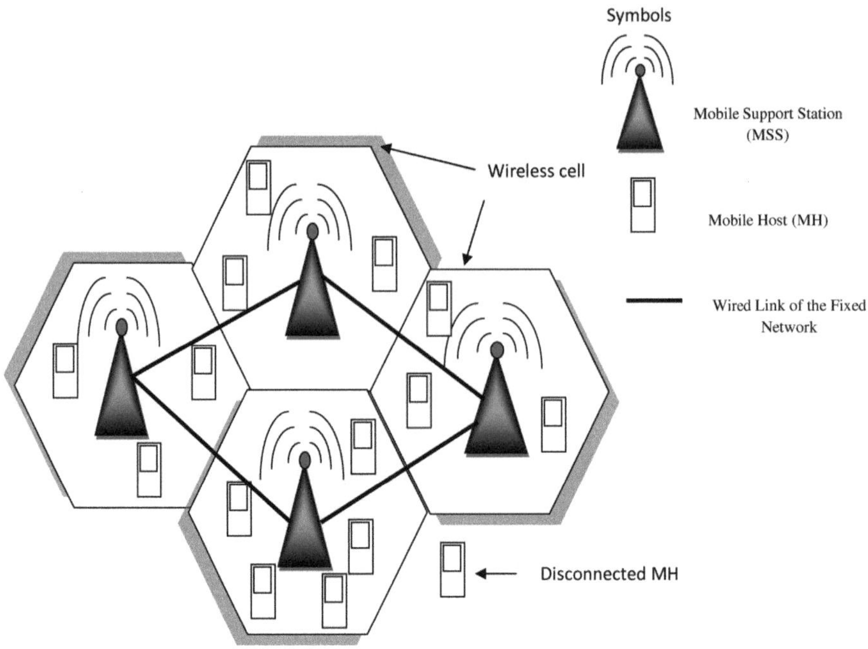

Fig. 10.1 (Color online) The model for mobile computing

Security is a pre-requisite for every network especially for mobile hosts (WLAN or AD-Hoc network) due to the following listed issues:

- Constrained network bandwidth
- Energy constraint
- Decentralized Infrastructure
- Node selfishness

A distributed mobile system used in [11, 17, 30] has been adopted for our scheme. Figure 10.1 shows an illustration of mobile computing/communication scheme. A mobile computing environment consists of a set of static Mobile Support Stations (MSSs) and Mobile hosts (MH). A set of dynamic and wireless communication links can be established between a MH and a MSS; and a set of high speed static and wired communication links is assumed between MSSs. A geographical region covered by a MSS is called a cell. Each MSS is responsible for managing a number of MHs in its cell. A MH on entering its cell informs the MSS about its presence. The MSS servicing the MHs is known as the Base station or the Base MSS. Each MH must be connected to one Base MSS. MH's communicate only with its Base MSS. In order to communicate with another MH which is located in another cell, it has to communicate through its local/base MSS. It is assumed that the links in the dynamic network support FIFO communication in both directions, however, there is no assumption on the message delivery order by the static links. This chapter presents a novel communication scheme proposed for mobile computing which handles: (a) logging and checkpointing; (b) security through chaos based cryptography using optical chaotic system which is used to address the Authentication and key management processes. The proposed method is computationally inexpensive and a fast mode for secure communication since chaos cryptography operates on the physical level of the transmission system. Rest of the chapter is organized as: In Sect. 10.2 introduces the concept of achieving fault tolerance through checkpointing and message logging. Section 10.3 presents various literature review on the checkpointing and logging of messages. Section 10.4 handles the security aspect in mobile communication and introduces chaos based cryptography. The subsection presents the system used in rendering security to the scheme by a coupled chaotic laser system. Section 10.5 introduces the proposed system with the Algorithm. In Sect. 10.6, the security analysis is performed with a colored image to test the efficiency of the proposed secure communication technique followed by the conclusion of our work.

10.2 Checkpointing and Message Logging: A Feasible Solution

Checkpointing is a technique for inserting Fault tolerance [11, 17] into computing systems. It consists of storing a snapshot of the current application state and uses it for restarting the execution in case of failures. Its main property lies in preserving the system consistency for it saves the status information. According to Elnozahy et al. [17] and Alvisi et al. [3] checkpointing for mobile computing can be categorized into three classes:

- Independent or uncoordinated (asynchronous) checkpointing: This scheme was presented first by Acharya and Badrinath [1] for mobile systems. Here each process can independently take the decision when to take its own checkpoints. The primary advantage is that each process of a MH may take a checkpoint in accordance to its convenience. But there are certain disadvantages. Firstly, it is unsuitable for applications which commit their output frequently as these require global coordination to compute the recovery line. Secondly, a process may take an unnecessary checkpoint which may never be a part of a global consistent state. These are undesirable for they incur overhead. Thirdly, uncoordinated checkpointing forces each process to maintain multiple checkpoints, and to invoke periodically a garbage collection algorithm to reclaim the outdated checkpoints. Lastly, there is always a possibility of domino effect and inconsistent states.
- Co-ordinated checkpointing or synchronous checkpointing: Here processes are forced to organize their checkpoints in order to generate a single, consistent application checkpoint. The main advantage is that there is no domino effect. This is achieved using synchronized clock or system messages.

Whenever a process changes its communication status from the sending to the receiving mode, each process independently takes checkpoint. Although there is no overhead related to the coordination, the frequency of checkpointing cannot be controlled by each MH. Therefore, MH may have to transfer a checkpoint with every outgoing message to MSS, which is not feasible and impractical for a mobile network with limited bandwidth. Message logging is a way to recover from failures which keeps a history of all the communications. In message logging, processes log their received messages and checkpoint their local state. A log-based rollback recovery makes use of deterministic and nondeterministic events in a computation. A nondeterministic event can be a receipt of a message from another process or an event internal to the process. A message send event is a deterministic event. So a log-based rollback recovery views a process as a sequence of deterministic state intervals, each starting with the execution of a nondeterministic event. Message logging advantage lies in its ability to avoid synchronization of processes during checkpointing. Message logging can be grouped into protocols: Optimistic and Pessimistic message logging:

- Pessimistic Logging: This protocol assumes that a failure can occur after any nondeterministic event in the computation. This protocol logs to the stable storage the determinant if each nondeterministic event before the event affects the computation. It incurs a high performance overhead.
- Optimistic Logging: The processes buffer the received messages in volatile storage and asynchronously logs them to stable storage. This eliminates blocking of the processes. In contrast to pessimistic protocol, this assumes that logging will be complete before failure occurs and it allows a process to receive a message and to continue its execution before the message is logged to a stable storage. But this protocol creates orphan messages because a process which has failed and hence not logged some of the messages received may not be able to restore its last state before it failed.

10.3 Background and Motivation

This chapter concentrates on asynchronous checkpointing as it is suitable for mobile computing even-though it suffers from domino effect. Message logging is done at receivers end which minimizes the chances of orphan message creation and overcomes the other pitfalls of asynchronous checkpoint discussed in above section. We have also attempted to adopt secure checkpointing.

10.3.1 Literature Review on Logging and Checkpointing

Following are the related work which gives motivation to this work.

Chandy-Lamport [12] first stated how a process in a distributed system can attain a global state of the system using a special type of message. The underlying computation is not blocked even-though all processes participate in checkpoint.

Elnozahy et al. [17] introduced a new concept of incremental checkpoint. This will increase the speed of the distributed application. A non-blocking copy-on-write technique stores those pages of the address space of process which have been modified since the previous checkpoint. Thus this method reduces overhead on stable storage and the effect of checkpoint efficiency increases.

Asynchronous checkpointing is considered favorable for highly mobile systems as it is not possible for some MHs which are disconnected from the network to participate in coordination. MH takes a local checkpoint whenever message reception is preceded by that sent by that MH. Other work done [8] reports that every process may accumulate multiple local checkpoints and logs on stable storage during normal operation. Further a checkpoint taken may also be discarded later if it is found that it is not needed during recovery process. For this it is required that processes periodically broadcast the status of the logs taken.

In [11] Cao and Singhal developed a checkpointing algorithm for mobile environment which was based on coordinated blocking checkpointing. It avoided search cost introduced in traversing the dependency list and reduces the blocking time. The MSS were blocked for duration of $2\times\ T_{static}$ and no MSS could send message during this duration.

The authors in [35] proposed that the advantage of asynchronous checkpointing can be augmented with message logging feature to nullify the pitfalls of asynchronous checkpointing scheme thereby reducing failure-free overhead. Log-based rollback recovery assumes that all nondeterministic events can be identified and their corresponding determinants can be logged into stable storage. During failure-free operation, each process logs the determinants of all the nondeterministic events on the stable storage. Moreover, each process also takes checkpoint to reduce the amount of rollback. After a failure occurs, the failed process recovers by using the checkpoints and logged determinants to replay the corresponding nondeterministic events.

Optimistic message logging protocols [31] buffer the received message in volatile storage and logs them to stable storage asynchronously in order to recover. But if logging is not complete before a failure occurs then execution of a failed process may be lost. In optimistic message logging, message are logged asynchronously after receipt so no synchronizing delays are necessary during any logging [3].

Hyochang Nam et al. [30] have provided a strategy for secure checkpointing to address the information leakage through checkpointing data by cryptographic methods which ensures that only the checkpoint and recovery nodes recognize the contents of the checkpoint data. One of the very common attack in these systems is Packet sniffing and another one is "man-in-the-middle attack" which modifies data. They proposed a model for a secure checkpointing system which combines the key agreement model with SPC scheme and achieved an increased overhead of 1.57 when compared to conventional checkpointing schemes.

N. Asokan [4] also realized the importance of information protection about the movements, location and activities of mobile hosts from potential threats in the form of onlookers. He stated that any one of these methods-shared key cryptosystem (SKCS) or a public key cryptosystem (PKCS) or a hybrid system, can be a solution to provide anonymity to mobile devices. Thus protection of relevant information can be achieved to some extend from man-in-the-middle attack and other such third party attacks which are prevalent in distributed systems. In [13] an efficient region based group key management protocol was introduced based on a region-based approach by which group members are broken into region-based subgroups. They compared their system with non region-based key management to demonstrate the effectiveness. Vamsi et al. [23] developed a consensus protocol based on Byzantine fault detectors using which a group of servers can determine if some member of the group has been attacked and compromised. Secondly, a checkpointing protocol based on incremental checkpointing with compression techniques [33] was also developed. This protocol divided the checkpoint data into n chunks and dispersed securely among n recipients. This was done in such a manner that any k of those n recipients can get together and re-create the checkpoint information however, no $(k-1)$ or lesser number of recipients can. Moreover, no checkpoint information is dispersed to a server that has been determined as malicious. Their third contribution is the development of a simple yet elegant group encryption scheme which will allow any member to be thrown out if a quorum of the group determines that the member is malicious.

Bhargava and Lilien [9] identified two kinds of vulnerabilities in distributed systems – Operational such as unexpected broken communication link and Information based such as unauthorized access, unauthorized modification, traffic analysis and Byzantine input in a distributed system. Following were the proposed mechanisms for reducing vulnerabilities and threats:

- Using reliability and fault tolerant principles to security – a security function R(e) was proposed which was analogous to reliability function
- Using trust in role- based access control – This states that authorization and credential (in form of digital certificates) is more secure
- Privacy-preserving data dissemination
- Fraud countermeasure mechanism [10]

Mostafa and his team [16] developed a hybrid scheme with a strong combination of pessimistic and the optimistic replication schemes with an added feature of balancing the delay incurred by the pessimistic and the high memory requirements of the optimistic schemes.

10.4 Security Aspect in Checkpointing and Message Logging: Motivation

Security vulnerabilities exist when during recovery process the checkpointing data is transferred from a local host to the stable storage when a checkpoint occurs and from the stable storage to a recovery host. To inject security into any system there are two categories of classical encryption modes – Symmetric (also known as Private key encryption) and Asymmetric (or Public key encryption).

Generally, public key cryptography is used only for key exchange or for digital signatures due to slow computational speed. Moreover, it finds its importance in message integrity during key exchange session and authentication of communicatingparties. However, problem lies in devising a method to securely transfer the keys and to prevent illegitimate access to the information. Common protocols for over coming this issue is Diffie Hellman key exchange protocol for securely exchanging the keys over the network together with digital signature which takes care of the authentication and data integrity aspects. A digital signature is an electronic signature that can be used to authenticate the identity of the sender of a message or the signer of a document and possibly to ensure that the original content of the message or document that has been sent is unchanged. The authors in [30] attempted to make the log system secure using a proxy server to store the sessional private keys generated using DHKE protocol together with digital signature algorithm (DSA) for authentication. DHKE is a method of digital encryption that uses large prime numbers raised to specific powers to produce decryption keys on the basis of components that are never directly transmitted, making the task of a hacker/"man-in-the-middle" impossible to break the mathematical code. In [14], the authors devised an efficient secret sharing mechanism for mobile ad-hoc networks. The most common approach to tackle security in a static environment is through Shared Key Cryptosystem (SKCS) or public key cryptosystems (PKCS). Shared key Cryptography uses one or more authentication servers (AS) for each communicating party (say A and B) and Public Key Cryptography uses a Certificate issuing Authority (CA). In the former method, AS serves as a third party and generates a random session key K_{AB} shared publicly by A and B. Apart from this, each one of them has its own private key K_{AS} and K_{BS} respectively. Then the sender encrypts the data with the shared public key. The receiver recovers the message by decrypting it with its private key. The vital issue in this *asymmetric* way of communication is correct authentication of the sender. In the second method,the CA generates several public keys for all the entities in the network. Then, any entity which wants to communicate selects a public key and registers itself with the CA. In this way, CA issues a certificate as a guaranteer and any other entity can trust the certificate and hence also the transmitter since it also has the shared public key.

In a wireless network with dynamically changing configuration including location of the MH, security can be compromised by an unauthorized user eavesdropping over a communication. A cryptographic scheme is secure if its key generation system, key exchange mechanism and authentication process is robust and resistant to attacks. However, mobile computing is essentially a dynamic system. In [36] it was pointed out by Shannon that "In a good mixing transformation functions are complicated, involving all variables in a sensitive way. A small variation of any one variable changes the outputs considerably." This implies that small changes may result into large changes and if this be mapped to chaos theory it will mean that chaotic behavior will continue to manifest the effects of any minuscule disturbances. It is in this aspect that the well tested properties of ergodicity and randomness of yet another dynamic system adds sufficient confusion and diffusion properties required for secure systems. DHKE is a method of digital encryption that uses large prime numbers raised to specific powers to produce decryption keys on the basis of components that are never directly transmitted, making the task of a hacker/"man-in-the-middle" impossible to break the mathematical code. The computational overhead incurred in the exponential DHKE and the complex calculations in digitally signing the message is too taxing for a distributed dynamic system with limited resources. Hence, alternative non traditional ways of cryptographic solution has been explored which fulfills the dual purpose of secure key exchange and authentication. Successful authentication also overcomes the task of data integrity and non repudiation problem which are equally important in secure communications. The next section uncovers the proposed solution for secure message logging and authentication.

It is assumed that the MSS's are stable and the probability of their failure-free operation is more as compared to the MHs.

10.4.1 Chaos Based Technique for Secure Communication

Chaos theory from the area of non linear dynamics has proved to play a vital role in cryptography. Currently, a lot of developments have led to the fruitful implementation of chaotic sequences as subtle replacement for the random numbers for several application areas [28]. These sequences are totally random and depended on the system parameters and hence can be substituted as random number generators. The most attractive feature of chaotic system that makes it ideal for novel engineering applications is its unpredictable and random nature. The uncertain yet deterministic features of the highly fluctuating chaotic orbits are the key properties which are explored in securely transmitting the data. [2, 5, 6, 25] which exploits the properties of Chaotic dynamics viz. ergodicity sensitivity to initial condition, randomness and mixing. Chaos based cryptography can be categorized into analog and digital cryptosystem. The former is based on the techniques of control [32] and synchronization [21, 26] of chaos. Whereas digital chaos based cryptosystems can be realized by: block ciphers based on forward and/or reverse iterations of chaotic maps [37], block ciphers based on chaotic round functions [22], stream ciphers implementing chaos based pseudo random bit generators (PRBG) [29]. There are several ways through which analog chaos based cryptosystems can be realized such as: chaotic masking [15], chaotic modulation [18] chaotic switching [38] inverse system approach. On the contrary, in digital chaos based cryptosystems the chaotic discrete dynamical systems are implemented in finite computing precision. The primary requisite for a secure system is that the keys should be secure. Chaotic cryptography possesses reliability and added security due to the inherent randomness. The idea for opting chaos based cryptosystems is obtained from theoretical and practical efforts of the research in the past few years that this new breed of cryptography technique performs similar to the classical methods based on computational techniques.

Cryptography with chaos falls under the category of private key cryptography mode. This approach relies on the properties that chaotic signals are usually noise like and very sensitive to initial conditions and fluctuations. Therefore the secret keys usually contain the system parameters and the initial condition. Due to the nonlinear mechanisms that lead to a chaotic regime, the system properties are too difficult to be predicted by analytical methods without the *a priori* knowledge of the secret key,initial conditions and/or parameters. This would reduce a potential attack to one category – that of a brute force attack, in which any attempt to crack the key depends directly upon how long (or complicated) the key is. It is well known that traditional cryptography works with discrete values and in discrete time in contrast to chaotic cryptography which utilizes continuous-value systems that may operate in continuous or discrete time. As a result, in order to ensure an appropriate design and analysis methods for chaotic cryptographic systems, these crucial points must be taken into account. Chaotic maps and cryptographic algorithms (or more generally maps defined on finite fields) have also some similar properties: sensitivity to initial conditions and parameters, random like behavior and unstable orbits with long periods.

Encryption rounds of a cryptographic algorithm render the desired diffusion and confusion properties of the algorithm. In a similar manner, iterations of the chaotic map spread the initial region over the entire phase space while the parameters of the chaotic map may represent the key of the encryption algorithm. An important difference between chaos and cryptography is the fact that the encryption transformations are defined on finite fields, while chaos has meaning only on real numbers [20] and cryptography deals with integer numbers [34].

In this scenario, the chapter presents a cryptographic technique to inject a secure logging process with the modified logging and asynchronous checkpointing scheme. The keys are exchanged over a secure communication channel. In the original DHKE protocol, the short lived sessional keys were generated randomly. But, in our proposed method the keys used are continuously generated by a coupled chaotic system placed at the transmitter and receiver section of the two communicating parties. The driver system acts as the transmitter and the slave or the response system acts as the receiver. Thus the entire arrangement forms a secure communication channel [24]. Further, the chaos based keys are never revealed over the network. This will further enrich the security because no information regarding the manner in which these random keys are generated need to be transmitted over the channel. These are then used to mask data using any traditional symmetric key encryption technique like Vernam cipher. The chaotic stream of data which forms the private keys are used in encoding the message. By virtue of the unpredictable and stochastic properties of such dynamic system, the sequences are absolutely randomized in nature. Since, the chaotic sequences are used only once with an XOR mode for cryptography so this organization forms a one-time pad which is theoretically unbreakable [27].

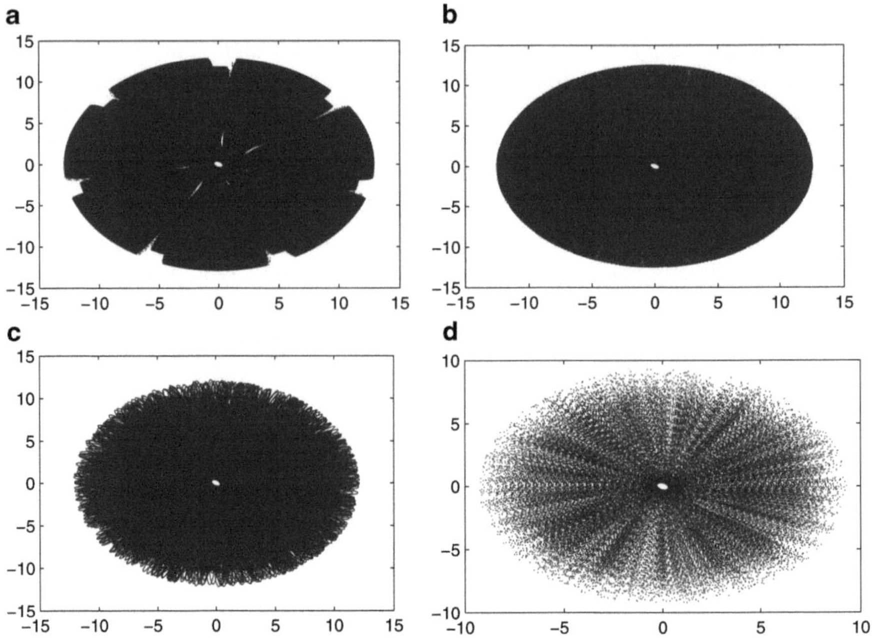

Fig. 10.2 (Color online) Phase space diagram for $a = 2.0$, $b = 0.8$, $c = 70.0$, $d = 0.01$, $q_1 = 0.005$ (**a**) $q_2 = 0.001$; (**b**) $q_2 = 0.005$; (**c**) $q_2 = 0.01$; (**d**) $q_2 = 0.05$

10.4.2 Optical Chaos: The System for Rendering Security in the Proposed Model

This section deals with the synchronization of a chaotic laser system from the field of optical chaos.

The Maxwell-Bloch equations are used to describe the dynamics of the propagation of the electromagnetic fields in an optical fiber which is doped with rare earth ions. From the governing PDE system we apply a set of discrete fourier expansions to convert them to a set of coupled nonlinear ODE given by [7]

$$\dot{x_1} = f_1(x_1,x_2,x_3,x_4,x_5) = -a(x_1 - x_3) - (x_1^2 + x_2^2)(q_2x_1 + q_1x_2) \tag{10.1}$$

$$\dot{x_2} = f_2(x_1,x_2,x_3,x_4,x_5) = -a(x_2 - x_4) + (x_1^2 + x_2^2)(q_1x_1 - q_2x_2) \tag{10.2}$$

$$\dot{x_3} = f_3(x_1,x_2,x_3,x_4,x_5) = -(x_3 - x_4d) + (c - x_5)x_1 \tag{10.3}$$

$$\dot{x_4} = f_4(x_1,x_2,x_3,x_4,x_5) = -(x_4 - x_3d) + (c - x_5)x_2 \tag{10.4}$$

$$\dot{x_5} = f_5(x_1,x_2,x_3,x_4,x_5) = -bx_5 + (x_1x_3 + x_2x_4) \tag{10.5}$$

In the above set x_1,x_2,x_3,x_4,x_5 are the state variables and a,b,c,d,q_1,q_2 are the system parameters. The system (10.1)–(10.5) produces multi periodic orbits and also chaotic attractor for different choice of the parameters is illustrated in Fig. 10.2.

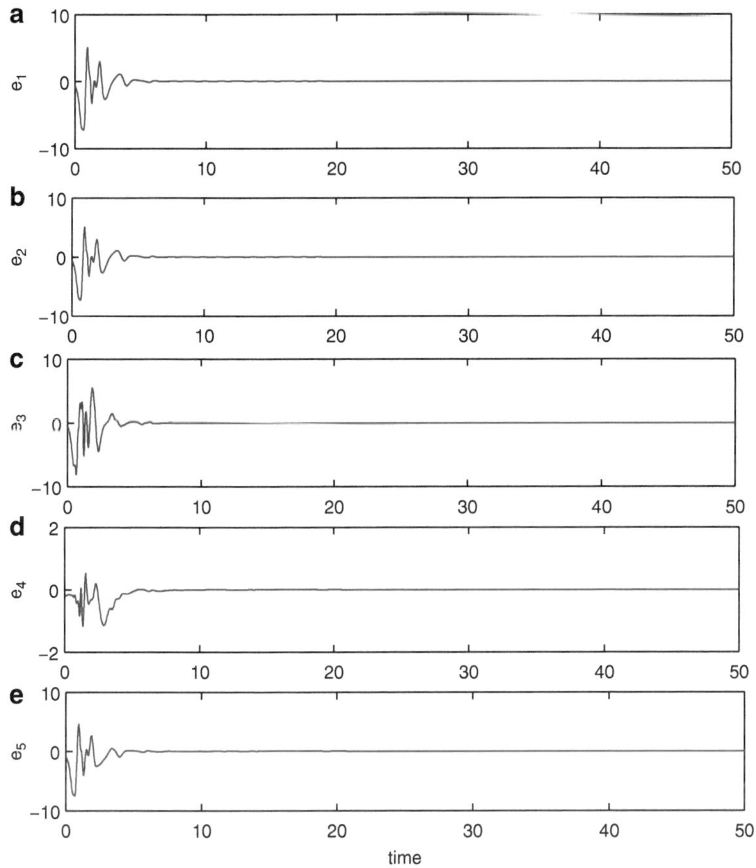

Fig. 10.3 (Color online) Synchronization error for the variables

Now we consider a synchronization scheme between two identical systems. Let us define the system (10.1–10.5) by $\dot{x} = f(x)$ which can be considered as a driving system and also identical system $\dot{y} = f(y)$ as a response system. The synchronization occurs by adding both way linear feedback coupling term $K_1(y-x)$ to the driving and $K_2(x-y)$ to the response system. The synchronized system is a two way coupled system. We choose all other parameters in the chaotic state, corresponding to the Fig. 10.2d. Figure 10.3 represents the synchronization error for the two electromagnetic systems for $K_1 = 3.5$ and $K_2 = 4$.

System (10.1) is the key generating system. The key sets are processed by the simple rule

$$keys_i \leftarrow integer(abs(10^3 \times |x_{1i}|))Mod(256) \qquad (10.6)$$

10.5 Proposed Scheme: Chaos Key Exchange Protocol for Mobile Computing and Secure Logging and Checkpointing

The proposed Chaos based Key Exchange technique is explained in this section for secure logging and transmission of messages. We propose: (a) A hybrid message logging and asynchronous checkpointing scheme at receiver's end with an improved performance where messages exchanged by Mobile Hosts (MH) are logged at the stable base station; (b) The proposed message logging is made secure by introducing cryptography devised from a coupled chaotic laser system derived form Maxwell Bloch's equations. The entire process serves a dual purpose. Firstly, it eliminates the computational overhead incurred during the complex way of exchanging keys in Diffie Hellman Key Exchange (DHKE) protocol. Also, synchronization of the two communicating devices would form a secure channel through which physical transfer of the private keys would be completely eliminated as otherwise present in the case of DHKE. In DHKE, there is an additional task of authentication and message integrity for which adopting digital signature becomes imperative. However, the method of chaos based cryptography with synchronized chaotic carrier wave at the physical layer is synonymous to exchanging the keys through a secure network. The proposed method eliminates the complex computation otherwise performed in DHKE protocol thereby achieving faster communication. Secondly, the inherent randomness and deterministic properties of chaotic system also renders the scheme immune to attacks with a computational edge. The entire scheme has been termed as Chaos Key Exchange (CKE) which uses the chaotic sequences generated by a synchronized chaotic laser system as the private keys in the cryptography. The proposed logging process as well as the communication of messages has been made secure by applying CKE method. If the MH is authenticated by the assumed stable MSS then only its log is recorded by the MSS. Similarly, only a legitimate MH can communicate with other legitimate MH in its same or different cell. This is possible if the receiver section is synchronized with the MH which is initiating the logging process or communicating at that session. CKE has been demonstrated by encoding a text as well as an image which are transmitted over a secure channel. The ergodicity of the chaotic system adds the necessary confusion and diffusion properties required for efficient cryptography.

This is illustrated schematically in Fig. 10.4. Whenever, a new MH first enters a cell, it registers itself with the Mobile Support Station (MSS). This MSS becomes its local/base MSS (MSS_b). The registration process involves the use of the synchronized system (10.1) and (10.2). In this phase, MSS_b becomes the driver and the MH becomes the response system. The MH and its MSS_b has its unique set of processed chaotic sequences obtained from (10.6) denoted as K_{mh} and K_b respectively. Both of them use the first key from the processed chaotic sequences generated through synchronization as the private session key for registration denoted as K_{reg}. Due to limited storage capacity of MH, the chaotic sequences for MH (K_{mh}) are stored at its MSS_b corresponding to the id of the MH (MH_{id}). By virtue

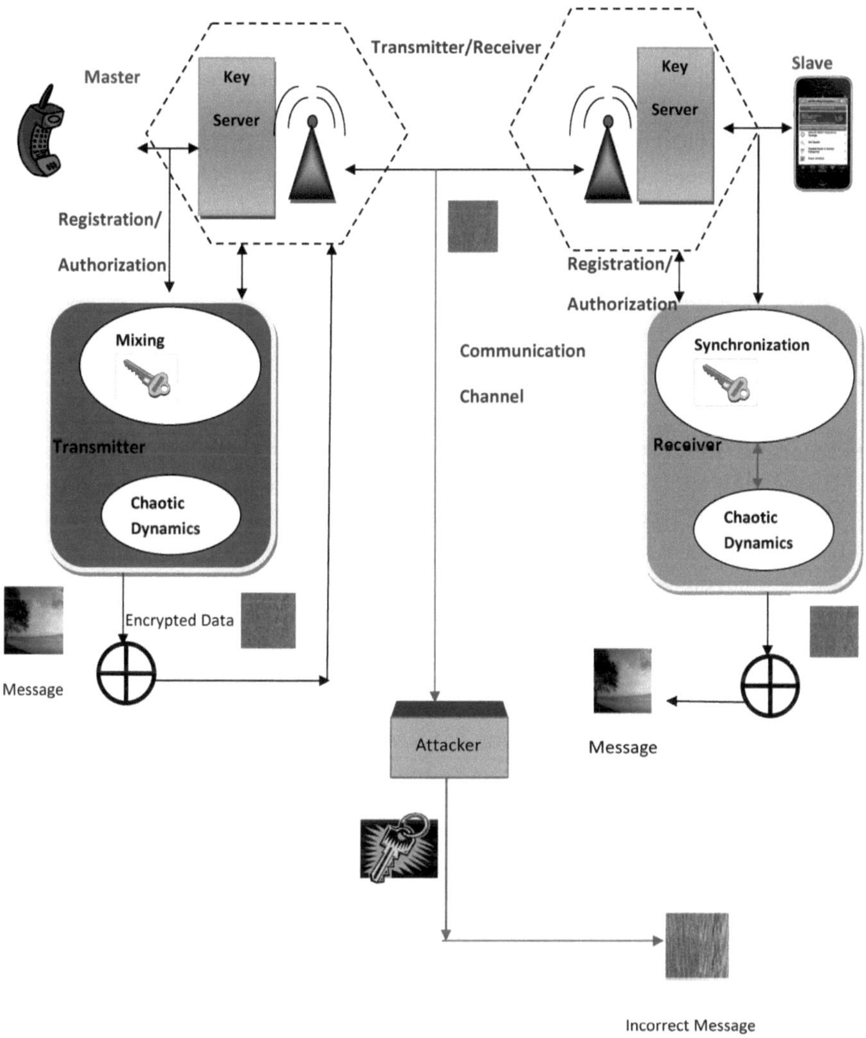

Fig. 10.4 (Color online) Illustration of Secure Communication of a message through Chaos Key Exchange Protocol

of chaos synchronization, $K_{mh} = K_b$ so this does not require a separate storage of another key set and any one set of keys can be recorded in the permanent storage space. This eliminates duplicate storage and reduces the storage space. This is the Authentication process. Next, if this MH wants to log its message in its MSS_b, its authentication is verified by checking the entry of K_{reg} corresponding to MH_{id}. If the entry matches, then MSS_b sends an acknowledgement ($Flag$) to MH_{id} after which the logging process is initiated ($Take_{log}()$). This secure mechanism of Registration and Authentication is explained in the Algorithm procedure $Authenticate_{protocol}$. The logging process is explained in the Algorithm procedure *Logging*.

Now, suppose two MH's, MH_s and MH_r under the same or different cell MSS_{rs}, wish to initiate a communication of a message *msg*. MH_s transmits msg to MH_r. The base station of MH_s authenticates it by its K_{reg}. If it is authenticated, then MH_s transmits msg by XORing msg with the rest of the keys in K_{mh} to produce the cipher message, CM. Then CM is transmitted to the base station of the intended receiver mobile MH_r. There also the same process of identification of MH_r is instantiated by its base MSS. On successful authentication by their pair of K_{reg}, CM is decoded by the MSS and transmitted to the receiver MH_r. This procedure is explained in detail under the procedure $Authenticate_{protocol}$.

10.5.1 Mobility Management

In apropos to the above discussions, the present work combines a secure optimistic checkpointing technique with receiver-based message logging strategy enriched with the sessional keys generated by system (10.1)–(10.5). It is a non-blocking algorithm since the underlying computations by the MSS are not stalled when it does the authentication and secure logging of the message. Checkpoints are taken after the handoff parameter for a MH exceeds a certain threshold which can be a function of failure rate, log arrival rate and the mobility rate of the MH.

Mobility management is handled by sending the message *leave* when a handoff occurs indicating the sequence number of the message received from MSS. The MH can establish a connection with MSS when it enters a new cell by sending the message join (MH_{id}, $prevMSS_{id}$) to this new MSS. The id of MHs which are currently connected to a MSS are recorded in a list known as $Active_{MHlist}$ stored at the MSS and an entry is made when a MH joins the MSS, whereas an entry is deleted when a MH leaves that MSS. The keys used by the proposed arrangement of secure communication is shared by the MH and the MSS. With this the logged message is encrypted and stored at MSS. During checkpointing, this agreement is again used to encrypt the checkpoint data. The MSS and MH communicate privately over a public medium via a proxy key server like interface implementation with the encryption and authentication processes so that the MSS is not kept busy and other computations are not blocked. Once the MH is authenticated by the proxy server for the local MSS, encrypted data is transferred seamlessly between MH and MSS which buffers it. The private keys are not transmitted over the network thereby overcoming "man-in-the-middle" attack. $AUTHENTICATE_{PROTOCOL}$ (function signature) is used to securely log the messages communicated.

10.5.2 Data Structure

We have used the following variables in the implementation of our scheme:
MH_{id} = MH identity number(id) communicating with a MSS.
MH_s = represents the transmitting MH which is initiating the log or sending a message.

MH_r = represents the MH intended to receive a communication message.
MSS_b = stands for the base station.
MSS_s = represents the MSS transmitting the message.
$Active_{MHlist}$ = This contains all the MH id's of the connected MH's.

$Disconnected_{MHlist}$ = It contains all the MH id's which are currently disconnected or are in sleep mode. This is also stored at the MSS_b.

leave($prev_{MSS}$, $leaving_{MH}$) = When a MH leaves a cell, it sends this message to the MSS by informing the MSS about the sequence number of the message it received from the MSS.

$Chkpt_{seqnum}$ = It keeps a record of the number of checkpoints of MH.

msg = is the message. Consists of non negative integer source and the data which may be a string or an image.

Log_{table} = This structure contains the log of the MH_{id} and stores checkpoint data which is also logged along with it. It contains a non negative integers.

$Dependency_{Loglist}$, $Chkpt_{seqnum}$, $Chkpt_{MSSid}$ all initialized to zero which stores the mentioned information:

- id's of the MSS where the log is taken,
- Sequence order of taking the checkpoint,
- id MSS where the checkpoint has been taken,
- It also contains variable msg_{source} indicating the MH which has send the message.

The log table (Log_{table}) logs the messages received by the logging MH and the process which took the checkpoint in the buffer $msg_{received}$ and $Chkpt_{process}$ respectively. Mobility history of the logging MH is thus maintained by keeping the record of $msg_{received}$, $Chkpt_{seqnum}$ and $Dependency_{Loglist}$. When a MH takes a checkpoint, the $Chkpt_{seqnum}$ variable is updated and the identification of the MSS under which it took checkpoint is put in $Chkpt_{MSSid}$. In order to retrieve the logged messages after the latest checkpoint, the $Dependency_{Loglist}$ can be used which stores the id of the last MSS where the log of the message for this failed MH was stored. The order in which checkpoint was taken can be easily found out by looking at the log table entries which are recorded in LIFO format.

$mobility_{rate}$ = It is a Counter which keeps the record of the number of times a MH leaves its cell.

K_{mh1} = One time sessional Private key of MH generated by (10.1).

K_{b1} = One time sessional Private key of proxy server generated by (10.2).

K_{reg} = variable denoting the $K_{mh1} - K_{b1}$ pair.

Flag = represents the acknowledgement status and it is in either of the state: high or low. If Flag is high then it implies that the MH has been authenticated and if its is low then it means otherwise.

$MH_{AUTHENTICATEFLAG}$ = It is set to one if the MH is authenticated by the MSS else it is reset to zero.

CM = The encrypted message.

10.5.3 Algorithm

Logging of Messages

Case: $MH_s \rightarrow MSS_b$

When MH_s sends Message msg to its MSS_b which serves as the MSS_s:
if(msg ε join, leave, disconnect, reconnect) then
{
if (msg ε leave, disconnect) then
{
leave MH_s from MSS_b
Reconnect (MH_s) to other MSS chosen randomly, say MSS_x
MH_s.handoff = MH_s.handoff + 1;
Send [msg, $mobility_{history}$] to MSS_b
}
Else
{
Join(MSS_b, MH_s)
}
Else
{
/∗ *If msg is a communication message* ∗ /
Flag = $AUTHENTICATE_{PROTOCOL}$(MSS_b, MH_s, msg, $process_{id}$);
}

/∗ *When* MSS_b *receives Message from* MH_s ∗/

if(msg ε join, leave, disconnect, reconnect) then
{
if (msg ε leave, disconnect) then
{
search if($MH_s \varepsilon Active_{MHlist}$). If true then
{

Update $Active_{MHlist}$ stored at MSS_b
Update $Disconnected_{MHlist}$ of MSS_b
}
Else MH_s is already disconnected from MSS_b
}
Else /* *If msg is a join message, then this new MSS becomes its base or home MSS**
{
Join(MSS_b, MH_s)
Update $Active_{mhlist}$ stored at MSS_b
Fetch Log_{table} from $MSS_{previous}$ to this MSS_b
}
Else /* *If msg is a communication message* /*
{

If (Flag is true) then
{
MSS_b accepts msg from MH_s
msg.source = MH_s;
msg.data = $process_{id}$;
Receive [msg, $Mobility_{history}$] from MH_s;
MSS_b transmits msg to MSS_r of the destination MH;

}
Else
msg is discarded by the MSS_b
}

Case: $MSS_r \rightarrow MH_r$

if ($MH_r \varepsilon$ MSS_r, $Active_{MHlist}$)
for all msg ε communication message
do
{
Flag = $AUTHENTICATE_{PROTOCOL}$(MSS_r, MH_r, msg, $process_{id}$)
$Take_{log}$ (MH_s, msg, MH_r, MSS_r, $process_{id}$)

if (Flag is true) then
{
MH_r receives mg
$Insert_{log}$(MH_s, msg, MSS_r, MH_r, process) into MH_r volatile Log_{table};
log the checkpoint data if any
Periodically flush the log into log space of MSS_r
}

Else
discard msg
}

Checkpointing

Procedure: Checkpoint(MH)
{
if (MH.handoff > max ($Mobility_{rate}$) then
{
Take checkpoint
MH.$Chkpt_{MSSid} = Chkpt_{MSSid}$;
MH.$Chkpt_{seqnum} =$ MH.$Chkpt_{seqnum} + 1$;
Insert(MH.$Chkpt_{MSSid}$, MH.$Chkpt_{seqnum}$, $Chkpt_{processid}$) into Log_{table} of the MSS under which checkpoint was taken;
Reset the handoff counter;
}
Else
no checkpoint
resume computation
}

$AUTHENTICATE_{PROTOCOL}$ *(MSS_{id},MH_{id},msg,$process_{id}$)*. This block is executed between any communicating MH, be it a sender (MH_s) or a receiver (MH_r) and its base MSS (MSS_b). The subscript *id* in the function signature is a unique identification number assigned to MSS and MH in communication protocol.

1. *Registration Process*: Whenever, the MH first enters its base MSS (MSS_b), it registers itself through the process of chaos synchronization. MH uses system (10.1) and MSS_b uses system (10.2).

1.1 *Selection of session key by the Sender Mobile Host*: Each MH obtains a unique private key k_{mh1}, which is the first key from the processed key set obtained in (10.6) as $K_{MH} = \{k_{mh1}, k_{mh2}, \ldots, k_{mh\infty}\}$.

1.2 *Selection of session key by the MSS*: MSS_b also has its private key k_b1 from the set $K_b = \{k_{b1}, k_{b2}, \ldots, k_{b\infty}\}$. During this phase, the connection between the two are assumed to be that of a widely deployed wired-line. Privacy issues and authentication between the MH and the base station are then treated seamlessly, authenticating MH with the base station and establishing a secure channel at their respective link layer. By virtue of chaos synchronization, $k_{mh1} = k_{b1}$. We have denoted this unique key used for registration as K_{reg}.

2. *Authentication*: MSS_b will first verify MH_s with itself since this MSS contains the credentials of the MH under it and MSS_r possesses no knowledge or secret information about MH_s. To proceed further, it is assumed that every pair of static MSS, say MSS_b and MSS_x shares a secret key K_{bx} ε $keys_i$ chosen similarly by each MSS as in Step 1.2.

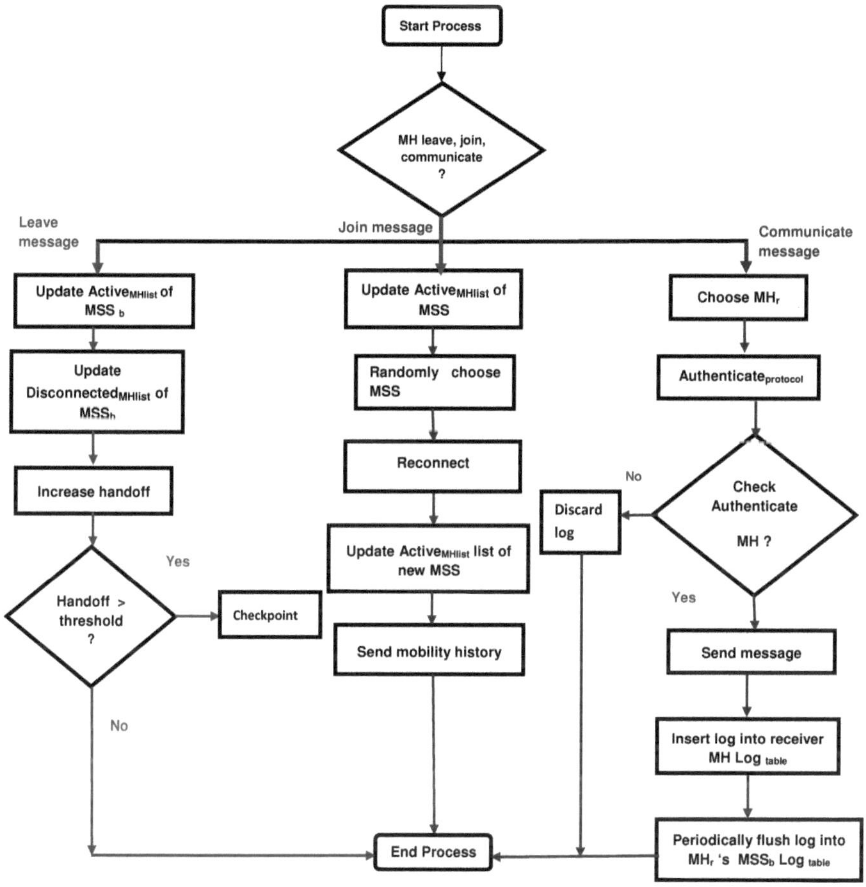

Fig. 10.5 (Color online) The Flowchart illustrating the activity of different processes in the scheme

2.1 Verification: By virtue of the properties of chaos synchronization, for a MH_y, $K_{by} = K_y$. If MSS_b is satisfied with the recent time stamp (the time stamp should be fresh time stamp so as to prevent message replay attack), then MSS_b confirms MSS_r about the authenticity of MH_s. MSS_b sends K_{bs} to MSS_r after MSS_b verifies the credentials sent by MH_s through MSS_r.

2.2 Communication: MSS_r and MH_s establish a secure session. MH_s then transfers msg of length say l encrypting with the key sets $K_s - K_{bs}$ by XOR operation forming a stream cipher CM

$$CM \leftarrow msg \bigoplus K_s - K_{bs}$$

In the meantime, MSS_b relays the synchronized key set between MH_s and itself to MSS_r relevant for decrypting over its secure channel to MSS_r. MSS_r correctly recovers msg

$$msg \leftarrow CM \bigoplus K_b - K_{bs}$$

The same steps discussed above occur between the intended receiver MH (MH_r) and its local MSS. During decryption, the MSS of the MH_r performs anti-XOR operation to obtain the original message.

A flowchart illustrating the above working of the various activities involved is shown in Fig. 10.5.

10.6 Security Analysis

The proposed CKE method and the subsequent encoding of the message *msg* to Cipher Message (*CM*) has been illustrated with a colored 256×256 sized image which is the message transmitted. It is represented as a $M \times N$ dimension RGB image with M rows and N columns and the three color channels red,green and blue are represented by symbols R, G, B respectively. Figure 10.6 illustrates the encoding and decoding scheme through the set of equations (10.1)–(10.6) representing the system.

10.6.1 Key Sensitivity Analysis Against Noise

The scheme correctly recovers the original image and this can be shown by a metric known as Mean Square Error $MSE = [MSE^R\ MSE^G\ MSE^B] = [0\,0\,0]$ calculated as

$$MSE = \frac{1}{M \times N} \sum_{r=1,c=1}^{M,N} [P(r,c) - CI(r,c)]^2$$

The effect of changing the key set through the operation

$$keys_i \leftarrow integer(abs(10^3 \times |x_{1i}|))Mod(256) \tag{10.7}$$

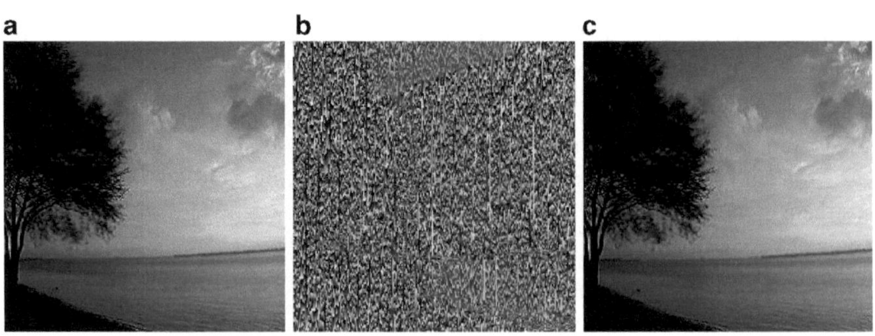

Fig. 10.6 (Color online) Cryptography Illustration. (**a**) Original multimedia picture message; (**b**) Encoded transmitted message; (**c**) Decoded message

Fig. 10.7 (Color online) Statistical test result due to tampering of the key set. (**a**) Correct key set; (**b**) Correct decryption; (**c**) Incorrect key set; (**d**) Incorrect recovery of the message

Table 10.1 Mean square error for altered key set

MSE	R	G	B
Test result	1.0e+004 ×1.0844	1.0e+004 ×1.0340	1.0e+004 ×1.0048

is demonstrated in Fig. 10.7. On changing the key set through the (10.7), it is observed that the recovered image highly differs from the original. So, an attacker will never be able to decipher the original message transmitted. The MSE also changes drastically for the incorrectly recovered message as tabulated in Table 10.1.

Table 10.2 Result of differential attack: NPCR and UACI performance metric

	R	G	B
NPCR	0.9958	0.9958	0.9958
UACI	0.0223	0.0219	0.0587

Also, the key space is vast since each key is composed of 8 bits having a total of 2^8 combinations, which is a good exhaustive key set. In general this encryption technique is impressive due to the sensitivity of the cipher image/message to the changes in its keys and the combined effect of these together with a huge key space renders the system immune to major statistical brute force attacks. These are discussed subsequently through statistical tests conducted.

10.6.2 Differential Attack Analysis

A good cipher should be immune to the influence of a change in a pixel. The change in a few voxel values are listed below to demonstrate the dispersion of voxels through the scheme.

The rate of change of the number of pixels of the cipher image when the plain image is altered by just one pixel is quantified by a metric known as NPCR. This also measures the efficacy of the diffusion phase. For each r ε M and c ε N, NPCR is given by

$$NPCR = \frac{\sum_{r=1,c=1}^{M,N} D(i,j)}{M \times N} \times 100$$

Here, $D(r,c)$ represents the change in the picture element due to the encryption of a monochrome image and

$$D(r,c) = 0 \text{ when } msg(r,c) = CM(r,c) else$$
$$D(r,c) = 1 \text{ when } msg(r,c) \neq CM(r,c)$$

Next quantifiable measure of diffusion properties is the Unified Average Changing Intensity (UACI) quantifies the average intensity of the differences between the pixel values of the original and encrypted image. It is represented by

$$UACI = \frac{1}{M \times N} \sum_{r=1,c=1}^{M,N} \frac{|msg(r,c) - CM(r,c)|}{255} \times 100$$

Table 10.2 shows the effective performance of the scheme with a high value of NPCR. It is evident from the result, that higher the value of NPCR, more will be the tolerance of the system to differential attacks.

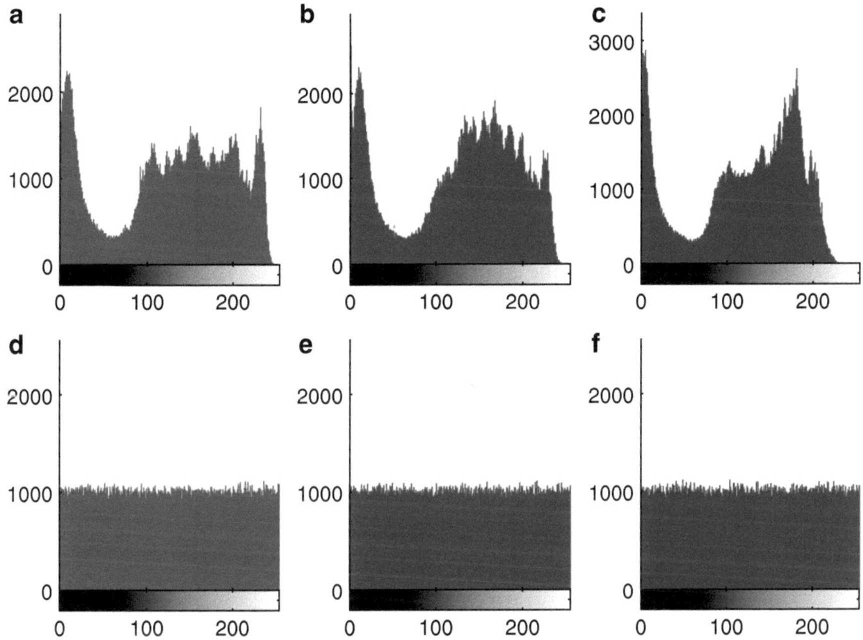

Fig. 10.8 (Color online) Histogram: (**a**)–(**c**) original image, (**d**)–(**e**) cipher image

10.6.3 Histogram Analysis

The distribution of the pixels after being distorted can be analyzed from the intensity plot of pixels in the range 0-255 known as the histogram. A monochrome image with gray level r_k is represented by a function $h(l_k) = n_k$, where l_k denotes the kth gray level and n_k are the number of pixels in an image.

As seen from Fig. 10.8, the intensity plot of the encrypted image is uniformly distributed and completely differs from the original image. So, from the histogram it can be proved that an attacker/unauthorized MH cannot deduce any statistical information about the original image.

10.6.4 Statistical Analysis: Correlation

Correlation between the secret key and the output of the cryptosystem is the main source of information to the cryptanalyst. In the easiest case, the information about the secret key is directly leaked by the cryptosystem. More complicated cases require studying the correlation (basically, any relation that would not be expected on the basis of chance alone) between the observed (or measured) information

Table 10.3 Correlation coefficient for each color band

	Original			Cipher		
	R	G	B	R	G	B
Horizontal	0.9764	0.9737	0.9706	0.0091	−0.0206	0.0001
Vertical	0.9707	0.9677	0.9645	−0.4476	−0.4179	−0.4053
Diagonal	0.9609	0.9571	0.9531	0.0023	0.0077	0.0003

about the cryptosystem and the guessed key information. This shows if there is any occurrence of an association between the guessed secret key and the observed information of the cryptosystem and forms the core information for cryptanalysis. Mathematically, covariance $cov(x,y)$ between a pair of pixel values x and y in a gray image is formulated as

$$cov(x,y) = E(x - E(x))(y - E(y))$$

Then, the correlation coefficient r_{xy} is given by

$$r_{xy} = \frac{Cov(x,y)}{\sqrt{D(x)}\sqrt{D(y)}}$$

where $E(x), E(y)$ denotes the mean; $D(x), D(y)$ stands for the variance between the pixels. Correlation coefficient values of a random sample of 2,000 voxels placed adjacently along the horizontal,vertical and diagonal directions have been computed for each of the three color channels and enumerated in Table 10.3 from which it can be observed that the correlation of cipher image is almost zero. For an efficient encryption methodology, it is imperative that the correlation values of adjacent pixels be minimal for the encoded image. Our scheme fulfills this criteria. Figure 10.9 displays the spatial distribution of the randomly selected voxels from the three color planes placed adjacently along the horizontal direction.The graphical result emphasizes that there is hardly any correlation between the original and the distorted version of the image.

10.6.5 Conclusion

The present work proposes a secure communication technique and efficient key management method for mobile computing environment based on the chaotic sequences generated by an optical chaos system with high fluctuating chaotic orbits. Such a system possess high dimensionality and uncertainty which assists in making the system immune to attacks. The ergodicity of the chaotic system adds the necessary confusion and diffusion properties required to render the scheme secure. Exhaustive statistical test have been conducted to show the merit of the proposed scheme. Moreover, we have also attempted to minimize the need for taking

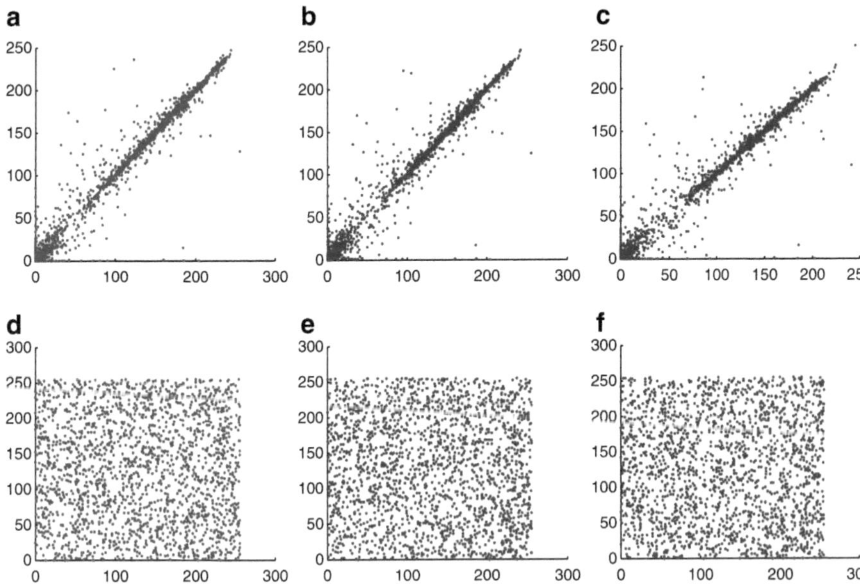

Fig. 10.9 Scatter plot of Correlation Coefficients of adjacent pairs of pixels along the horizontal direction: (**a**) Correlation coefficient of the original image in the *red* color plane; (**b**) Correlation coefficient of the original image in the *green* color plane; (**c**) Correlation coefficient of the original image in the *blue* color plane

checkpoints by using secure message logging at receiver's end. Also, the log is only stored if the communicating MH is authenticated by the sender and receiver MSS. In effect, this arrangement is ideal for a device with low computing power in a dynamically changing configuration.

References

1. Acharya, A., Badrinath, B.R.: Checkpointing distributed applications on mobile computers. In: Proceedings of the Third International Conference on Parallel and Distributed Information Systems, pp. 73–80 (1994)
2. Alvarez, G., Monotoya, G., Pastor, F., et al.: Chaotic cryptosystems. In: Proc. IEEE Int. Carnahan Conf. Security Technology, pp. 332–338 (1999)
3. Alvisi, L., Hoppe, B., Marzullo, K.: Nonblocking and orphan free message logging protocols. In: Proceedings of IEEE 23rd Int. Symp. Fault Tolerant Computing, pp. 145–154 (1993)
4. Asokan, N.: Anonymity in a mobile computing environment. In: Proceedings of Workshop in Mobile Computing Systems and Applications, Santa Cruz, California, pp. 200–204 (1994)
5. Banerjee, S., Ghosh, D., Ray, A., et al.: Synchronization between two different time-delayed systems and image encryption. Europhys. Lett. **81**(2), 20006 (2008)
6. Banerjee, S., Rondoni, L., Mukhopadhyay, S., et al.: Synchronization of spatio-temporal semiconductor lasers and its application in color image encryption. Optic. Comm. **284**(9), 2278–2291 (2011)

7. Banerjee, S., Saha, P., Chowdhury, A.R.: Chaotic aspects of lasers with host-induced nonlinearity and its control. Phys. Lett. A **291**, 103–114 (2001)
8. Bhargava, B., Lian, S.R.: Independent checkpointing and concurrent rollback for recovery in distributed systems. In: Proceedings of Seventh Symposium Reliable Distributed Systems, p. 312 (1988)
9. Bhargava, B., Lilien, L.: Vulnerabilities and threats in distributed systems. Distributed computing and internet technology, p. 146. Springer (2004)
10. Bhargava, B., Zhong, Y., Lu, Y.: Fraud formalization and detection. In: Proc. Intl Conf on Data Warehousing and Knowledge Discovery DaWak (2002)
11. Cao, G., Singhal, M.: On the Impossibility of Min-Process Non-Blocking Checkpointing and an Efficient Checkpointing Algorithm for Mobile Computing Systems. In: Proceeding of the 27th International Conference on Parallel Processing, 37–44 (1998)
12. Chandy, M., Lamport, L.: Distributed snapshots: Determine global states of distributed. In: ACM Transactions on Computing Systems, **3**(1), 63–75 (1985)
13. Cho, J.H., Chen, I.R., Wang, D.C.: Performance optimization of region-based group key management in mobile ad hoc networks, Perform. Eval. **65**(5), 319–344 (2008)
14. Daza, V., Herranz, J., Morillo, P., et al.: Cryptographic techniques for mobile ad-hoc networks. Comput. Network. **51**(18), 4938–4950 (2007)
15. Dedieu, H., Kennedy, M.P., Hasler, M.: Chaos shift keying: Modulation and demodulation of a chaotic carrier using self-synchronizing. IEEE Trans. Circuits Syst. II **40**, 634–641 (1993)
16. El-Barrv, M.I.H.A., Khan, S.A.: Design and analysis of a fault tolerant hybrid mobile scheme. Inform. Sci. **177**(12), 2602–2620 (2007)
17. Elnozahy, E.N., Johnson, D.B., Zwaenpoel, W.: The performance of consistent checkpointing. In: Proceedings of the 11th IEEE Symposium on Reliable Distributed Systems, pp. 39–47. Houston, TX (1992)
18. Feldmann, U., Hasler, M., Schwarz, W.: Communication by chaotic signals: The inverse system approach. Int. J. Circuit Theor. Appl. **24**, 551–579 (1996)
19. Gerla, M., Kleinrock, L.: Vehicular networks and the future of the mobile internet. Comput. Network. **55**(2), 457–469 (2011)
20. Gickenheimer, J., Holmes, P.: Nonlinear oscillations. Dynamical systems and bifurcations of vector fields, p. 453. Springer, Berlin, Germany (1983)
21. Habutsu, T., Nishio, Y., Sasase, I., et al.: A secret key cryptosystem by iterating a chaotic map, in Advances in Cryptology – EUROCRYPT'91. Lecture Notes in Computer Science, vol. 547, pp. 127–140. Spinger (1991)
22. Halle, K.S., Wu, C.W., Itoh, M., et al.: Spread spectrum communication through modulation of chaos in Chua's circuit. Int. J. Bifurc. Chaos **3**, 469–477 (1993)
23. Kambhampati, V., Ray, I., Kim, E.: A secure checkpointing protocol for survivable server design, distributed computing and internet technology. Lecture Notes in Computer Science, pp. 430–440 (2005)
24. Klein, E., Mislovaty, R., Kanter, I., et al.: Public-channel cryptography using chaos synchronization. Phys. Rev. E **72**, 016214 (2005)
25. Kocarev, L.: Chaos-based cryptography: A brief overview. IEEE Circ. Syst. Mag. **1**(3), 6–21 (2001)
26. Li, Y.X., Tang, W.K.S., Chen, G.: Generating hyperchaos via state feedback control. Int. J. Bifurc. Chaos, **15**(10), 3367–3375 (2005)
27. Menezes, A.J., Oorschot, P.V., Vanstone, S.: Handbook of Applied Cryptography. Boca Raton: CRC Press (1997)
28. Mukhopadhyay, S., Banerjee, S.: Cooperating swarms: A paradigm for collective intelligence and its application in finance. Int. J. Comput. Appl. **6**(10), 31–41 (2010a)
29. Mukhopadhyay, S., Mitra, M., Banerjee, S.: Chaos Synchronization with Genetic Engineering Algorithm in secure communication. In: Banerjee, S. (ed.), Chaos Synchronization and Cryptography for Secure Communications: Applications for Encryption, IGI Global Publishers: U.S.A., pp. 476–509 (2010b)

30. Nam, H., Kim, J., Hong, S.J., et al.: Secure checkpointing. J. Syst. Architect. **48**, 237–254 (2003)
31. Park, T., Woo, N., Yeom, H.Y.: An Efficient Optimistic Message Logging Scheme for Recoverable Mobile Computing Systems. IEEE Transactions on Mobile Computing **1**(4), 265–277 (2002)
32. Pecora, L.M., Carroll, T.L.: Synchronization in chaotic systems. Phys. Rev. Lett. **64**, 821–824 (1990)
33. Plank, J.X., Netzer, R.: Compressed differences: an algorithm for fast incremental checkpointing. Technical Report CS, University of Tennesse, pp. 95–302 (1995)
34. Schneier, B.: Applied cryptography: protocols, algorithm, and source code in C. Wiley, New York (1996)
35. Strom, R., Yemini, S.: Optimistic Recovery in Distributed Systems. ACM Trans. Comput. Syst. **3**, 204–226 (1985)
36. Shannon, C.E.: Communication theory of secrecy systems. Bell Syst. Tech. J. **28**, 656–715 (1949)
37. Tang, G., Liao, X., Chen, Y.: A novel method for designing S-boxes based on chaotic maps. Chaos, Solitons, Fractals, **23**, 413–419 (2005)
38. Zhou, H., Ling, X.: Problems with the chaotic inverse system encryption approach. IEEE Trans. Circuits Syst. I. **44**, 268–271 (1997)

MIX
Papier aus verantwortungsvollen Quellen
Paper from responsible sources
FSC® C105338

If you have any concerns about our products,
you can contact us on
ProductSafety@springernature.com

In case Publisher is established outside the EU,
the EU authorized representative is:
Springer Nature Customer Service Center GmbH
Europaplatz 3, 69115 Heidelberg, Germany

Printed by Libri Plureos GmbH
in Hamburg, Germany